I0930012

PROTEINS
structure and function

PROTEINS
structure and function

Vol. 1

Edited by

M. FUNATSU
K. HIROMI
K. IMAHORI
T. MURACHI
K. NARITA

A Halsted Press Book

KODANSHA LTD.
Tokyo

JOHN WILEY & SONS
New York-London-Sydney-Toronto

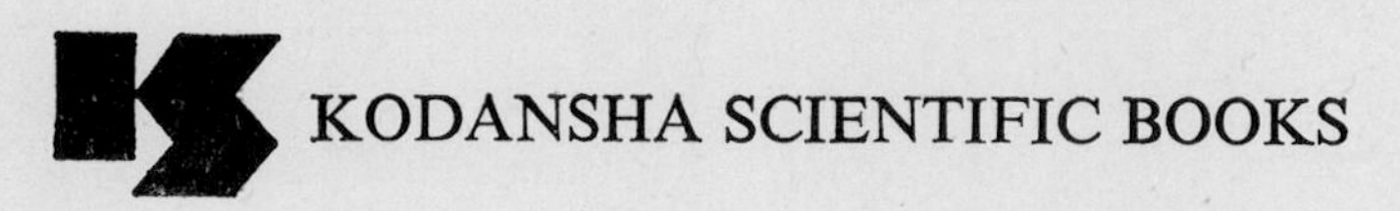

KODANSHA EDP NO.: 3047-247382-2253 (0)

Library of Congress Cataloging in Publication Data
Main entry under title:

Proteins: structure and function.

"A Halsted Press book."
1. Proteins. I. Funatsu, Masaru, 1913- ed.
QP551.P6978 574.1′9245 72-10197
ISBN 0-470-28770-5 (v. 1)
ISBN 0-470-28771-3 (v. 2)

Published in Japan by

KODANSHA LTD.

12-21 Otowa 2-chome, Bunkyo-ku, Tokyo 112, Japan

Published by

HALSTED PRESS

a division of John Wiley & Sons, Inc., New York

PRINTED IN JAPAN

Contributors

Kozo HAMAGUCHI, Dept. of Biology, Faculty of Science, Osaka University, Osaka, Japan

Katsuya HAYASHI, Dept. of Agricultural Chemistry, Faculty of Agriculture, Kyushu University, Fukuoka, Japan

Yoshiko IKEYA-OCADA, Dept. of Biology, Medical School, University of North Carolina, Chapel Hill, North Carolina, U.S.A.

Kazutomo IMAHORI, Dept. of Agricultural Chemistry, Faculty of Agriculture, University of Tokyo, Tokyo, Japan

Tadashi INAGAMI, Dept. of Biochemistry, Vanderbilt University School of Medicine, Nashville, Tennessee, U.S.A.

Tairo OSHIMA, Dept. of Agricultural Chemistry, Faculty of Agriculture, University of Tokyo, Tokyo, Japan

Kenji TAKAHASHI, Dept. of Biophysics and Biochemistry, Faculty of Science, University of Tokyo, Tokyo, Japan

Akira TSUGITA, Laboratory of Molecular Genetics, Medical School, Osaka University, Osaka, Japan

Contents

Preface

Each chapter of the present book consists of a review covering a specific topic within the broad field suggested by the title, *Proteins: structure and function.* The editors have encouraged the authors to deal primarily with their own work, referring to the work of others only where it is relevant in this context, and the reviews are therefore not intended to be fully comprehensive. It is hoped that this editorial policy has retained the distinctive style of reasoning and *modus operandi* of each of the authors.

Protein chemists in this country usually publish most of their work in domestic periodicals, including those in the Japanese language, and although various reports have appeared in journals published in Western countries, much important work by Japanese protein chemists may still be ignored or misunderstood where it is available only in summary form in English. This is clearly unfavorable for the worldwide development of protein science, and the present book is intended to make the activities of Japanese protein chemists more widely known abroad. In publishing these contributions by leading Japanese researchers the editors therefore hope to bridge the gap between protein chemists in Japan and elsewhere.

Proteins: structure and function is published in two volumes, of which volume 1 contains chapters on hydrolytic enzymes whose primary structures have been elucidated. Thus a comparison of the various chapters may prove useful. For example, one can compare work on the three dimensional structure of egg-white lysozyme with the present state of knowledge concerning T2 and T4 phage lysozymes, where elegant structural studies have been carried out using a variety of mutant phages. In addition, two chapters are devoted to ribonuclease T_1, concentrating on the chemical and physicochemical facets of the problem, respectively. A comparison between work on glycosidases, ribonucleases and proteinases serves to illustrate both the uniformity and variability of the techniques used to study hydrolytic enzymes, and it is from broad considerations of this type that further progress may be made in uncovering the general mechanism of enzymatic hydrolysis.

The editors wish to express their appreciation and thanks to the authors for their contributions to these volumes and would also like to take this opportunity to thank the staff of Kodansha, and Dr. G. Shaw and other foreign scientists, for invaluable assistance in the preparation of the English manuscripts comprising this book.

November 1971 The Editors

Trypsin

Tadashi INAGAMI
Department of Biochemistry, Vanderbilt University School of Medicine
Nashville, Tennessee, U.S.A.

1. Introduction

The study of trypsin (EC 3.4.4.4) has benefited greatly from its close resemblance to chymotrypsin (EC 3.4.4.5), both in mode of action and in various molecular properties. Many of the important results obtained from the work on chymotrypsin have been applicable to trypsin directly or with slight modification. On the other hand, the two enzymes differ markedly in their patterns of specificity, trypsin being specific to bonds involving the carboxyl groups of lysine and arginine whilst chymotrypsin is specific to those involving aromatic amino acids and some others with aliphatic side chains. Comparative studies of the catalytic action of the two enzymes have produced valuable information about their mechanisms of action. For these reasons the present account is largely written as a comparative study of the two enzymes, with the emphasis on trypsin.

Numerous reviews on trypsin have been published either as parts of more comprehensive discussions of proteolytic enzymes in general or as single articles. Recent reviews include those of Cunningham,[1] Bender and Kezdy,[2] and Desnuelle.[3] The present article is mainly concerned with findings which have appeared since existing reviews were published, although a brief discussion of basic, classical facts necessary for a full understanding of these recent developments is included. More recently Walsh[4] and Keil[5] have also published articles on trypsin.

A few preliminary remarks on the terminology to be used will serve to avoid the possibility of later confusion or ambiguity. The term "active site" is used to designate the entire area involved in the interaction of the enzyme with the substrate. The use of the word "specificity" is limited to specificity with respect to the side chain of amino acid residues in the substrate. Since the primary substrates of trypsin and chymotrypsin are derivatives of L-α-amino acids, specificity for this common structure is taken for granted unless otherwise stated. Therefore the term "specificity site" designates the area which binds or interacts with the side chain of the amino acid substrate or a competitive inhibitor. Also "recognition site" or "specificity determining site" are used synonymously with the phrase "specificity site". "Catalysis site" refers to the area or to functional groups which directly take part in the actual catalysis. A topologically clear-cut division of the active site into the two sites may not be possible but their individual function can be observed separately as will be discussed later.

2. Primary Structure

2.1. Bovine Trypsinogen

The amino acid sequence of bovine trypsinogen was determined by Walsh *et al.*[6,7] and Mikeš *et al.*[8] The poor solubility of reduced and carboxylmethylated trypsinogen and some of the peptides obtained by its enzymatic digestion presented considerable difficulty to the sequence study. Special techniques, such as cleavage of methionine peptide bonds by cyanogen bromide,[9] redirection of the site of tryptic hydrolysis to cysteine peptide bonds by masking the ε-amino groups of lysine residues, and reduction followed by aminoethylation of disulfide bridges,[6] were employed. There are some minor differences in the sequences published by the two groups. Glu-58 and 68 and Asp-67 and 151 determined by Walsh *et al.*[7] were reported by Mikeš *et al.*[8] to be glutamine and asparagine residues, respectively. Asn-177 has recently been corrected to Asp by several groups[10–12] working on the specificity determin-

Trypsinogen Val–Asp–Asp–
Chymotrypsinogen Cys–Gly–Val–Pro–Ala–Ile–Gln–Pro–Val–Leu–Ser–Gly–
5 10

5 ⇓ 10 15 20
Asp–Asp–Lys–*Ile–Val*–Gly–*Gly*–Tyr–Thr–Cys–Gly–Ala–Asp–Thr–Val–*Pro*–Tyr–
Leu–Ser–Arg–*Ile–Val*–Asn–*Gly*–Glu–Glu–Ala–Val–Pro–Gly–Ser–Trp–*Pro*–Trp–
15⇑ 20 25

25 30 35
Gln–Val–Ser–Leu–Asn– –Ser–*Gly*–Tyr–*His–Phe–Cys–Gly–Gly–Ser–Leu*–
Gln–Val–Ser–Leu–Gln–Asp–Lys–Thr–*Gly*–Phe–*His–Phe–Cys–Gly–Gly–Ser–Leu*–
30 35 40 45

40 45 50
Ile–Asn–Ser–Gln–*Trp–Val–Val*–Ser–*Ala–Ala–His–Cys*–Tyr–Lys–Ser–Gly–Ile–
Ile–Asn–Glu–Asn–*Trp–Val–Val*–Thr–*Ala–Ala–His–Cys*–Gly–Val–Thr–Thr–Ser–
50 55 60

55 60 65
Gln–*Val*–Arg–Leu–Gly–Gln–Asp–Asn–Ile–Asn–Val–Val–Glu–Gly–Asn–Gln–
Asp–*Val*–Val–Val–Ala–Gly–Glu–Phe–Asp–Gln–Gly–Ser–Ser–Ser–Glu–Lys–
65 70 75

70 75 80
Gln–Phe–Ile–Ser–Ala–Ser–Lys–Ser–Ile–Val–His–Pro–*Ser*– –*Tyr–Asn–Ser*–
Ile–Gln–Lys–Leu–Lys–Ile–Ala–Lys–Val–Phe–Lys–Asn–*Ser*–Lys–*Tyr–Asn–Ser*–
80 85 90 95

85 90 95 100
Asn–*Thr*–Leu–*Asn–Asn–Asp–Ile*–Met–*Leu*–Ile–*Lys–Leu*–Lys–Ser–*Ala–Ala–Ser*–
Leu–*Thr*–Ile–*Asn–Asn–Asp–Ile*–Thr–*Leu*–Leu–*Lys–Leu*–Ser–Thr–*Ala–Ala–Ser*–
100 105 110

105 110 115
Leu–Asn–Ser–Arg–*Val*–Ala–Ser–Ile–Ser–*Leu–Pro*–Thr–Ser–Cys– –Ala–
Phe–Ser–Gln–Thr–*Val*–Ser–Ala–Val–Cys–*Leu–Pro*–Ser–Ala–Ser–Asp–Asp–Phe–
115 120 125 130

120 125 130
Ser–*Ala–Gly–Thr*–Gln–*Cys*–Leu–Ile–Ser–*Gly–Trp–Gly*–Asn–*Thr*–Lys–Ser–Ser–
Ala–*Ala–Gly–Thr*–Thr–*Cys*–Val–Thr–Thr–*Gly–Trp–Gly*–Leu–*Thr*–Arg–Tyr–Thr–
135 140 145

135 140 145 150
Gly–Thr–Ser–Tyr–*Pro–Asp*–Val–*Leu*–Lys–Cys–Leu–Lys–Ala–*Pro*–Ile–*Leu–Ser*–
Asn–Ala–Asn–Thr–*Pro–Asp*–Arg–*Leu*–Gln–Gln–Ala–Ser–Leu–*Pro*–Leu–*Leu–Ser*–
150 155 160

155 160 165
Asp–Ser–Ser–*Cys–Lys*–Ser–Ala–Tyr–Pro–Gly–Gln–*Ile*–Thr–Ser–Asn–*Met*–Phe–
Asn–Thr–Asn–*Cys–Lys*–Lys–Tyr–Trp–Gly–Thr–Lys–*Ile*–Lys–Asp–Ala–*Met*–Ile–
165 170 175 180

170 175 180
Cys–Ala–Gly–Tyr–Leu–Glu–Gly–Gly–Lys–Asp–*Ser–Cys*–Gln–*Gly–Asp–Ser–Gly*–
Cys–Ala–Gly–Ala–Ser–Gly–Val–Ser– –*Ser–Cys*–Met–*Gly–Asp–Ser–Gly*–
185 190 195

185 190 195
Gly–Pro–Val–*Val–Cys*–Ser–Gly–Lys–Leu–Gln– *Gly–Ile–Val*–
Gly–Pro–Leu–*Val–Cys*–Lys–Lys–Asn–Gly–Ala–Trp–Thr–Leu–Val–*Gly–Ile–Val*–
200 205 210

200 205 210
–*Ser–Trp–Gly–Ser*–Gly–Cys–Ala–Gln–Lys–Asn–Lys–*Pro–Gly–Val–Tyr*–Thr–
–*Ser–Trp–Gly–Ser*–Ser–Thr–Cys–Ser–Thr–Ser–Thr–*Pro–Gly–Val–Tyr*–Ala–
215 220 225

215 220 225
Lys–*Val*–Cys–Asn–Tyr–*Val*–Ser–*Trp*–Ile–Lys–*Gln–Thr*–Ile–*Ala*–Ser–*Asn*
Arg–*Val*–Thr–Ala–Leu–*Val*–Asn–*Trp*–Val–Gln–*Gln–Thr*–Leu–*Ala*–Ala–*Asn*
230 235 240 245

Fig. 1. Amino acid sequences of bovine trypsinogen[8,10–12] and bovine chymotrypsinogen A.[15,16] Residues indicated in *italic* letters are homologous in the two enzymes. Those in roman letters are non-homologous.

ing site. The sequence involving residue 84 through 87, which was determined as Ser–Asn–Thr–Leu by Mikeš *et al.*,[8] has been reported to contain (Pro,Leu,Thr,Asn) by Walsh, Neurath and their co-workers.[6,7] Based on these sequences the molecular weight of trypsinogen is calculated as 23,985 for the structure of Mikeš *et al.*, and 23,991 for that of Walsh *et al.* For trypsin it is 23,279 and 23,285, respectively. The primary structure of trypsinogen was found to have as much as 40% of its amino acid sequences homologous to those of chymotrypsinogen A.[7,13,14] Furthermore, if analogous amino acid residues are taken into consideration the similarity between the two pancreatic endopeptidases is strikingly high. The sequences of bovine trypsinogen and chymotrypsinogen A[15,16] are shown in Fig. 1, in which amino acid residues in *italic* letters indicate the sequences which are common to both. The positions of the disulfide bridges in the two zymogens, determined by Kaufman,[17] Brown and Hartley,[18] and Keil *et al.*[19] are compared in Table 1. It will be noted that three disulfide bridges of trypsinogen and

TABLE 1
Disulfide Bridges of Bovine Trypsinogen and Chymotrypsinogen

S-S Bridge†1	Cysteine residues involved		Homology
	Trypsinogen†2	Chymotrypsinogen†3	
A	31–47	42–58	homologous
B	154–168	169–182	homologous
C	179–203	191–220	Cys-179 (trypsin) and Cys-191 (chymotrypsin) are homologous Cys-203 (trypsin) and Cys-221 (chymotrypsin) are out of position by 1 residue
D	122–189	136–201	homologous
E	–	1–123	not homologous
F	115–216	–	not homologous
G	13–143	–	not homologous

†1 Alphabetical notation by ref. 14.
†2 Numbering of trypsinogen shown in Fig. 1 (ref. 17).
†3 Numbering of chymotrypsinogen shown in Fig. 1 (ref. 18 and 19).

chymotrypsinogen are in identical positions and another bridge is almost identical. Since the catalytic mechanisms of the two enzymes are very similar it may be inferred that the homologous sequences will be important for the catalysis, whereas some of the dissimilar sequences may form the specificity determining site structure which ought to be different in the two enzymes.

2.2. Trypsinogen and Trypsin from Other Sources

Desnuelle and his co-workers[20,21] have isolated trypsinogen from porcine pancreas, studied the mechanism of its activation to trypsin and isolated the trypsin. Liener and his co-workers[22,23] and Van Melle *et al.*[24] have purified and crystallized the porcine trypsin. Liener's group also determined the amino acid sequence of peptides in the active site area.[25,26] Florkin and his co-workers have isolated trypsinogen from sheep and goat and studied the mechanism of activation of these zymogens.[27,28] Bier and Nord and their associates have compared molecular and catalytic properties of trypsins from different animals and shown that their catalytic properties are similar[29] but porcine trypsin was found to be more stable in alkaline solution.[30] Human trypsin has also been isolated, purified and characterized by several groups of investigators.[31–3]

2.3. Activation of Trypsinogen to Trypsin and Further Autolysis of Trypsin

Trypsinogen is activated to trypsin by autocatalytic cleavage of a single peptide bond. In the case of bovine trypsinogen the bond between Lys-6 and Ile-7, indicated by an arrow in Fig. 1, is cleaved to release the *N*-terminal hexapeptide.[34] Activation of trypsinogens from goat,[35] pig,[21] and sheep[27,28] proceeds by a similar mechanism releasing short peptides from the *N*-terminal portion of the trypsinogens. These peptides are listed in Table 2. It will be noted that the new *N*-terminal sequence Ile–Val is homologous with that of chymotrypsin. Such homology suggests that the amino acid residues in the *N*-terminal position of trypsin may be essential for activity. Evidence for the special importance of the *N*-terminal α-amino groups of trypsin[36,37] and chymotrypsin[38,39] has been obtained and is discussed in sections 3.3.2 and 4.4.2.

TABLE 2
N-Terminal Peptides of Trypsinogen Which Are Cleaved off during Activation to Trypsin

Animal	Sequence	Ref.
Pig	Phe–Pro–Thr–Asp–Asp–Asp–Asp–Lys	21
Sheep 1	Phe–Pro–Val–Asp–Asp–Asp–Asp–Lys	27
Sheep 2	Val–Asp–Asp–Asp–Asp–Lys	27
Goat	Val–Asp–Asp–Asp–Asp–Lys	35
Ox	Val–Asp–Asp–Asp–Asp–Lys	34

Although cleavage of the above mentioned bond is sufficient for activation, cleavage of additional bonds which are susceptible to tryptic digestion takes place even under the mildest laboratory conditions. Thus Schroeder and Shaw[40] were able to fractionate the product of activation, by ion-exchange chromatography on Sulfoethyl-Sephadex, into trypsin (β-trypsin), trypsin with one additional lysine peptide bond cleaved at Lys_{131}–Ser_{132} (α-trypsin) and minor inactive components. Maroux *et al.*,[41,42] reported that autolysis proceeded beyond the cleavage of the hexapeptide from trypsinogen. In the presence of $CaCl_2$ at 4°C the bond Arg_{105}–Val_{106} was cleaved without loss of enzyme activity. The bond Lys_{131}–Ser_{132} was also substantially autolyzed but the enzyme activity was not affected. Cleavage of the bond Lys_{49}–Ser_{50}, however, seemed to result in the formation of an inactive enzyme. Smith and Shaw[43] found that autolytic cleavage of the bond Lys_{176}–Asp_{177} caused loss of ability to hydrolyze the specific substrate N^{α}-benzoyl-L-arginine ethyl ester (BAEE) but activity towards a nonspecific substrate *N*-acetyl-L-tyrosine ethyl ester (ATEE) was not lost, nor was susceptibility to inactivation by diisopropyl fluorophosphate (DFP) (I) lost. The cleavage should have destroyed the specificity site.

$$(i\text{-}C_3H_7O)_2{=}\overset{\overset{\large O}{\|}}{P}\text{—}F \quad \text{(I)}$$

2.4. Amino Acid Residues in the Active Site

2.4.1. Ser-183

Trypsin, chymotrypsin and other proteases, called serine proteases, were found to be rapidly inactivated by diisopropyl fluorophosphate and other phosphate derivatives.[44,45] The inhibited enzyme contained a stoichiometric amount of the diisopropylphosphoryl group which was bound to the hydroxyl group of a particular serine residue by a phosphate ester linkage. A peptide containing this phosphorylated serine residue was isolated by Dixon *et al.*[46] The phosphorylated serine residue was found to be residue 183 using the numbering of trypsinogen. The sequence

–Gly–Asp–Ser–Gly–

is the same as that containing the active serine residue of chymotrypsin.[47,48] The same sequence has also been found in all other mammalian serine proteases inhibited by DFP such as thrombin[49] and elastase,[50,51] whereas bacterial serine proteases such as subtilisins[52,53] and aspergillopeptidases[54] seem to possess the sequence

–Thr–Ser–Met–Ala–

By analogy with chymotrypsin the hydroxyl group of the active serine residue can be considered as the site of acylation by *p*-nitrophenyl acetate,[55,56] *N*-*trans*-cinnamoyl imidazole,[57] *p*-guanidinobenzoyl esters,[58,59] a carbamyl chloride,[60] and various sulfonyl fluorides,[61] and is discussed in section 3.1.4.

2.4.2. His-46

Gutfreund[62] and Inagami and Sturtevant[63] showed that the catalytic rate constant, k_{cat}, for the hydrolysis of a specific substrate, N^{α}-benzoyl-L-arginine ethyl ester (BAEE), was controlled by a single ionizing group in trypsin. The pK_a value of the ionizing group was 6.1–6.3 at 25°C, and studies at different temperatures gave the ΔH^0 of ionization of the group as 7 kcal/mole. These values suggested that the imidazole group of histidine was implicated in the catalytic mechanism. A similar situation was observed with chymotrypsin. Hammond and Gutfreund[64] obtained a pK_a of 6.85 for the hydrolysis of *N*-acetyl-L-phenylalanine ethyl ester at 25°C. Cunningham and Brown[65] obtained a pK_a of 6.7 for N^{α}-acetyl-L-tryptophan ethyl ester (ATrEE) and *N*-acetyl-L-tyrosine ethyl ester (ATEE), and ΔH^0 of ionization as 11 kcal/mole.

An unequivocal proof of the presence of a histidine residue in the active site was obtained by Shaw and his co-workers using a chemical method. A bifunctional alkylating reagent L-1-chloro-3-tosylamido-7-amino-2-heptanone (the chloromethyl ketone derivative of N^{α}-tosyl-L-lysine, TLCK) (II) was designed so that it would specifically bind to the active site of the enzyme by virtue of its lysine-like structure before alkylating a functional group by the chloroketo moiety.[66] The reagent inhibited trypsin stoichiometrically.

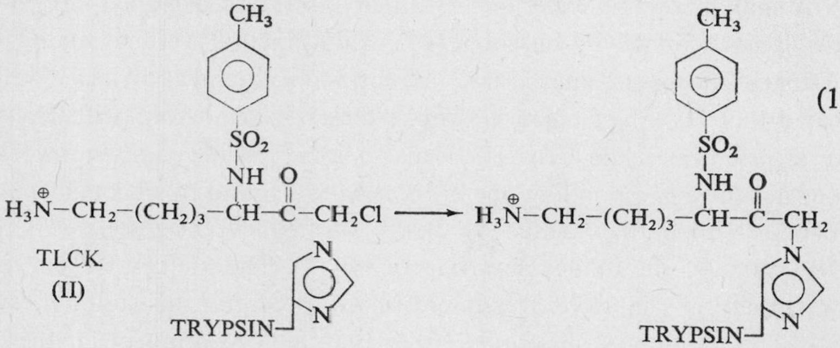

$$\overset{\oplus}{H_3N}-CH_2-(CH_2)_3-\underset{}{CH}(NH-SO_2-C_6H_4-CH_3)-\overset{O}{\overset{\|}{C}}-CH_2Cl \longrightarrow \overset{\oplus}{H_3N}-CH_2-(CH_2)_3-CH(NH-SO_2-C_6H_4-CH_3)-\overset{O}{\overset{\|}{C}}-CH_2-N(\text{imidazole})-\text{TRYPSIN} \quad (1)$$

TLCK
(II)

TRYPSIN (imidazole N, N)

$$C_6H_5-CH_2-CH(NH-SO_2-C_6H_4-CH_3)-\overset{O}{\overset{\|}{C}}-CH_2Cl \longrightarrow C_6H_5-CH_2-CH(NH-SO_2-C_6H_4-CH_3)-\overset{O}{\overset{\|}{C}}-CH_2-N(\text{imidazole})-\text{CHYMOTRYPSIN} \quad (2)$$

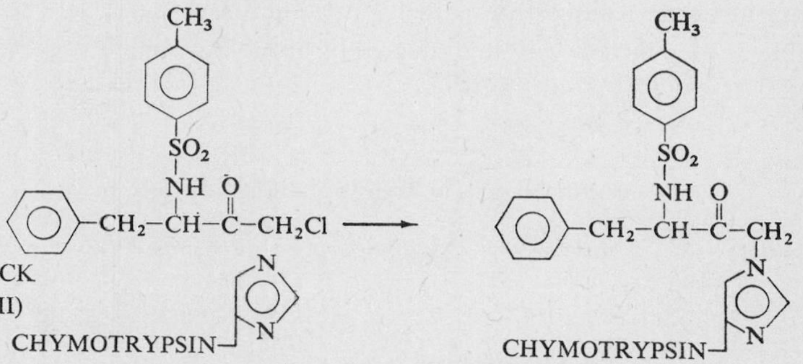

TPCK
(III)

CHYMOTRYPSIN (imidazole N, N)

The inhibition was irreversible and was retarded by competitive inhibitors. Amino acid analysis of the inhibited enzyme showed loss of only one histidine residue per molecule of enzyme.[66,67] After reduction of the keto group by sodium borohydride to prevent destruction of the inhibitor, a peptide containing the modified histidine residue was isolated. The site of alkylation by TLCK was identified with His-46.[68,69] Oxidation at the keto group of the inhibited enzyme by performic acid followed by hydrolysis produced 3-carboxymethyl histidine, indicating that the inhibition by TLCK was due to an exclusive alkylation of the N-3 atom of the imidazole moiety of His-46.[70] Beeley and Neurath[71] found that the same site was most rapidly alkylated by bromoacetone and that the alkylation resulted in irreversible inhibition of the enzyme.

Iodoacetamide was also found to alkylate the N-3 atom of a histidine residue preferentially in the presence of methylguanidine.[72,73] Thus it is clear that His-46 is at the catalysis site and most probably is the catalytically important histidine residue.

A similar labeling experiment with chymotrypsin using L-1-chloro-3-tosylamido-4-phenyl-2-butanone (the chloromethyl ketone analog of *N*-tosyl-L-phenylalanine, TPCK) (III) by Schoellmann and Shaw[74] indicated that His-57 of chymotrypsin, which is homologous with His-46 of trypsin, was at the active site. This histidine residue of chymotrypsin, but not the corresponding one of trypsin, was found by Nakagawa and Bender[75] to be methylated by methyl *p*-nitrobenzenesulfonate at the N-3 atom of the imidazole group of His-57. The methylated enzyme was practically inactive. It is interesting to note that this sulfonyl ester alkylates the imidazole group rather than sulfonylating the hydroxyl group of the active serine residue. The long sequences which involve these histidine residues are called "histidine loops" and, as shown in Table 3, they have a very high degree of homology which suggests that

TABLE 3

Histidine Loops of Trypsin and Chymotrypsin

Trypsin†1	29 His–30 Phe–Cys(S)–Gly–Gly–Ser–Leu–Ile–Asn–Ser–Gln–40 Trp–Val–
	Val–Ser–Ala–Ala–46 His–Cys(S)
Chymotrypsin†2	40 His–Phe–Cys(S)–Gly–Gly–Ser–Leu–Ile–Asn–Gln–50 Asn–Trp–Val–
	Val–Thr–Ala–Ala–57 His–Cys(S)

†1 Ref. 7 and 8. †2 Ref. 15.

these structures are important for catalysis. A second histidine residue (His-29) of trypsin also occurs at a position homologous with His-40 of chymotrypsin. However, this residue has never been detected by chemical methods. Recent reports of X-ray crystallographic work on chymotrypsin indicated that this histidine residue is located deep inside the enzyme molecule but not in the area of the active site.[16,76] In view of the high degree of homology in the sequence of this area, it may not be too far amiss to assume that the His-29 of trypsin may not directly participate in the catalysis.

Another approach for affinity labeling by Lawson *et al.*[77] led to the alkylation of the active serine hydroxyl group instead of a histidine residue. These workers synthesized a series of bromoacetamidoalkylamines $BrCH_2\text{–}CONH\text{–}(CH_2)_n\text{–}NH_3^{\oplus}\cdot Cl^-$ and bromoacetamidoalkylguanidines $BrCH_2\text{–}CONH\text{–}(CH_2)_n\text{–}NHC(=NH)NH_3^{\oplus}\cdot Cl^-$, where $n = 2,3,4,5$. All of these compounds slowly inactivated trypsin at rates much slower than TLCK. The inactivation was thought to be due to alkylation of an essential group by the bromoacetyl group. Analysis of acid hydrolyzates of trypsin inactivated by bromoacetamidopropylguanidine, which was the most effective inhibitor, revealed that alkylation had taken place at the hydroxyl group of the active serine residue. No histidine residue was alkylated.

2.4.3. Asp-90

A third sequence in the vicinity of Asp-90 in trypsin highly homologous to the corresponding sequence in chymotrypsin is outlined in Table 4. Asp-102 of the latter, which is in the position equivalent to Asp-90 of the former, was originally reported as asparagine. The X-ray crystallographic study of chymotrypsin by Blow *et al.*[16] revealed that the β-carboxyl group of this residue was in a position which allowed it to be hydrogen-bonded to an imidazole N-atom of the catalytically important His-57 of chymotrypsin. Since four other mammalian serine

TABLE 4

Amino Acid Sequences Involving Asp-90 of Trypsin and Asp-102 of Chymotrypsin

	82 ... 90
Trypsinogen†1	Tyr–Asn–Ser–Asn–Thr–Leu–Asn–Asn–Asp–Ile–Met–
	96
	Leu–Ile–Lys–Leu
Chymotrypsinogen†2	Tyr–Asn–Ser–Leu–Thr–Ile–Asn–Asn–Asp–Ile–Thr–
	94 ... 100
	Leu–Leu–Lys–Leu
	108

†1 Ref. 7 and 8. †2 Ref. 16.

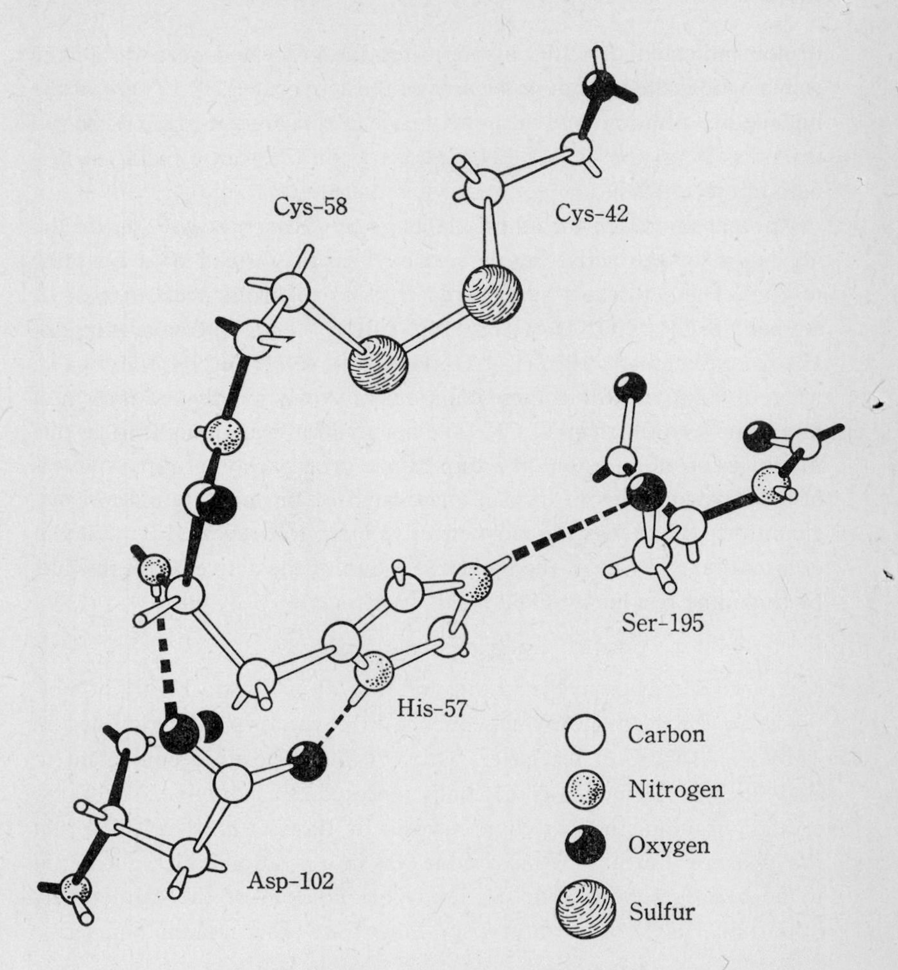

Fig. 2. Three dimensional structure of the catalysis site of chymotrypsin.

TABLE 5
Interatomic Distances of Some Functional Groups in the Catalysis Site of α-Chymotrypsin[†1]

Bonding Atoms		Distance (Å)
His-57 Im-N[1]	Asp-102 β-Carboxyl-O[1]	2.8
His-57 Im-N[3]	Ser-195 Hydroxyl-O	3.0
Asp-102 β-Carboxyl-O[2]	Ser-214 Hydroxyl-O[†2]	2.6

[†1] Ref. 16. [†2] Not shown in Fig. 2.

proteinases, chymotrypsin B, trypsin, elastase and thrombin, have been found to have Asp instead of Asn in the corresponding position, this residue of chymotrypsin was reinvestigated and subsequently corrected to Asp.[16]

Fig. 2 shows the three dimensional arrangement of the catalysis site of chymotrypsin thus determined. It has been reported that Asp-102 is inside the molecule and shielded from contact with water. As indicated in Table 5, one of the oxygen atoms of the β-carboxyl group of Asp-102 is within a 2.8 Å distance from the N-1 atom of the imidazole group of His-57, which is clearly within a hydrogen-bonding distance. The N-3 atom of the imidazole group of the same histidine residue is only 3.0 Å away from the hydroxyl oxygen atom of Ser-195, again a hydrogen-bonding distance. In view of the high degree of homology between trypsin and chymotrypsin it is highly probable that a similar three dimensional arrangement exists in the catalysis site of trypsin.

Stroud *et al.*[78] have very recently obtained an electron density map of trypsin inhibited by DFP (DIP–trypsin) at 2.7 Å resolution.[8] The presence of a similar structure in trypsin involving Ser-183, His-46 and Asp-90 has been confirmed. Also another hydroxyl group of Ser-198 was found to be within a hydrogen-bonding distance of the β-carboxyl oxygen atom of Asp-90, involved in the hydrogen-bonding with His-46, in a manner similar to Ser-214 of chymotrypsin.

Thus far we have seen that homologous sequences are involved in the formation of the catalytic site, and in view of the great similarity in the mode of catalysis of the two enzymes this is to be expected. Dissimilar sequences, then, should represent that part of the enzyme structure responsible for dissimilar action, namely, the different specificity. Comparison of the homologous sequences in relation to function, evolution and phylogeny has been made by Neurath *et al.*,[13] Hartley *et al.*,[14] Smillie *et al.*,[79] and Šorm *et al.*[80]

2.4.4. Asp-177 at the Specificity Site

A detailed discussion of the nature of the specificity site of trypsin will be presented later in section 4.4.1. Since trypsin is specific for derivatives of lysine and arginine, whose side chains have a positive electrical charge, the site of binding for the side chains was expected to contain a carboxylate anion. The β-carboxyl group of Asp-177 was found to exist in the site and was masked with a potent competitive inhibitor, benzamidine, by Eyl and Inagami[10] (see section 4.4.1.). It is interesting to note that this residue occurs in a sequence which as expected was not homologous with chymotrypsin.

	171	177 178
Trypsin	*Gly*–Trp–Leu–Glu–Gly–Gly–Lys–Asp–*Ser*	
Chymotrypsin	*Gly*–Ala–Ser–Gly–Val–Ser– –*Ser*	
	184	190

Moreover the nonhomologous sequence in trypsin is longer than the corresponding sequence of chymotrypsin by two amino acid residues, which indicates that insertion of the two amino acid residues may have occurred in the course of development of trypsin from chymotrypsin to create diversity in the specificity of pancreatic proteases. Alternatively, if chymotrypsin has developed from trypsin, deletion of two residues may have occurred.

Recent crystallographic work on DIP–trypsin by Stroud *et al.*[78] shows that the β-carboxyl group of Asp–177 is at the back of a pocket which presumably is the site of substrate binding. Interestingly enough this carboxyl group is within a hydrogen-bonding distance (3 Å) of the amide N-atom of Gln-205 (which is not homologous with chymotrypsin). The back of the pocket is bound by Val-197, Tyr-212 and Ser-178, which are homologous to chymotrypsin.

3. Catalytic Mechanism

Trypsin-catalyzed hydrolysis of peptide bonds or proteins is affected not only by the nature of bonds attacked but also by many complicating factors such as the state of denaturation of the substrate protein, accessibility of the bonds cleaved and so forth. Consequently most clear-cut information concerning the catalytic mechanism has come from studies with synthetic substrates of low molecular weight.

3.1. Acyl-enzyme Intermediate

The first evidence for the involvement of an acyl-enzyme intermediate in the course of catalysis came from Hartley and Kilby's work[81] on chymotrypsin. Since then many publications have appeared on this subject and generally confirm the occurrence of an intermediate in which the acyl group of a substrate is covalently bound to the hydroxyl group of the active serine residue with formation of an ester. Most of the initial work which suggested involvement of an acyl-enzyme intermediate was first carried out with chymotrypsin, but similar experiments with trypsin subsequently led to essentially similar conclusions. In this article experimental results obtained from studies with chymotrypsin will be reviewed, then compared with corresponding work on trypsin. For details of the mechanism of chymotrypsin catalysis readers may find a chapter by Bruice and Benkovic[82] of interest.

3.1.1. Acetyl-Enzyme Formation with *p*-Nitrophenyl Acetate as Substrate

Hartley and Kilby[81] discovered that *p*-nitrophenyl acetate (NPA) was hydrolyzed by chymotrypsin at neutral pH and the course of the catalysis was conveniently followed by measurement of the increase of optical density at 400 nm due to the *p*-nitrophenylate ion formed. The reaction, however, was found to proceed in two distinct phases, wherein an initial very rapid reaction was followed by a much slower release of the product. The stoichiometry of the first reaction was almost one mole of product to one mole of enzyme. These results have been interpreted as follows: an initial rapid acetylation at the active site of the enzyme resulted in an equally rapid formation of *p*-nitrophenylate ion followed by a second phase in which slow turnover of the acetyl group occurs. When Balls and his co-workers[83,84] carried out the same reaction at pH 5, the second phase of the reaction was arrested and monoacetyl-chymotrypsin was isolated. The acetyl group was lost as the pH was raised to a neutral value.[85,86] However, Dixon *et al.*[87] observed that if the acetyl-enzyme was first denatured in 8 M urea at pH 5, then the pH was increased to near neutrality in the same solution, no deacetylation took place. The acetyl-enzyme was inactive towards a more specific substrate such as *N*-acetyl-L-tyrosine ethyl ester (ATEE), strongly indicating that monoacetylation occurred by a mechanism closely related to the normal catalytic mechanism. This hypothesis was strengthened when Oosterbaan *et al.*[88] isolated from ^{14}C-acetylchymotrypsin a peptide which contained the radioactive acetyl group attached to the active serine residue which earlier had been shown to be phosphorylated by DFP.

It is thus clear that chymotrypsin catalysis of the NPA hydrolysis involves an acyl-enzyme intermediate which may be formed during a three-step sequence of the following type:

$$E + S \underset{k_{-1}}{\overset{k_1}{\rightleftharpoons}} ES' \xrightarrow{k_2} \underset{\substack{+ \\ P'}}{ES''} \xrightarrow{k_3} E + P'' , \tag{3}$$

proposed by Gutfreund and Sturtevant[89] and Wilson and Gabib[90] where E represents free enzyme, S substrate, ES′ the initial Michaelis complex, ES″ the acyl-enzyme intermediate, P′ the leaving group and P″ the acyl portion of the substrate hydrolyzed from the intermediate. Detailed kinetic studies were made to determine the rate constant for each step. Gutfreund and Sturtevant[89,91] have obtained rate equation (4) by applying the steady state approximation to ES′ but not to ES″

$$[E] = \frac{k_3\,K_m\,[E]_0 + \dfrac{k_2\,K_m\,[S]_0\,[E]_0}{K_m + [S]_0} \cdot \exp\left\{-\dfrac{(k_2+k_3)\,[S]_0 + k_3\,K_m}{K_m + [S_0]}\,t\right\}}{(k_2+k_3)\,[S]_0 + k_3\,K_m} \tag{4}$$

where $K_m = (k_{-1} + k_2)/k_1$ is the Michaelis constant, [E] and $[E]_0$ are concentrations of free and total enzyme, respectively. $[S]_0$ is the initial substrate concentration and t is time. Using [E] the rates of P′ and P″ formations are evaluated by Eq. (5) and (6).

$$\frac{d[P']}{dt} = \frac{k_2}{K_m}\,[E]\,[S]_0 \tag{5}$$

$$\frac{d[P'']}{dt} = k_3\,[E]_0 - k_3\,\frac{K_m + [S]_0}{K_m}\,[E] \tag{6}$$

At small values of t, when the initial burst of P′ formation is observed, the rate constant $k = \{(k_2 + k_3)\ [S]_0 + k_3K_m\}\ /\ \{K_m + [S]_0\}$ can be obtained, and when it is assumed that $(k_2 + k_3)\ [S]_0 \gg k_3K_m$ for high $[S]_0$, $(k_2 + k_3)$ and K_m can be evaluated by an Eadie[92] or Lineweaver-Burk[93] plot. At large t values the exponential term becomes negligibly small and the rate equation reduces to

$$v = \frac{d[P']}{dt} = \frac{d[P'']}{dt} = \frac{k_{cat}\,[E]_0\,[S]_0}{[S]_0 + K_m\,(app)}\,, \tag{7}$$

where

$$k_{cat} = \frac{k_2\,k_3}{k_2 + k_3} \tag{8}$$

$$K_m\,(app) = \frac{k_3}{k_2 + k_3}\,K_m \tag{9}$$

For cases in which $k_3 \gg k_2$, these constants reduce to $k_{cat} = k_2$ and $K_m(\text{app}) = K_m$, whereas if $k_3 \ll k_2$, $k_{cat} = k_3$ and $K_m(\text{app}) = (k_3/k_2)K_m$. From a study of the sequence of these reactions with a stopped flow apparatus, and making use of the above mentioned assumptions, k_2, k_3, K_m were obtained as shown in Table 6. The finding by Spencer and Sturtevant[94] that the NPA hydrolysis by chymotrypsin was inhibited by a specific substrate, *N*-acetyl-L-tyrosine ethyl ester (ATEE), and that the ATEE hydrolysis was inhibited by NPA was interpreted as indicating that the NPA hydrolysis by way of the acetyl-enzyme intermediate occurred by a mechanism similar to that of the true catalysis by the enzyme. Recently, however, Faller and Sturtevant[95] pointed out that the assumption $(k_2 + k_3)\,[S]_0 \gg k_3\, K_m$ was not correct and without this assumption their study did not indicate the presence of the Michaelis complex, ES′.

Trypsin was found by Stewart and Ouellet[56] to catalyze the hydrolysis of NPA in a similar fashion, as shown in Fig. 3. Application of techniques similar to those used by Gutfreund and Sturtevant and accepting the above mentioned assumption gave the kinetic constants shown in Table 6.

Since NPA has no structural feature required of a specific substrate for either chymotrypsin or trypsin, studies of its hydrolysis by these enzymes have often been considered not to be relevant to the elucidation of the true catalytic mechanism. However, Gutfreund and Hammond[96] observed that the chymotrypsin-catalyzed hydrolysis of *N*-benzyloxycarbonyl-L-tyrosine *p*-nitrophenyl ester at pH 7.2 proceeded with an instantaneous release of *p*-nitrophenol followed by a slow turnover. Kezdy *et al.*[97] were able to study the initial release of *p*-nitrophenol from a specific labile ester substrate of chymotrypsin, *N*-acetyl-L-tryptophan *p*-nitrophenyl ester, at pH values between 2 and 4. Thus

TABLE 6
Kinetic Constants for the Hydrolysis of *p*-Nitrophenyl Acetate Catalyzed by Trypsin and Chymotrypsin

	Trypsin†1	Chymotrypsin†2
k_2 (sec^{-1})	1.5	3.1
k_3 (sec^{-1})	0.013	0.025
K_m (mM)	21	6.8

†1 Ref. 56. †2 Ref. 91.

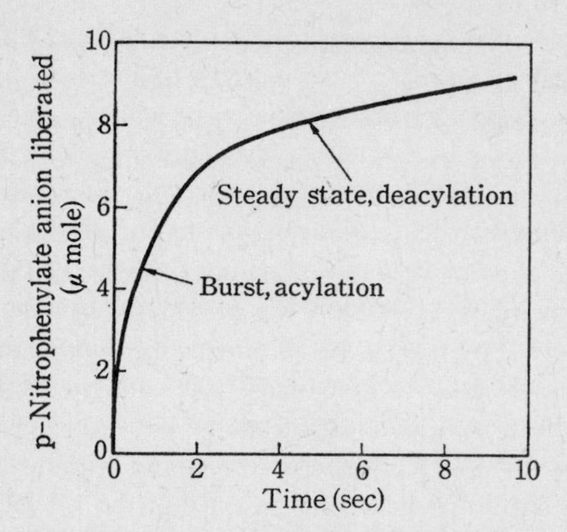

Fig. 3. Hydrolysis of *p*-nitrophenyl acetate by trypsin with an initial burst of *p*-nitrophenylate anion liberation.

acylation of the enzyme by a *p*-nitrophenyl ester seems to be part of a general mechanism for trypsin and chymotrypsin catalysis.

3.1.2. Kinetic Studies of the Hydrolysis of Specific Substrates

Further evidence for the involvement of an acyl-enzyme intermediate in trypsin and chymotrypsin catalysis was obtained by Bender and his co-workers,[98-102] who investigated the effect of leaving groups on the steady state kinetics. Their results for c. ym trypsin, listed in Table 7,

TABLE 7

The Chymotrypsin-catalyzed Hydrolys s of Derivatives of *N*-Acetyl-L-tryptophan and *N*-Acetyl-L-phenylalanine†

Substrate		k_{cat} (sec^{-1})	K_m(app) (M)
N-Acetyl-L-tryptophan	amide	0.026	7.3
	ethyl ester	26.9	0.097
	methyl ester	27.7	0.095
	p-nitrophenyl ester	30.5	0.002
N-Acetyl-L-phenylalanine	amide	0.039	37
	ethyl ester	63.1	0.88
	methyl ester	57.5	1.50
	p-nitrophenyl ester	77	0.024

† Ref. 102. Conditions: pH 7.0, 25°C, 3.17% v/v acetonitrile-water.

indicate that the labile *p*-nitrophenyl ester and the stable alkyl esters have practically identical catalytic rate constants. These results have been interpreted to indicate that an acyl-enzyme intermediate, which is identical for the three different esters, is formed, and its breakdown is the rate limiting step. Therefore k_{cat} must be equal to k_3. It has been suggested that the incomparably lower rate constant for amide substrates may be due to differences in the rate limiting step, the acylation reaction being the one for amides. Therefore k_{cat} for the amides is equal to k_2. Various esters of *N*-benzoylglycine or *N*-acetyl-L-valine, which are poor nonspecific substrates of chymotrypsin, did not have such uniform k_{cat} among themselves, indicating that neither acylation nor deacylation was a dominant rate limiting step.[102]

A similar kinetic method applied to trypsin has shown the involvement of an acyl-enzyme intermediate. Baines *et al.*[103] using four different esters of N^{α}-tosyl-L-arginine (IV) as substrates demonstrated that the rate of hydrolysis was identical for all the esters which had a common acyl moiety, as shown in Table 8. Similar results have been obtained with esters of N^{α}-tosyl-L-homoarginine (V),[103] N^{α}-benzoyl-L-lysine, N^{α}-tosyl-L-lysine (VI), N^{α}-benzoyl-L-thialysine [*S*-(β-aminoethyl)-*N*-benzoyl-L-cysteine] and N^{α}-tosyl-L-thialysine [*S*-(β-aminoethyl)-*N*-tosyl-L-cysteine] (VII).[104]

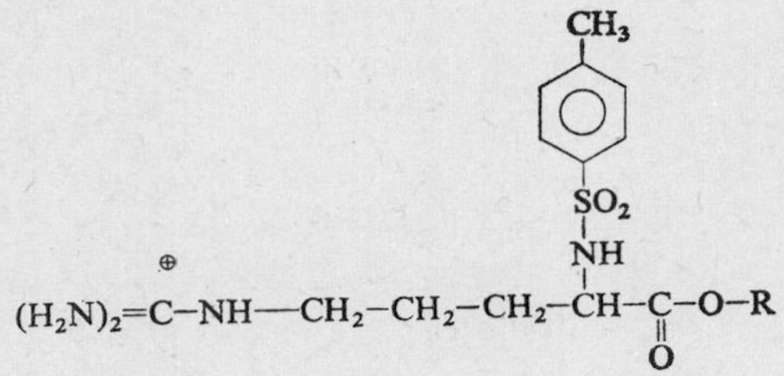

CH_3–C_6H_4–SO_2–NH–
$(H_2N)_2^{\oplus}$=C–NH—CH_2–CH_2–CH_2–CH–C(=O)–O–R

(IV)

CH_3–C_6H_4–SO_2–NH–
$(H_2N)_2^{\oplus}$=C–NH—CH_2–CH_2–CH_2–CH_2–CH–C(=O)–O–R

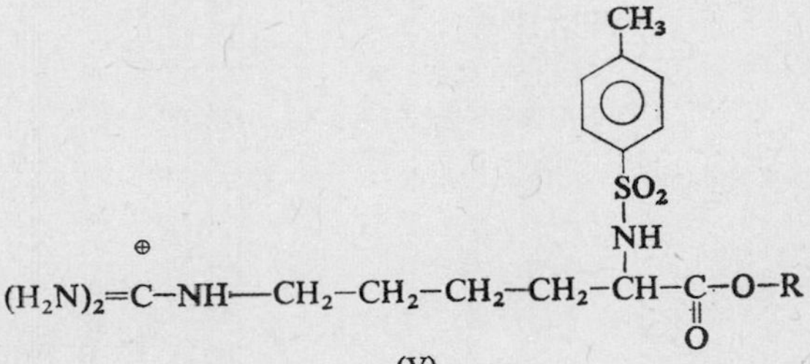

(V)

TABLE 8

Kinetic Parameters of Trypsin-catalyzed Hydrolysis of Various Ester Substrates

Substrates		k_{cat} (sec^{-1})	K_m(app) (μM)
N^{α}-Tosyl-L-arginine†[1]	methyl ester	75.0	6.4
	ethyl ester	74.1	—
	1-propyl ester	76.6	6.6
	cyclohexyl ester	76.9	8.2
N^{α}-Tosyl-L-homoarginine†[1]	methyl ester	4.04	330
	ethyl ester	4.08	290
	1-propyl ester	4.26	260
	cyclohexyl ester	3.46	240
N^{α}-Benzyloxy-carbonyl-L-lysine†[2]	*p*-nitrophenyl ester	6.79	10
	methyl ester	6.47	290
	benzyl ester	6.18	100
N^{α}-Benzoyl-L-lysine†[3]	methyl ester	16.6	17
	cyclohexyl ester	16.5	0.66
N^{α}-Tosyl-L-lysine†[3]	methyl ester	67.2	42
	ethyl ester	67.8	36
	1-propyl ester	69.2	22
	β-naphthyl ester	67.6	7.8
N^{α}-Benzoyl-L-thialysine†[3]	methyl ester	20.6	94
	cyclohexyl ester	22.0	29
N^{α}-Tosyl-L-thialysine†[3]	methyl ester	54.8	220
	ethyl ester	55.6	370
	1-propyl ester	56.7	200

†[1] pH 8.4, 25 °C (ref. 103). †[2] pH 5.8, 25 °C (ref. 105). †[3] pH 8.0, 25 °C (ref. 104).

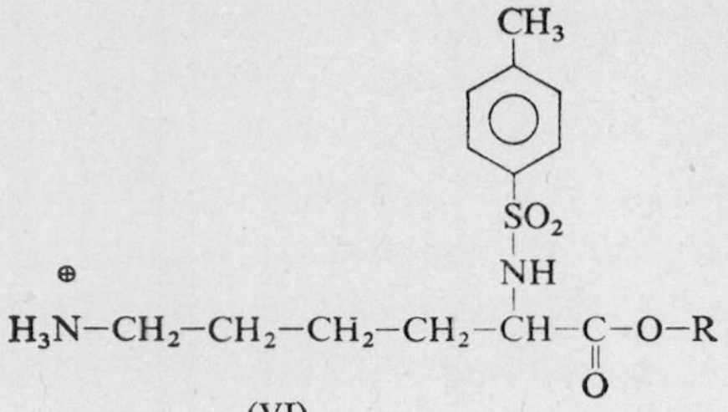

$$\overset{\oplus}{H_3N}-CH_2-CH_2-CH_2-CH_2-\underset{}{CH}(NH-SO_2-C_6H_4-CH_3)-\overset{O}{\overset{\|}{C}}-O-R$$

(VI)

$$\overset{\oplus}{H_3N}-CH_2-CH_2-S-CH_2-CH(NH-SO_2-C_6H_4-CH_3)-\overset{O}{\overset{\|}{C}}-O-R$$

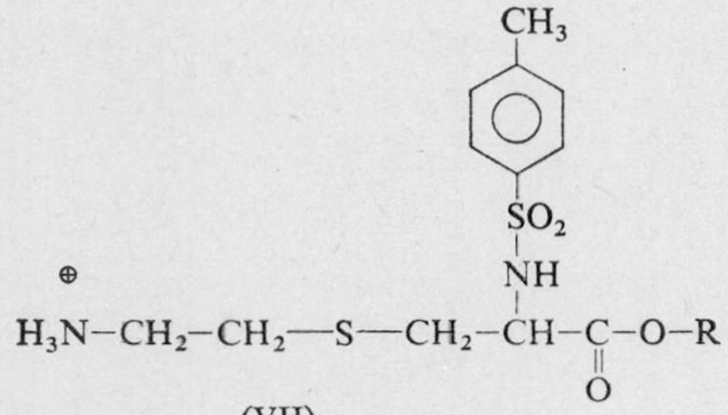

(VII)

Furthermore, Bender and Kezdy[105] showed that the rate of hydrolysis of the labile *p*-nitrophenyl ester of N^{α}-benzyloxycarbonyl-L-lysine was almost identical with those of various stable alkyl esters (Table 8). These results were interpreted as for chymotrypsin, namely that a common acyl-enzyme intermediate was formed and that its subsequent breakdown was the rate limiting step.

3.1.3. Spectrophotometric Observation of Acyl-enzyme Intermediates

In addition to a labile ester, NPA, numerous reactive derivatives of carboxylic acids including other nitrophenyl esters of various acids[106,107] anhydrides,[86] acid chloride,[86] and *N*-acylimidazoles[108] were shown to acylate the active site of chymotrypsin. Furthermore, Bender and Zerner[109] were able to demonstrate the formation of an acyl-chymotrypsin intermediate from a nonlabile ester, *trans*-methyl cinnamate (VIII). The formation of *trans*-cinnamoyl-chymotrypsin could be followed spectrophotometrically by measuring the rapid

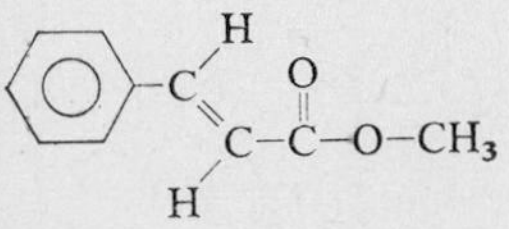

$$C_6H_5-CH=CH-\overset{O}{\overset{\|}{C}}-O-CH_3$$

(VIII)

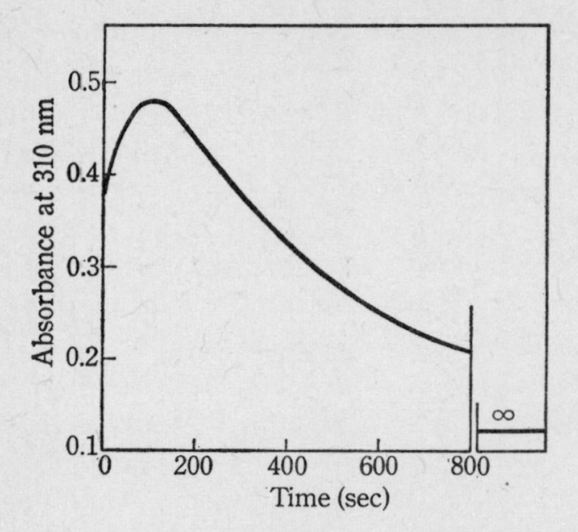

Fig. 4. The chymotrypsin-catalyzed hydrolysis of methyl *trans*-cinnamate. $[E]_0$ = 1.33 mM; $[S]_0$ = 0.12 mM; pH 7.8, 25°C, 3.2% CH_3CN.

increase in absorbance at 310 nm, and, as shown in Fig. 4, this was followed by a slow decrease of absorbance interpreted as a result of the deacylation reaction. This conclusion was substantiated by the fact that the first order rate constants for the decay of the absorbance obtained using methyl, *o*-, *m*-, and *p*-nitrophenyl cinnamates, as well as *trans*-cinnamoylimidazole, were practically identical[110] (see Table 9).

TABLE 9

Chymotrypsin and Trypsin Catalyses of Hydrolysis of *trans*-Cinnamic Acid Derivatives

$$\phi\text{-CH=CH-}\overset{\displaystyle O}{\overset{\|}{C}}\text{-X}$$

X refers to	Chymotrypsin		Trypsin
	$k_2(sec^{-1})$	$k_3(sec^{-1})$	$k_3(sec^{-1})$
Methyl ester	0.024	0.0125†1	
o-Nitrophenyl ester		0.0125†2	
m-Nitrophenyl ester		0.0125†2	
p-Nitrophenyl ester		0.0125†2	
Imidazole		0.011†2	0.014†3

†1 pH 7.9, 25°C, 3.2% acetonitrile, ref. 109.

†2 pH 9, 25°C, 1.6–3.2% acetonitrile, ref. 110.

†3 pH 9.0, 25°C, 1.6% acetonitrile, ref. 111.

This may be considered as additional evidence for the occurrence of a common rate limiting step and for the acyl-enzyme intermediate.

Bender and Kaiser found that trypsin also reacted with *N-trans*-cinnamoylimidazole in a fashion closely resembling chymotrypsin.[111] The acylation reaction was considerably slower than in the case of chymotrypsin, but, as shown in Table 9, the rate of the deacylation reaction was practically identical with that produced using chymotrypsin, which suggests that the catalytic mechanism of the two enzymes may be closely related although the mechanism of formation of the initial enzyme-substrate complex may be somewhat different.

The most interesting part of the work carried out by Bender and his co-workers with *trans*-cinnamoyl derivatives as substrate was the fact that the acyl-enzyme intermediates could be observed by virtue of their characteristic ultraviolet absorption peaks. The ultraviolet absorption spectrum of the phenyl ring is sensitively affected by the electronic state of the bond involving the carboxyl group due to the double bond which conjugates the phenyl and the carboxyl groups. By a difference spectrophotometric technique in which enzyme and substrate or enzyme and product were placed in separate tandem cells in the reference beam, the ultraviolet absorption spectrum of the acyl-enzyme intermediate was obtained. These curves were of almost symmetrical bell shape and their spectral characteristics are outlined in Table 10,[110,111] together with spectral characteristics of model compounds. The difference spectra were the same whether determined at pH 4 with the mixture of enzyme and substrate or after isolation of the acyl-enzyme intermediate. Degradative experiments with acylated or phosphorylated enzyme have always shown the presence of the acyl or DIP group on the hydroxyl group of the one particular active serine group. There was no definite evidence that acylation of the hydroxyl group was involved in the normal course of catalysis and that acylation of imidazole did not occur. In fact, there has always been a residual suspicion that one of the imidazole nitrogen atoms is first acylated and that the labile acyl group is subsequently transferred to the hydroxyl group during the degradative manipulation. It was hoped, therefore, that the spectrophotometric work on the cinnamoyl enzymes by Bender and his co-workers would resolve this important problem.

The data of Table 10 indicate that the wave lengths of the absorption maxima of the cinnamoyl enzymes are not closely similar to the values

TABLE 10
Absorption Spectra of *trans*-Cinnamoyl-Trypsin, Chymotrypsin and Model Compounds†

Compounds	Water		10 M LiCl	Isooctane
	λ_{max} (nm)	ε_{max}	λ_{max} (nm)	λ_{max} (nm)
Cinnamoyl trypsin	296	19,300		
Cinnamoyl chymotrypsin	292	17,700		
N-Cinnamoylimidazole	307	25,200	318	295
O-Cinnamoyl-*N*-acetyl-serinamide	281.5	24,300	286	275
O-Cinnamoyl-*N*-acetyl-tyrosinamide	285	27,900	290	
Methyl cinnamate	279.5	22,100		
Cinnamate ion	269.5	20,300		
Cinnamic acid	278	21,500		
Cinnamaldehyde	292	24,400		278

† Ref. 110 and 111.

for cinnamoyl imidazole or *O*-cinnamoyl serine, and consequently the site of the acyl group remained inconclusive. However, if cinnamoyl chymotrypsin was first denatured in 7.74 M urea at a low pH where it was stable and the solution then adjusted to pH 12–13, the rate of the subsequent hydroxide ion-catalyzed reaction was almost indistinguishable from that of a model compound *O*-cinnamoyl-*N*-acetylserinamide.[110] If the cinnamoyl group was attached to the imidazole nitrogen it should have been lost very rapidly. Thus, although definite information about the transient state during the actual catalysis has not yet been obtained, it is more likely that the cinnamoyl group is attached to the serine hydroxyl group. This group, however, may be in a perturbed state due to interaction with neighboring groups, and such perturbation may explain the absorption maxima occurring at wave lengths considerably higher than those of model esters.

3.1.4. Inhibition by Esterification of the Active Serine Hydroxyl Group

As discussed in section 3.1.1, the active serine hydroxyl group is acetylated in the course of hydrolysis of NPA, then deacylated slowly in the subsequent step, and so the enzyme turns over to serve as catalyst.

Methyl cinnamate is hydrolyzed in a similar way. On the other hand, DFP reacts very rapidly with the hydroxyl group to form a phosphate ester but the subsequent hydrolysis step is too slow to be observed, and consequently DFP is considered to be an inhibitor. Cunningham[112] was able to show that the DIP group could be removed from the inhibited chymotrypsin, and Cohen and Erlanger's detailed studies[113,114] indicated that the inactivated chymotrypsin and trypsin could be dephosphorylated by standing in aqueous solution. The difference between substrate and inhibitor may therefore be considered in terms of a quantitative difference in the rate of deacylation, whereas both may have reasonably high rates of acylation. In other words, the inactivated enzyme is a frozen state of the acyl-enzyme intermediate. A number of inhibitors which behave like DFP have been discovered in addition to various phosphoryl halides analogous to DFP.

Methane sulfonyl fluoride is known to be a potent inhibitor of acetylcholinesterase,[115] and 1-dimethylaminonaphthalene-5-sulfonyl chloride (IX) has been shown by Massey *et al.*[116] to inhibit chymotrypsin. Fahrney and Gold[61,117] demonstrated that various sulfonyl fluorides such as phenylmethane- (X), 2-phenylethane-1- (XI), benzene- (XII), 2-methylpropane-1- (XIII), and methane- (XIV) sulfonyl fluorides rapidly inhibited chymotrypsin and trypsin by a stoichiometric sulfonylation of the enzymes.

$(CH_3)_2N-C_{10}H_6-SO_2Cl$ (IX)

$R-SO_2-F$

R= $C_6H_5-CH_2-$ (X), $C_6H_5-CH_2-CH_2-$ (XI), C_6H_5- (XII), $(CH_3)_2CH-CH_2-$ (XIII), CH_3- (XIV)

The site of sulfonylation was thought to be the hydroxyl group of the active serine residue. The second-order rate constants of the sulfonylation are compared for the two enzymes in Table 11. Aryl sulfonyl halides reacted more favorably with chymotrypsin than trypsin probably due to selective affinity. The rapid sulfonylation did not occur in 8 M urea and was inhibited by competitive inhibitors. Interestingly

TABLE 11

Second Order Rate Constant, k'', for Inactivation of Trypsin and Chymotrypsin by Sulfonylation, Phosphorylation, and Carbamylation

Inhibitor	k'' ($M^{-1}sec^{-1}$) Trypsin	Chymotrypsin
Sulfonylfluoride[†1]		
Phenylmethane- (X)	271	14,900
2-Phenylethane-1- (XI)	120	6780
Benzene- (XII)	14.7	238
2-Methylpropane- (XIII)	4.6	72
Methane- (XIV)	0.75	1.3
Diisopropylfluorophosphate[†1] (I)	300	2700
Diphenylcarbamyl chloride[†2] (XV)	8.2	610
Ethyl *p*-guanidinobenzoate[†3] (XVIII)	6.93	—

[†1] Ref. 117. Conditions: 25°C, pH 7.2 for trypsin, pH 7.0 for chymotrypsin.
[†2] Ref. 118. Conditions: 25°C, pH 7.8. [†3] Ref. 99. Conditions: 20°C, pH 8.15.

enough the sulfonyl group was remarkably stable at neutral pH, whereas in 8 M urea desulfonylation readily took place. Obviously the rapid sulfonylation required the integrity of the catalysis site. The desulfonylation was infinitely slower than the decarboxylation reaction in the course of catalysis of a normal substrate, probably due to an extensive perturbation of the intricate conformation of the catalysis site by the sulfonyl group.

Erlanger and his co-workers[118,119] reported that diphenylcarbamyl chloride (DPCC) (XV) inhibited trypsin and chymotrypsin by forming stable diphenylcarbamyl enzymes by stoichiometric reactions. Again the hydroxyl group of the active serine residue seemed to be the most probable site of the carbamylation. The rates of this acylation for the two enzymes are given in Table 11. Faster inactivation of chymotrypsin than trypsin was explained by the aromatic structure of DPCC. The deacylation was negligible at neutral pH. The acylation, therefore, leads to inactivation of the enzymes.

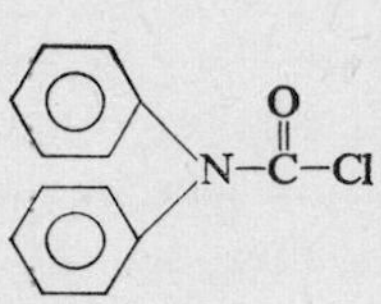

(XV) 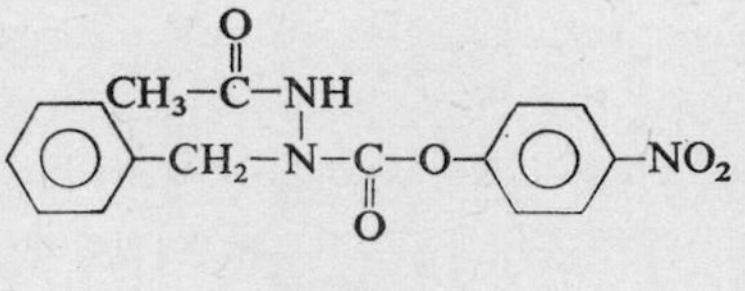 (XVI)

Elmore and Smyth[120] reported another acylating inhibitor, *p*-nitrophenyl N^2-acetyl-N^1-benzyl carbazate (NPABC) (XVI). It inhibited chymotrypsin almost instantaneously at pH 7.0 and 25°C and reacted with trypsin somewhat more slowly. In both reactions stoichiometric amounts of *p*-nitrophenol were released, thus making it possible to titrate the active site of the enzymes.

Recently Shaw and his co-workers synthesized *p*-nitrophenyl *p*-guanidinobenzoate (XVII)[59] and ethyl *p*-guanidinobenzoate (XVIII).[121] These compounds were subsequently found to inhibit trypsin specifically instead of reacting as substrates.

$$(NH_2)_2C^{\oplus}-NH-C_6H_4-C(=O)-O-R \qquad R = -C_6H_4-NO_2 \ \text{(XVII)} \qquad -C_2H_5 \ \text{(XVIII)}$$

$$(NH_2)_2C^{\oplus}-NH-C_6H_4-CH_2-C(=O)-O-C_2H_5 \quad \text{(XIX)}$$

$$(NH_2)_2C^{\oplus}-C_6H_4-C(=O)-O-CH_3 \quad \text{(XX)}$$

Since both compounds contain a positively charged guanidino group, and the distance between the guanidino group and the ester linkage is approximately equal to that in the synthetic arginine substrate, they were expected to behave as substrates. However, they inhibited trypsin. Inhibition by the *p*-nitrophenyl ester was accompanied by an instantaneous, stoichiometric release of *p*-nitrophenylate ion with concomitant *p*-guanidinobenzoylation of the catalysis site, most probably the serine hydroxyl group. The deacylation was extremely slow, but could be facilitated by hydroxylamine, and this reaction was utilized for titrating the active site of trypsin. Such diminished reactivity in the deacylation step may be due to perturbation of the catalysis site caused by interaction of the specificity site with the benzene ring of the inhibitor. Slight changes in its structure, however, were found to convert the inhibitor to substrates. Thus Mares-Guia *et al.*[122] found that ethyl *p*-guanidinophenylacetate (XIX) and methyl *p*-amidinobenzoate (XX) were continuously hydrolyzed by trypsin. Also Tanizawa *et al.*[123] discovered that the ethyl- and *p*-nitrophenyl esters of *p*-amidinobenzoic acid were substrates of trypsin. After a rapid acylation, the acyl-enzyme intermediate was hydrolyzed at the considerable rate of $k_{cat} = 0.04$ sec^{-1} at pH 8.2 and 25°C. The increase of approximately 1.5 Å between the guanidino and ester groups due to the additional nitrogen atom in the *p*-guanidinobenzoate derivatives caused a conspicuous difference in the pattern or strength of interaction with the enzyme. Mix *et al.*[124] reported syntheses and inhibitory effects of benzyl and *p*-nitrobenzyl esters of *p*-guanidinobenzoic acid. An analogous situation was encountered by Bender and Hamilton[125] when *p*-nitrophenyl trimethylacetate was reacted with chymotrypsin. The trimethylacetyl group attached to chymotrypsin was extremely stable.

The relationship between the rate of acylation and deacylation depends on the type of substrate and varies over a wide range as discussed in section 3.1.2. In the case of a normal amide substrate the rate constant for acylation, k_2, is smaller than that for deacylation, k_3, whereas for a specific ester substrate k_2 may be higher than k_3 by 1 to 3 orders of magnitude. The inhibition discussed in this section may be understood as an extreme case of such a relationship where k_3 is incomparably lower than k_2. Thus the inhibited enzyme may be considered as a model of the acyl-enzyme intermediate in the normal catalysis.

3.1.5. Effect of Hydroxylamine on the Rate of Catalysis

Hydroxylamine was used by Bernhard *et al.*[126] as a nucleophile to compete with water in the chymotrypsin-catalyzed decomposition of methyl hippurate. In such concomitant hydroxylaminolysis and hydrolysis, part of the acyl-enzyme is converted to the hydroxamic acid of the acyl moiety by the following sequence of reactions:

$$E + S \underset{k_{-1}}{\overset{k_1}{\rightleftharpoons}} ES' \xrightarrow{k_2} \begin{matrix} ES'' \\ + \\ P' \end{matrix} \left\{ \begin{array}{l} \xrightarrow[H_2O]{k_3^{H_2O}} E + RCOOH \\ \xrightarrow[NH_2OH]{k_3^{NH_2OH}} E + RCONHOH \end{array} \right. \qquad (10)$$

where RCOOH is the product of hydrolysis and RCONHOH is the hydroxamic acid. Epand and Wilson[127] were able to show that the ratio of the rates of the hydroxylaminolysis to that of the hydrolysis were identical with ten different esters of hippuric acid. Methyl, ethyl, isopropyl, isobutyl, choline, homocholine, (4-pyridyl) methyl, isoamyl, benzyl, and glyceryl esters were used. Of all these substrates 36–38% were converted to hippuryl hydroxamate by 0.1 M hydroxylamine at pH 6.6 and 25°C. This observation could not be explained unless a common acyl-enzyme intermediate was postulated. Inagami and Sturtevant[128] made use of the fact that the chymotrypsin-catalyzed hydrolysis of an amide or anilide substrate had a rate limiting step at the acylation reaction and examined the effect of hydroxylamine on the hydrolysis of *N*-acetyl-L-tyrosine *p*-nitroanilide by chymotrypsin. Hydroxylamine to 1.6 M did not affect the overall rate of disappearance of substrate although almost half the product was converted to the hydroxamic acid of *N*-acetyl-L-tyrosine. This result could again be explained only when formation of the acyl-enzyme intermediate was postulated. Caplow and Jencks[129,130] obtained a result which did not seem to be compatible with the presence of an acyl-enzyme intermediate from the study of chymotrypsin-catalyzed concurrent hydrolysis and hydroxylaminolysis. However, this could be explained by the fact that the ratio $k_3^{NH_2OH} / k_3^{H_2O}$ depends on the concentration of enzyme.[131]

Inagami[132] applied the hydroxylaminolysis technique to the trypsin-catalyzed hydrolysis of an anilide substrate, N^α-benzyloxycarbonyl-L-arginine *p*-toluidide. The rate of disappearance of the substrate due to concurrent hydrolysis and hydroxylaminolysis, determined by the rate of release of free *p*-toluidine, was found to be unaffected by the presence and absence of hydroxylamine, again supporting the acyl-enzyme hypothesis for trypsin. Since most of the convincing evidence for the acyl-enzyme intermediate has been obtained with nonspecific or abnormal ester substrates, there has always been a suspicion that hydrolysis of specific peptide or amide substrates may not proceed by way of the acyl-enzyme intermediate, formed by esterification of the active serine hydroxyl group. Therefore the results obtained with the specific substrates for both trypsin and chymotrypsin are significant for generalization of the mechanism.

Alcohols of small molecular size such as methanol have also been used, like hydroxylamine, as nucleophiles to compete with water. Kinetic analyses of the chymotrypsin-catalyzed hydrolyses of methyl and ethyl hippurate in the presence of methanol made by Bender *et al.*[133] also suggested the presence of an acyl-enzyme intermediate. Similar acyl transfer from the acyl-enzyme to alcohol by trypsin was demonstrated by Glazer.[134] Lysine methyl ester was partially converted to ethyl, propyl, butyl and ethylene glycol esters when the substrate was incubated with trypsin in the presence of the respective alcohols. Although this observation of trans-esterification may not alone be convincing evidence for an acyl-enzyme intermediate, it strongly suggests the presence of such an intermediate.

3.1.6. Comments

As can be seen from the preceding discussion, a large amount of evidence, both direct and suggestive, has accumulated which points to the acyl-enzyme intermediate. Nevertheless, doubts regarding the generality of the hypothesis, and especially the exact nature of the bond formed in the intermediate, have often been expressed. Since much of the experimental evidence has been derived from results obtained with either labile or nonspecific substrates, or after denaturation of the acyl-enzyme, it has been frequently argued that in the actual catalysis the acyl group in the intermediate may not be localized on the hydroxyl group of the active serine residue but may be in a more labile state. The fact that the spectral characteristics of the cinnamoyl-enzyme do not agree either with those of typical aliphatic cinnamoyl esters or *N*-cinnamoylimidazole is one point frequently employed in such arguments, although it is recognized that the shift in the wave length of the absorption maximum may have been caused possibly by non-bonding interaction between the chromophore of the cinnamoyl group and the specificity site of the enzyme. Poor design of experiments or misinterpretation of results has occasionally resulted in dismissal of the intermediate concept but in any case the precise chemical nature of the acyl-enzyme bond is as yet unknown.

Conclusion of this discussion would not be complete without mentioning the studies on ^{18}O-exchange of virtual substrates catalyzed by chymotrypsin. Although a corresponding experiment with trypsin has not to the author's knowledge been reported, the result obtained with chymotrypsin may be directly applied to trypsin, and will provide important information about the nature of the acyl-enzyme intermediate. Sprinson and Rittenberg[135] demonstrated chymotrypsin-catalyzed incorporation of ^{18}O from $H_2{}^{18}O$ into the carbonyl oxygen atoms of virtual substrates, *N*-benzyloxycarbonyl-L-phenylalanine, *N*-acetyldibromo-L-tyrosine and N^{α}-acetyl-L-tryptophan. Subsequent studies[136-9] indicated that the exchange reaction proceeded by Michaelis-Menten type kinetics with only L-optical isomers and that the K_m value was identical with the dissociation constant of the compounds when determined by equilibrium dialysis or by competitive inhibition against hydrolytic reaction. Since methylphenylpropionate carbonyl-^{18}O and *N*-benzoyl-L-phenylalanine ethyl ester carbonyl-^{18}O did not exchange carbonyl oxygens with the media in the chymotrypsin-catalyzed hydrolysis, as shown by Eq. (11), a mechanism

$$R-C(={}^{18}O)-O-X \xrightarrow{Enz.} R-C(={}^{18}O)-Enz. \xrightarrow{H_2O} R-C\langle^{O\frac{1}{2}{}^{18}O}_{O\frac{1}{2}{}^{18}O}\ominus \tag{11}$$

involving a tetrahedral intermediate such as Eq. (12) was not a plausible explanation.

$$R-C\langle^{O}_{O}\ominus + H_2{}^{18}O \xrightarrow{Enz.} \left[R-C(OH)(^{18}OH)-OH \cdot Enz.\right] \longrightarrow R-C\langle^{O\frac{1}{2}{}^{18}O}_{O\frac{1}{2}{}^{18}O}\ominus \tag{12}$$

Therefore, even in the enzyme-catalyzed exchange reaction of the carbonyl group, a mechanism involving the acyl-enzyme intermediate such as Eq. (13) was the only reasonable explanation of the exchange reaction.

$$R-C\langle^{O}_{O}\ominus \xrightarrow{Enz.} R-C(=O)-Enz. \xrightarrow{H_2{}^{18}O} R-C\langle^{O\frac{1}{2}{}^{18}O}_{O\frac{1}{2}{}^{18}O}\ominus \tag{13}$$

Another result with chymotrypsin but not extended to trypsin was obtained by Kezdy *et al.*[97] who were able to demonstrate spectrophotometrically the formation of an acyl-chymotrypsin with N^α-acetyl-L-tryptophan (note, not the ester) at pH 2.4, where the carboxyl group of tryptophan was in the un-ionized form. The spectrum obtained after acylation of chymotrypsin by N^α-acetyl-L-tryptophan methyl ester in the presence of excess enzyme, was found to be identical with that obtained after the enzyme had reacted with N^α-acetyl-L-tryptophan. Since it was reported that the reaction of the methyl ester led to the formation of an acyl-enzyme intermediate which remained unhydrolyzed at the acid pH, it was concluded that the free acid also formed the same intermediate.

3.2. Initial Michaelis-complex Formation

Thus far we have exhaustively examined evidence for the formation of an acyl-enzyme intermediate. Although the precise nature of the bond between enzyme and substrate in this intermediate has yet to be carefully examined, there seems to be sufficient evidence for the existence of the intermediate. Consequently the three-step mechanism of Eq. (3) will be taken as a basis for subsequent discussion in which studies on the first, second, and third steps in this order, will be surveyed. Gutfreund[139] was the first worker to apply a stopped flow technique to the study of trypsin catalysis in an attempt to detect the enzyme-substrate complex. Using BAEE as substrate, a 40-fold excess of trypsin, and *p*-nitrophenol as indicator to detect the reaction product, N^{α}-benzoyl-L-arginine, he was unable to observe a transient pre-steady state reaction, although such a reaction was observed with ficin. He explained this by postulating a fast reaction which was completed in a period shorter than the resolution time of his instrument (5 msec). From this resolution time the minimum value of k_1 was estimated to be 4×10^6 sec^{-1}M^{-1}.

Johannin and Yon[140] applied a similar stopped flow technique to the hydrolysis of a chromogenic substrate, N^{α}-benzoyl-D,L-arginine *p*-nitroanilide (BANA).[141–3] BANA and its hydrolysis product, *p*-nitroaniline plus N^{α}-benzoylarginine exhibited an isosbestic point at 341 nm. When absorbance at this isosbestic wave length was observed with a mixture containing a 3-fold excess of trypsin over the substrate, a biphasic change shown by a solid line in Fig. 5 was observed. The rate constant

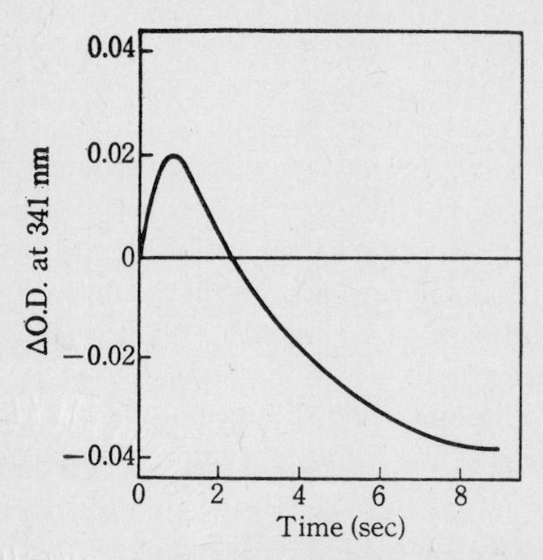

Fig. 5. Optical density change at 341 nm during the hydrolysis of N^{α}-benzoyl-D,L-arginine *p*-nitroanilide.

for the slower phase agreed with that of acylation. The formation of the initial Michaelis complex is usually diffusion limited and is much faster than the resolution time of a stopped flow apparatus as reported by Gutfruend.[139] Therefore the faster phase which peaked at about $t=1$ sec should represent the formation of an additional intermediate X in the following reaction sequence.

$$E + S \rightleftharpoons ES' \rightleftharpoons X \longrightarrow \underset{\displaystyle P'}{\overset{\displaystyle ES''}{+}} \longrightarrow E + P'' \tag{14}$$

It was suggested that the spectral difference between ES' and X might be due to conformational changes in the enzyme caused by substrate binding.

A further new technique, which allowed individual fast reaction steps in the hydrolysis of specific substrates to be followed, was developed by Bernhard, Gutfreund and co-workers. Proflavin (XXI) had been known to bind to trypsin and chymotrypsin with a considerable spectral change

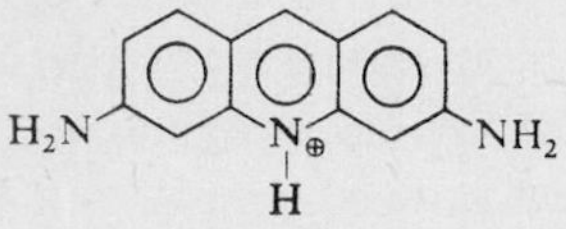

H_2N — $N^{\oplus}$—H — NH_2

(XXI)

in the visible wave length region. Since the dye could be displaced from these enzymes by their specific substrates, the binding was thought to occur in the area of the active site. Binding to the enzymes and release could be conveniently followed by the change in absorbance at 480 nm. When the reaction between equimolar amounts of BAEE and trypsin in the presence of proflavin was studied by a stopped flow apparatus, a rapid release of proflavin with a first-order rate constant of 150~200 sec^{-1} was observed.[146] In the meantime Barman and Gutfreund[147] had devised a technique which enabled a reaction to be quenched rapidly a few milliseconds after mixing the two reactant solutions. This method was applied to trypsin catalysis whereby equimolar quantities of trypsin and BAEE were mixed and quenched, after 3 to 70 msec of reaction, by a buffer of pH 2. When the quantity of ethyl alcohol formed by tryptic hydrolysis of BAEE in the quenched mixture was determined, it was not large enough to account for the amount of proflavin released, indicating that an additional intermediate such as X in Eq. (14) might have been formed.[148] Further investigation of the relationship between the kinetics of the proflavin release and the formation of ethyl alcohol, revealed that the reaction preceding the acylation step was of a complex nature and suggested involvement of more than one intermediate between the Michaelis complex and the acyl-enzyme intermediate.[149]

3.3. Mechanism of the Acylation Step

3.3.1. Proton Transfer Mechanism

The numerous studies on the acyl-enzyme intermediate reviewed in sections 3.1.1 through 3.1.6 indicated that the hydroxyl group of Ser-183 of trypsin, or Ser-195 of chymotrypsin was the site of acylation. One might question the role played by other functional groups such as His-46 of trypsin or His-57 of chymotrypsin. Valuable contributions to the elucidation of this central problem of chymotrypsin catalysis have come from X-ray crystallographic studies and also from investigations of the hydrolysis of specific anilide substrates. As outlined in sections 2.4.1 through 2.4.3 and 3.1.1 through 3.1.6, the structure of the catalysis sites of trypsin and chymotrypsin are almost identical as is the manner in which the acyl-enzyme intermediate is formed. Therefore, a review of the results obtained with chymotrypsin should contribute to understanding the catalytic mechanism of trypsin.

Study of the acylation by steady state kinetics required substrates whose rate of hydrolysis was limited at the acylation step. Many ester substrates such as ATEE whose hydrolysis could be followed easily and accurately by a pH-stat method were of limited value here since the hydrolysis of such specific ester substrates were rate-limited at the deacylation step. Amide substrates including N^{α}-acetyl-L-tyrosinamide (ATA) or N^{α}-acetyl-L-tryptophanamide (ATrA) were known to be hydrolyzed much slower than related esters,[98–100] and consequently these reactions were thought to be rate-limited at the acylation reaction. Since hydrolysis of an amide was not easy to follow with great accuracy, corresponding anilides were employed. Hydrolysis of an anilide can be followed titrimetrically using a pH-stat system, and in addition, a further advantage is that various substituents can be introduced into the aniline moiety and the electronic effect of these on the enzyme catalysis investigated.

Sager and Parks[150,151] using *meta*- and *para*-substituted anilides of *N*-benzoyl-L-tyrosine (XXII) as substrates for chymotrypsin discovered that the apparent second order rate constant, k_2/K_m, followed Hammett's equation[152] with a negative ρ value suggesting acid catalysis. The reaction could not be analyzed by the Michaelis-Menten kinetics due to poor solubility of the highly hydrophobic substrates. Consequently the first order rate constant, k_2, and the true Michaelis constant, K_m, could not be obtained separately. Using the more water-soluble *meta*- and *para*-substituted anilides of *N*-acetyl-L-tyrosine (XXIII),

$$HO-C_6H_4-CH_2-CH(NH-CO-C_6H_5)-CO-NH-C_6H_4-X$$

X: m–CH_3O–, p–CH_3O–, H–

(XXII)

$$HO-C_6H_4-CH_2-CH(NH-CO-CH_3)-CO-NH-C_6H_4-Y$$

Y: p–NO_2–, m–Cl–, p–Cl–, p–CH_3–, m–CH_3O–, p–CH_3O–

(XXIII)

Inagami *et al.*[153,154] were able to measure these constants. The k_2 value was found to be strongly affected by substituents on the leaving group aniline. This was strong evidence that the acylation, being slower than the deacylation, was the rate-limiting step in hydrolysis of the anilide substrates. The plot of $\log k_2$ against Hammett's σ[154] (a measure of the electron attracting power of the substituent) gave a straight line of slope (Hammett's ρ value) -2.0, as shown in Fig. 6. The negative ρ value indicated that the more strongly the electron was removed from the bond cleaved, by the electron-withdrawing increasing influence of the substituent, the slower the reaction became. This in turn indicated that acid catalysis by the enzyme played a dominant role in the acylation reaction. Proton transfer from the enzyme to either the anilide N-atom or the carbonyl O-atom should control the overall rate of catalysis. However, proton transfer along a hydrogen-bonded path has been shown by Eigen and De Maeyer[155] to proceed incomparably faster than a normal organic reaction. From this, Wang and Parker[156,157] estimated that the

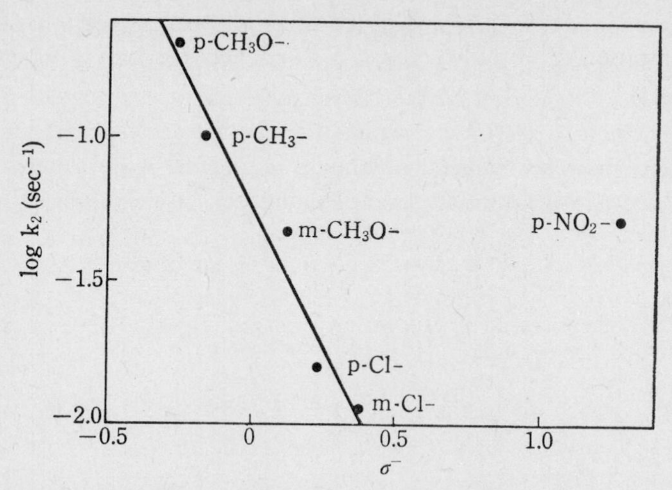

Fig. 6. Hammett's ρ-σ plot of the first-order rate constant, k_2, of the chymotrypsin-catalyzed hydrolysis of *meta*- and *para*-substituted anilides of *N*-acetyl-L-tyrosine at pH 8 and 25°C in 5% dimethylformamide.[153,154]

rate of the proton transfer was too fast to be rate-limiting in the transition state of the acylation reaction. In order to explain the dependence of the rate of acylation on the acidity of the anilides, they postulated that the proton transfer was a rapid equilibrium process established prior to and prerequisite to the transition state, with the substrate molecules in a Michaelis complex indicated by A and B in the following scheme.

His—57
H—N N···H—O—Ser—195
H\N—C(=O)—R / R′
A

$\underset{k_b}{\overset{k_f}{\rightleftharpoons}}$

His—57
H—N N ···H ⊖O—Ser—195
H\N⁺(R′)—C(=O)—R
B (15)

$\xrightarrow{k_a}$

His—57
H—N N
R′—NH₂
O—Ser—195
O=C—R
C

The subsequent acylation reaction leading to the acyl enzyme intermediate, C, was postulated as the rate-limiting step. Two basic assumptions were made: (a) the rate of the acylation was proportional to the concentration of B but k_a was a constant independent of the substituents:

$$v = k_a [B] \tag{16}$$

(b) the proton was distributed between the hydroxyl group of Ser-195 of chymotrypsin and the anilide N-atom of the substrates in proportion to the basicity of the two species, $\mathrm{R{-}\overset{\overset{\displaystyle O}{\|}}{C}{-}\overset{\overset{\displaystyle H}{|}}{N}{-}R'}$ and Ser-O$^{\ominus}$, in their free state. In other words, the equilibrium constant K was given by Eq. (17).

$$K = \frac{[\mathrm{R{-}\overset{\overset{\displaystyle O}{\|}}{C}{-}\overset{\overset{\displaystyle H_2}{|}}{N^{\oplus}}{-}R'}]\,[\text{Ser}{-}\text{O}^{\ominus}]}{[\mathrm{R{-}\overset{\overset{\displaystyle O}{\|}}{C}{-}\overset{\overset{\displaystyle H}{|}}{N}{-}R'}]\,[\text{Ser}{-}\text{OH}]} = \frac{k_f}{k_b} \tag{17}$$

where k_f and k_b are as in scheme (15). Therefore K can be represented by the following relationship

$$K = (K_a)_{\text{Ser}} \,/\, (K_a)_{\text{Substrate-H}^{\oplus}}\,, \tag{18}$$

where $(K_a)_{\text{ser}}$ and $(K_a)_{\text{Substrate-H}^{\oplus}}$ represent the acid dissociation constants for the hydroxyl group of Ser-195 and the protonated substrate (the conjugate acid of the substrate). The latter constant was estimated using substituted acetanilides as model compounds. It is important to note that no specific influence by the enzyme on this protonation equilibrium, such as strain of the substrate molecule, was postulated. The protonation equilibrium was the same as that existing between the two species in free solution. The steady state condition gave the following

$$[B] \,/\, [A] = k_f \,/\, (k_b + k_a); \tag{19}$$

since $k_a \ll k_b$ in this case, Eq. (19) simplifies to Eq. (20).

$$[B] \,/\, [A] = k_f \,/\, k_b \tag{20}$$

From Eq. (16) to (18)

$$v = [A]\, k_a\, (K_a)_{\text{Ser}} \,/\, (K_a)_{\text{Substrate-H}^{\oplus}} \tag{21}$$

$$= [A]\, k_2\,, \tag{22}$$

where k_2 was the experimentally obtained rate constant of the acylation, which can be related to the acid dissociation constant of the conjugate acid of the substrate by Eq. (23).

$$k_2 = k_a\, (K_a)_{\text{Ser}} \,/\, (K_a)_{\text{Substrate-H}^{\oplus}} \tag{23}$$

Since k_a and $(K_a)_{\text{Ser}}$ are constant and independent of substituents

$$k_2 = \text{Constant} \,/\, (K_a)_{\text{Substrate-H}^{\oplus}} \qquad (24)$$

From Eq. (24) the relative rate constants of acylation, k_2, were calculated by making the absolute value for the *m*-Cl compound equal to the experimentally obtained value (0.011 sec^{-1}) so that "Constant" in Eq. (24) could be estimated. The k_2 values so calculated for all the substrates and the experimental results are compared in Table 12 and the agreement seems to be satisfactory. The mechanistic basis for the calculation, therefore, seems to be justified. It was assumed that the hydrogen bond between the substrate and Ser-195 facilitated proton transfer, and as stated by Wang and Parker,[156-8] additional postulates such as the strain of the substrate molecule or a push-pull or acid-base catalysis were not necessary. Note, however, that the calculated rates were relative values and no attempt was made to calculate the absolute value of the rate of enzyme catalysis to explain its high efficiency compared with non-enzymatic catalysis.

Closer studies of the pH-profile for chymotrypsin-catalyzed hydrolysis of anilide substrates by Caplow[159] and Inagami *et al.*[154] produced data which are contradictory to the observations of Parker and Wang.[158] The pH-dependence of k_{cat} as well as K_m was found to depend strongly on the electronegativity of the substituent groups on the aniline ring. Based on these observations, Caplow[159] has proposed a mechanism, as shown below (Eq. 25), involving an additional tetrahedral intermediate, ES*, between the Michaelis complex, ES′, and the acyl-enzyme intermediate ES″. The breakdown of this intermediate to the acyl-enzyme

TABLE 12

Comparison of Observed and Calculated Values of the Rate Constant for Acylation, k_2, in the Chymotrypsin-catalyzed Hydrolysis of *meta*- and *para*-Substituted Anilides of *N*-Acetyl-L-tyrosine†1

Substituents	k_2 (observed)†2 (sec^{-1})	k_2 (calculated) (sec^{-1})
m-Cl	0.011	(0.011)†3
p-Cl	0.014	0.016
m-CH_3O	0.047	0.037
p-CH_3	0.087	0.073
p-CH_3O	0.21	0.11

†1 Ref. 156. †2 Ref. 153. †3 Set to be equal to the observed.

intermediate with concomitant transfer of proton from imidazole to the leaving N-atom of aniline is proposed to be the rate limiting step. This mechanism accounts for the strong dependence of the pH-profile and K_m on the electronegativity of the leaving group as well as for the negative ρ value in the log k_{cat} *vs.* σ plot.

ES′ ES* ES″

HN N--H O Ser His N-H C=O R ⇌ HN (+) NH ⊖O C N-H O R (25)

⇌ HN (+) NH N-H ⊖O C O R —k_2→ HN N O C O R

Cunningham[160] proposed that the hydroxyl group of Ser-195 was made extraordinarily reactive by being hydrogen-bonded to an N-atom of a histidine residue. A number of similar mechanisms with hydrogen-bonded serine and histidine residues have been proposed to explain chymotrypsin catalysis.[86,94,157,161-4] Recent X-ray crystallographic work by Blow *et al.*[16] indicated that not only are the hydroxyl group of Ser-195 and the N-3 atom of His-57 hydrogen-bonded but the N-1 atom of His-57 and one of the β-carboxyl O-atoms of Asp-102 is also hydrogen bonded, as illustrated in Fig. 2. Using crystals of α-chymotrypsin and its monotosylated derivative equilibrated in a strong salt solution at pH 4.2, interatomic distances have been obtained to an accuracy of 0.4 Å, as presented in Table 5. At this resolution H-atoms are not visible in an electron density map. However, N- and O-atoms or two O-atoms separated by 3 Å or less can be considered to be linked by a hydrogen bond. It has also been found that Asp-102 and that half of the imidazole unit of His-57 closest to Asp-102 are made inaccessible to proton exchange from outside by adjacent hydrophobic side chains of Ala-55, Cys-58, and Ile-99, which fill up space within 4 Å of the β-carboxyl O-atom of Asp-102. Thus, as shown in scheme (26), the negative charge of Asp-102,

Asp–102
C–O
$O^{\ominus}$
H
N
His–57
N
H–O
Ser–195

pH8

Asp–102
C–O
O
H
N
His–57
N
H–O
$\ominus$
Ser–195

(26)

located inside the molecule, is channelled to the surface of the molecule at the O-atom of Ser-195. Incidentally such a negative charge, relayed to the surface, can repulse a negatively charged substrate or inhibitor as postulated by Johnson and Knowles.[164] By such a "charge relay mechanism" the oxygen atom of Ser-195 will become a much stronger nucleophile than when it is linked only to the imidazole of His-57. Since the proton of Ser-195 is less localized to Ser-195 in this system than in the Ser-His system, the proton transfer to the anilide or amide substrate as discussed above will be more facilitated since the existence of a significant proportion of Ser-195 in the form of $-CH_2O^{\ominus}$ can be justified by scheme (26).

The importance of histidine imidazole has been clearly demonstrated by the inactivation of chymotrypsin due to methylation of the imidazole nitrogen atom by methyl *p*-nitrobenzenesulfonate[75] as discussed above in section 2.4.2. Although the enzyme activity seemed to be almost completely destroyed by the methylation, a closer study by Henderson[165] revealed that the enzyme retained a residual activity 5–6 orders of magnitude less than that of native chymotrypsin. Moreover this activity was found to have a pH profile similar to that of the native enzyme. These observations are valuable in assessing the contribution of the imidazole group to the catalytic activity, and also provide an important clue to the origin of the sigmoidal pH-profile with a pK_a of about 7, which has simply been attributed to the ionization of the imidazole group.

From such considerations of the catalysis site, Blow *et al.*[16] presented a mechanism for the acylation step shown in scheme (27). Although the scheme looks like an ordinary acid-base catalysis in which Ser-$O^{\ominus}$ is a base catalyst and His-H is an acid catalyst, the above discussion of the recently proposed mechanisms makes it clear that the two catalytic species may not work in a simultaneous push-pull mechanism. However, it must be admitted

Asp–102 C–O O H N His–57 N Ser–195 H···O⊖ N–C R′ H O R $\xrightarrow{k_2}$ Asp–102 C–O O⊖ H N His–57 N Ser–195 H···O N C R′ H O R (27)

that the negative electrical charge on the hydroxyl group of the serine residue further facilitates the proton transfer.

A structure very closely resembling that in chymotrypsin exists in trypsin.[78] Not only is the charge relay system identical, but many groups in its periphery are quite similar. The hydroxyl group of Ser-198 (Ser-214 of chymotrypsin) is hydrogen-bonded to the β-oxygen atom involved in the relay of Asp-90 (Asp-102 in chymotrypsin). His-29 and the corresponding His-40 of chymotrypsin are in a similar environment. The *N*-terminal α-amino groups of isoleucine in the two enzymes appear to be in electrostatic interaction with the β-carboxyl group of Asp-182 of trypsin and the Asp-194 of chymotrypsin, respectively. Thus it is likely that the catalytic mechanisms of trypsin and chymotrypsin are almost identical, with minor quantitative differences.

3.3.2. The pH Profile of the Acylation Reaction

The above discussion of the acylation step in chymotrypsin catalysis indicated that an amide substrate is needed if steady state kinetics are to be used in the investigation of the acylation reaction, since the rate-limiting step of the amide hydrolysis occurs at the acylation step, which therefore produces the observed rate. An ester substrate is not useful in this case because the steady state rate of its hydrolysis is equal to the rate of deacylation (see section 3.1.1). The acylation reaction of trypsin has been studied with various amide substrates. Chevallier and Yon[166] followed the course of hydrolyses of N^{α}-benzoyl-L-argininamide (BAA) and N^{α}-*p*-toluenesulfonyl-L-argininamide (TAA) by absorbance change at wave lengths 257 nm and 229 nm, respectively, using a method similar

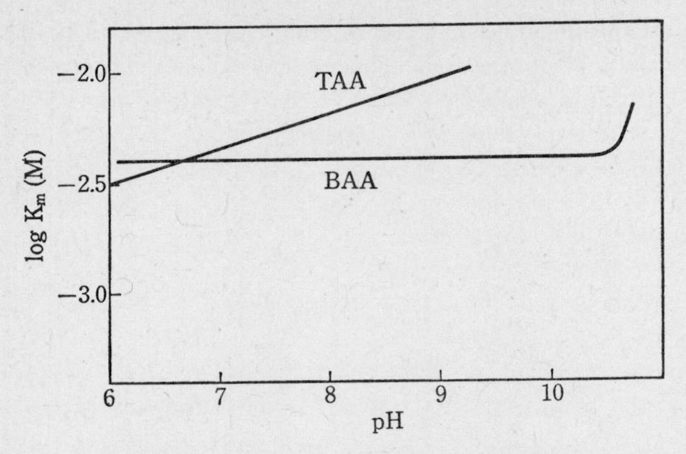

Fig. 7. Change in log K_m as a function of pH for the trypsin-catalyzed hydrolysis of N^{α}-benzoyl-L-argininamide (BAA) and N^{α}-tosyl-L-argininamide (TAA) at 35 °C.

to that of Schwert and Takenaka.[167] Changes in K_m(app) and k_{cat} with pH were followed at 35°C. (Note that since $k_2 \ll k_3$ in the amide hydrolysis, K_m(app) reduces to the true $K_m = (k_2 + k_{-1})/k_1$ as discussed in section 3.1.1, and that k_{cat} is k_2.) For BAA, K_m remains constant at 3.3 mM from pH 6 to 10, as shown in Fig. 7, in which log K_m is plotted against pH. It shows a sharp increase above pH 10.5 according to Wang and Carpenter.[168] The pH profiles of K_m of N^{α}-benzoyl-L-lysinamide (BLA)[168] and *N*-benzoyl-*S*-2-aminoethyl-L-cysteinamide (BAECA)[168] also showed similar behavior although the sharp rise occurred at pH values slightly below 10. Since the side chain amino groups of the last two substrates have pK_a values near 10, the pH profile of K_m discussed above may be explained by an ionic interaction between the positively charged side chain of the substrate molecule and the binding site of trypsin. However, the similar pH profile of K_m for BAA cannot be explained by the ionization of its guanidino group, whose pK_a is higher than 12. Since such interaction should require a certain conformation of the enzyme, it is more likely that the effect of the pH is due to a pH-dependent conformational change of the enzyme.

In the case of chymotrypsin, a similar sharp increase in K_m in alkaline solution was reported by Himoe and Hess.[169,170] This increase was found to occur at a pH considerably lower than that for trypsin and was attributed to disruption of the integrity of the active site due to loss of a proton from an ionizing group of pK_a 8.5, probably the α-amino group of the *N*-terminal isoleucine. X-ray crystallographic studies[76,171] revealed that the positively charged amino group of the isoleucine residue and the negatively charged β-carboxyl group of Asp-194 were in sufficiently close proximity to allow a strong coulombic interaction. Such an interaction seems to be essential for the maintenance of the active conformation of the enzyme protein. Scrimger and Hofmann's finding[36] that the chemical change of the *N*-terminal α-amino group of trypsin by nitrous acid completely destroyed the active conformation of the trypsin, renders support to the above argument.

As will be discussed in section 4.4.2, the denaturation was accompanied by the loss of ability to bind substrate and inhibitor. Such events were found to be under the control of an ionizing group with a pK_a of 10. Chevallier *et al.*[37] made observations which strongly suggested that the ionizing group was the α-amino group of the *N*-terminal isoleucine, and it is highly likely that the steep increase in K_m in the alkaline pH was due to denaturation controlled by deprotonation of this group. Binding of inhibitors or substrate analogs prevented the denaturation to a varying degree which depended on the structure and affinity of the inhibitor. Such variable influence of the enzyme-bound species may explain the somewhat different pH-profiles of K_m for different substrates. Assuming that the ionizing group is the α-amino group, the pK_a of this group in trypsin is approximately 10 compared with 8.5 in chymotrypsin (also see section 3.4). It is not understood why the α-amino group of the *N*-terminal isoleucine of trypsin should ionize at such a high pK_a value compared with the same group in chymotrypsin.

The pH-dependent change of k_2 for the amide substrates were also analyzed by Chevallier and Yon[166] and Wang and Carpenter.[168] Plots of log k_2 against pH according to Dixon[172] were made for the trypsin-catalyzed hydrolysis of BAA and TAA, as shown in Fig. 8. The intersection of the extrapolation of the acidic limb and the horizontal line from the plateau gave the pK_a value of the ionizing group responsible for the pH profile on the acid side, as shown in Table 13. The values for both BAA and TAA are not far from 7 and may be identified with the pK_a value of an imidazole group. The pK_a value of the ionizing group which determines the profile on the alkaline side is almost identical for BAA and TAA. It is noteworthy that the ionization of the sulfonamide group of TAA with a pK_a of 9.5 (ref. 168) does not affect the profile any

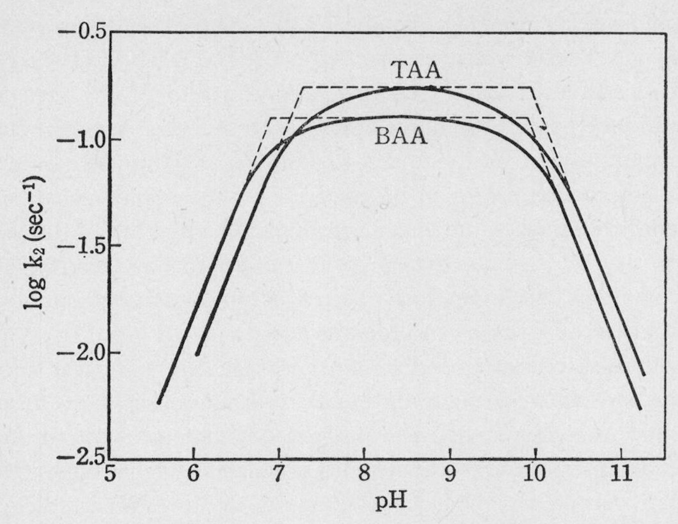

Fig. 8. Change in log k_2 as a function of pH for the trypsin-catalyzed hydrolysis of N^{α}-benzoyl-L-argininamide (BAA) and N^{α}-tosyl-L-argininamide (TAA) at 35 °C.

TABLE 13

pH Profiles of the Rate Constants of the Acylation and Deacylation in Trypsin Catalysis at 35 °C

Substrate	Acylation pK_a'	Acylation pK_a''	Deacylation pK_a'	Deacylation pK_a''	Ref.
BAA	6.9	10.1			166
TAA	7.2	10.0			166
BAEE			6.08	12†[1]	62, 174
TAME			7.3	10.1†[2]	173, 174
NPA	6.9		7.0		56

†[1] 15 °C. †[2] 25 °C.

more than the benzamide group of BAA which does not ionize in this pH region. The ionizing group of the enzyme which might be responsible for the pH profile in the alkaline region is not yet known, although it is possible that the group is the same as that which gives rise to the steep increase of K_m in this region.

3.4. Mechanism of the Deacylation Step

The deacylation reaction is the ratelimiting step in the hydrolysis of specific *ester* substrates (section 3.1.5). Thus data obtained from studies of the steady state kinetics of hydrolysis of BAEE or TAME (N^{α}-*p*-tosylarginine methyl ester) by trypsin pertain to the mechanism of the deacylation step. The pH-profile of this step has been studied by a number of investigators with the aim of identifying catalytically essential groups in the enzyme.[37,62,63,104,105,173,174] An example obtained for BAEE hydrolysis is shown in Fig. 9. Some of the results obtained at 35°C are listed in Table 13 together with corresponding values obtained for the acylation step with amide substrates. The pK_a' values between 6.0 and 7.5 can be attributed to the ionization of an imidazole group as discussed more thoroughly in section 2.4.2. The substrate with an N^{α}-tosyl group seems to shift this ionization by more than 1 pH unit presumably due to interaction of the easily ionizable tosylamide group with the imidazole. This conclusion was supported by similar studies with N^{α}-benzoyl-L-lysine esters and N^{α}-tosyl-L-lysine esters by Elmore *et al.*[104] The acyl group attached to the α-amino group of a substrate has an even stronger effect on the pH profile in the alkaline region. With the benzoylated substrate, BAEE,[62,63] or BAME,[174] no change in k_3 occurs up to pH 11.5, indicating that no group ionizing in this area is essential in the catalysis of deacylation, and provides an interesting contrast between the acylation and deacylation processes. It is likely that in the deacylation reaction the covalently bound acyl group prevents dissociation of the proton from the *N*-terminal α-amino

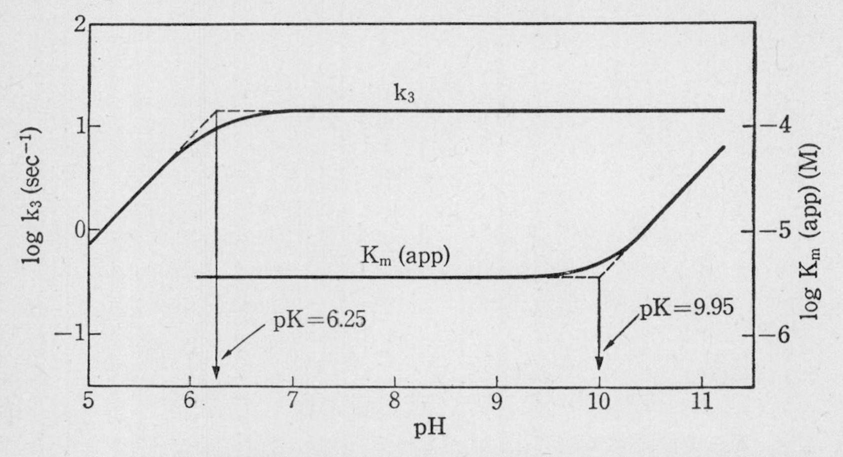

Fig. 9. Changes in log k_3 and log K_m(app) for the trypsin-catalyzed hydrolysis of BAEE at 25°C (composed from data in ref. 62, 175 and 176).

group and the ensuing denaturation. Such substrate influence should have been weaker in the acylation step where the substrate is present in the non-covalently bonded Michaelis complex. The pK_a'' of 10.1 for the TAME compared with the much higher value for BAEE seems to reflect the effect of deprotonation of the tosylamido group ($pK_a = 10$)[174] of the substrate molecule. This leaves a negative charge on the substrate molecule which will be repulsed by the negative charge in the electron relay system consisting of the hydrogen-bonded aspartic acid, histidine and serine residues at the catalysis site (see section 3.3).

The pH dependence of K_m(app) for BAEE is similar to that of BAA in solutions of neutral and alkaline pH; it remains constant between pH 6.5 and 10, then starts to rise steeply as the pH approaches 10.5.[175] An argument similar to that outlined in section 3.3 in connection with the pH profile of K_m for BAA may be equally applicable in this case. Again it is likely that ionization of the *N*-terminal α-amino group controls changes in the conformation which may be essential for substrate binding. At any rate in the hydrolysis of BAEE, K_m(app) and k_3 are controlled by two completely independent ionizing groups.

Analogous results derived from studies of chymotrypsin catalysis

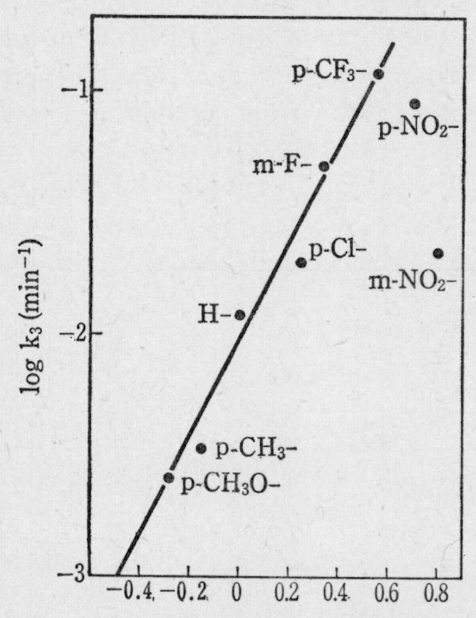

Fig. 10. Hammett's ρ-σ plot of the first-order rate constant, k_3, of deacylation of *meta*- and *para*-substituted benzoyl-chymotrypsin at pH 7 and 25°C.

were reported by Himoe *et al.*[169,170] and Bender *et al.*[175] The pH profiles of k_2 obtained with N^α-acetyl-L-tryptophan esters gave sigmoidal curves with flat plateaus extending to pH 11.5. A similar profile was obtained with N^α-acetyl-L-tryptophanamide with a plateau up to pH 10.5. The pH-dependent change of K_m(app) of chymotrypsin was also closely related to that of trypsin. Its change was independent of that of k_2 or k_3, and remained constant in the neutral pH region but sharply increased as the pH was raised.

The similar kinetics of trypsin and chymotrypsin catalyses justify discussion of an important result obtained by Caplow and Jencks[176] from a study of the electronic mechanism of chymotrypsin catalysis which is presumably applicable to trypsin catalysis. Benzoyl-chymotrypsins with various *meta* and *para* substituents on the benzoyl group were prepared from corresponding substituted *N*-benzoylimidazoles at pH 6.0. Rate constants of deacylation of these substituted benzoyl groups, which are equivalent to k_3 in the three-step mechanism, were determined at pH 7. The plot of log k_3 was made against the σ value according to Hammett,[152] as shown in Fig. 10. Points representing the various substituents fell on a straight line except for *m*- and *p*-nitro derivatives. The slope of the straight line, ρ, was found to be $+2.1$ in sharp contrast to the ρ value of -2.0 obtained for the acylation step and discussed in the preceding section.[153,154] The positive ρ value indicated that the lower electron density on the carbonyl carbon of the benzoyl group facilitated the reaction. This result indicated that general base catalysis must play a key role in the mechanism of deacylation. Recently, from a consideration of the three dimensional structure of the catalysis site, Blow *et al.*[16] proposed a mechanism for the deacylation reaction as shown below in scheme (28).

Asp–102 C–O O⊖ H N His–57 N H O Ser–195 O C H O R $\xrightarrow{k_3}$ Asp–102 C–O O H N His–57 N H O⊖ Ser–195 O–C H O R (28)

3.5. Substrate Activation

Trowbridge, Krehbiel and Laskowski Jr.[177] found that hydrolysis of N^{α}-*p*-tosyl-L-arginine methyl ester (L-TAME) did not fit simple Michaelis-Menten kinetics when the rate of the hydrolysis was examined over a wide concentration range (100,000-fold) of substrate. Similar observations with L-TAME were reported by Curragh and Elmore[178] and Bechet and Yon.[179] A plot of initial rate of catalysis $v_0/[E]_0$ against $v_0/[E]_0[S]_0$ (according to Eadie) obtained by Trowbridge *et al.*[177] is shown in Fig. 11. High concentrations of TAME caused a several-fold increase in rate over that predicted by the Michaelis-Menten equation, and shown by the broken line in Fig. 11. Similar results were obtained with the D-isomer of the substrate. D-TAME,[177] N^{α}-tosyl-L-lysine methyl ester (TLME),[178] and N^{α}-tosyl-*S*-amino ethyl-L-cysteine (tosylthialysine) esters.[178] All the N^{α}-tosylated derivatives exhibited extensive activation and an attractive explanation of the deviation involved the presence of a second species of the substrate, namely the one deprotonated at the sulfonamide group $-SO_2-NH-$ to a negatively charged state $-SO_2-N^{\ominus}-$. However, this explanation became untenable when

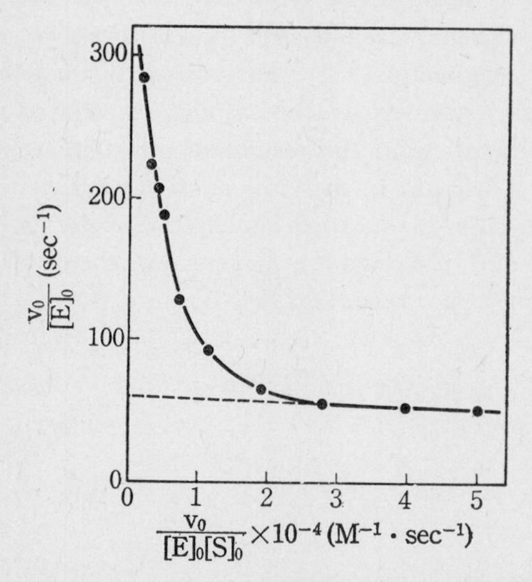

Fig. 11. The Eadie plot for L-TAME hydrolysis catalyzed by trypsin indicating deviation from the normal Michaelis-Menten kinetics (broken straight line) due to the substrate activation. Conditions: pH 8.0, 0.2 M KC1, 0.05 M $CaCl_2$ at 25°C.

Trenholm *et al.*[180] found a similar, though minor, substrate activation with BAEE, since its benzamide groups would not undergo ionization at neutral pH. Trowbridge, Krehbiel and Laskowski Jr.[177] showed that the complex kinetics could be fitted by the following scheme, which assumes the presence of two kinds of enzyme-substrate complexes, ES and ESS.

$$E + S \overset{K'_s}{\rightleftharpoons} ES \overset{k'_3}{\longrightarrow} E + P \tag{29}$$

$$ES + S \overset{K''_s}{\rightleftharpoons} ESS \overset{k''_3}{\longrightarrow} ES + P \tag{30}$$

ESS is a complex which contains two molecules of substrate, one of which is hydrolyzed. K'_s and K''_s are dissociation constants. This model does not presuppose any specific mechanism for the binding of the second substrate molecule whether it is allosteric, as proposed by Bechet and Yon,[179] or isosteric. For L-TAME the values of K_s's and k_3's listed in Table 14 were found to fit the observed data. An indication that the second substrate was not hydrolyzed but acted to modify the enzyme protein was obtained by Howard and Mehl.[181] Over a range of high concentrations of substrate where binding of the second molecule took place, part of TAME was replaced by its hydrolysis product, N^α-tosyl-L-arginine, and this had no effect on the extent of the substrate activation. For example, when 27.5 mM TAME was replaced by a mixture of 2.5 mM TAME and 25 mM N^α-tosyl-L-arginine, the activation remained constant at 82% more than that predicted by the Michaelis-Menten kinetics.

The substrate activation was not observed in the hydrolysis of amide substrates. Wang and Carpenter[168] failed to observe any activation in the hydrolysis of N^α-tosyl-L-arginamide or N^α-tosyl-L-lysinamide up to a substrate concentration of 50 mM. In fact, a slight substrate inhibition was observed with both amide substrates. Inagami[132] did not

TABLE 14

Kinetic Parameters for the Tryptic Hydrolysis of N^α-Tosyl-L-arginine Methyl Ester†

Complex	k_3 (sec^{-1})	K_s (mM)
ES	$k_3' = 60$	$K_s' = 0.0013$
ESS	$k_3'' = 340$	$K_s'' = 53$

† Conditions: pH 8.0, 25°C in 0.2 M KCl and 0.05 M $CaCl_2$ (ref. 177).

observe any activation in the hydrolysis of N^{α}-benzyloxycarbonyl-L-arginine *p*-toluidide up to a substrate concentration of 5 mM. Since the catalytic rate observed with an ester substrate is that of the deacylation reaction and for an amide substrate that of the acylation reaction, it is likely that substrate activation occurs only at the deacylation step. However, it must be noted that recent observations by Nakata and Ishii[182] of substrate activation in the hydrolysis of N^{α}-benzoyl-L-arginine *p*-nitroanilide do not support such a hypothesis, unless the rate limiting step for the nitroanilide is different from other amides and anilides.

3.6. Conformational Change Caused by Acylation

The pH profiles of the acylation and deacylation reactions can be compared from the pK_a values listed in Table 13. For example, the pK_a of acylation by BAA is 6.9 and deacylation of its acyl group, benzoylarginyl, has a pK_a of 6.08 as determined with BAEE. Such a shift in the pH dependence can be interpreted as involving a change in the environment of the catalytically important ionizing group by the acylation reaction. In some instances, the shift is small as in the case of the trypsin-catalyzed hydrolysis of TAA and NPA (Table 13), whereas the chymotrypsin-catalyzed hydrolysis of NPA has been reported[94] to cause a large shift of pK_a from 6.6 to 7.4. Acylation of chymotrypsin by various acylating inhibitors and substrates was found to cause some conformational change of the enzyme protein and such change has been extensively studied by a number of investigators.[39,183-96] In the case of trypsin, however, rapid autolysis makes similar studies less easy or even not feasible.

Hopkins and Spikes[195] circumvented this problem by comparing the speed of denaturation of native trypsin with trypsins whose active sites had been modified by diisopropylphosphorylation (DIP) of the Ser-183 by DFP or by alkylation of His-46 with TLCK as an alternative to attempting to detect difference in parameters which could be determined only under a static equilibrium condition. The denaturation in 8 M urea was followed by changes in tryptophan fluorescence at 340 nm, when the presence of the DIP group caused a 3-fold increase in the rate of denaturation. The alkylation by TLCK which resembled the specific substrate of trypsin more closely than DIP increased the rate by a factor of more than 100. Thus, these modifications of the active site of trypsin were shown to make the enzyme more susceptible to the urea denaturation, whereas the same investigators[196] found that similar modification of the active site of chymotrypsin increased its stability towards 8 M urea.

4. Specificity

Trypsin is well known for its restricted specificity to peptide bonds whose carbonyl groups are contributed by lysine and arginine. This clear-cut specificity has made trypsin the most widely used enzyme in the sequence determination of peptides and proteins. The term "specificity" in this discussion is used to signify the side chain specificity with respect to substrate amino acid residues. Consideration of the specificity in such a restricted sense is based on a tacit assumption that the active site of a protease consists of two sites, namely, the catalysis site and the specificity determining site or recognition site as it is often called. The latter serves to recognize and bind a specific substrate molecule to the site in order to facilitate the ensuing chemical reaction performed by the catalysis site. Since the normal reaction catalyzed by the protease is the cleavage of a peptide, ester or amide bond of an L-α-amino acid derivative, the catalysis itself is specific for the basic L-α-amino acid structure. In this sense the catalysis site itself plays a part in determining the specificity of the enzyme. But this contribution is taken for granted, and will not be discussed in connection with the specificity of trypsin. In addition the two sites may not be topographically separate areas, but, as the present discussion progresses it will be seen that their function can be separated and that such an approach is useful when studying the active site of trypsin.

In the preceding sections many observations have consistently suggested that a close similarity exists between trypsin and chymotrypsin, and information obtained with either of the enzymes has often been used interchangeably. Obviously such a relationship cannot be used in discussions of specificity, and attention will be confined here to the contrast in specificity which distinguishes the two enzymes. Careful comparative studies of the difference between the two enzymes will shed some light on the factors which differentiate their specificity patterns.

4.1. Trypsin Specificity in the Cleavage of Peptides and Proteins

It has generally been recognized that trypsin is highly specific to lysyl and arginyl peptide bonds. However, accumulated knowledge of the tryptic digestion process resulting from increasing efforts in sequence determination studies, has emphasized that even the recognized specificity of the process is not free from exceptions. It was noted that some tyrosyl peptide bonds of the oxidized B-chain of insulin[197-200] and of other proteins, for example ferredoxin,[201] were split by trypsin. It was suspected that this might be due to a small amount of chymotrypsin contaminating crystalline trypsin preparations. However, Inagami and Sturtevant[202] and Cole and Kinkade[203] demonstrated that trypsin preparations were able to hydrolyze a synthetic tyrosine ester substrate, *N*-acetyl-L-tyrosine ethyl ester (ATEE) (see Table 15), and that this activity was mostly due to the intrinsic action of trypsin rather than to that of contaminating chymotrypsin. Kostka and Carpenter[204] showed that the contaminating chymotrypsin in a trypsin preparation could be destroyed by specific alkylation at the active site of the chymotrypsin with TPCK. When this preparation was applied to the B-chain of insulin and to the heptapeptide from the B-chain, the hydrolysis of tyrosyl peptide bonds was negligible. However, the TPCK-treated trypsin retained its ability to hydrolyze ATEE at approximately 50% of its original level. Thus, trypsin seems to have a weak, but definite, specificity to the tyrosine ester, most tyrosyl peptide bonds are not hydrolyzed probably due to their high K_m values.

There have been reports that a number of arginyl or lysyl peptide bonds are either not cleaved by trypsin or only cleaved with difficulty, under the usual conditions for tryptic digestion. Examples include Lys–Pro[205] in bovine ribonuclease and bovine, orcine and porcine ACTH[206-8] and trypsin inhibitor of bovine pancreas,[209,210] which remained intact during the tryptic digestion. When cystine in a protein is reduced and aminoethylated with ethylenimine, *S*(β-aminoethyl)-L-cysteine (AECys) (XXIV) is formed.

$$NH_2\text{–}CH_2\text{–}CH_2\text{–}S\text{–}CH_2\text{–}CH(NH_2)\text{–}COOH$$
(XXIV)

Since the size and configuration of this molecule is similar to lysine, AECys peptide bonds are susceptible to tryptic digestion (see also Table 15), but its peptide with the imino group of proline in spinach ferredoxin was not cleaved by trypsin.[201]

The resistance of peptide bonds involving the imino group of proline nevertheless does not appear to be completely absolute. For example, Ando *et al.*[211] Ishii *et al.*[212] and Hayashida and Yamasaki[213] have observed the cleavage of arginyl proline bonds in protamine and synthetic peptides by slow reaction.

Concerning exceptions to recognized tryptic specificity, the following additional details should be given. The presence of basic or acidic amino acids in the immediate vicinity of potentially susceptible bonds seems to affect their hydrolysis rates. Using synthetic oligomers of lysyl as substrates, Waley and Watson[214] and Levin *et al.*[215] showed that lysine bonds penultimate to *C*-terminals were not cleaved. The splitting of an *N*-terminal lysine was also slower than normal. It was proposed that such effects were due to the negative electrical charge of the *C*-terminal carboxylate group and the positive charge of the *N*-terminal α-ammonium group, respectively. Data supporting such conclusions were obtained with oligomers of arginine[216] and with the *N*-terminal lysine residue of pancreatic ribonuclease.[217] When several basic amino acid residues are clustered in a consecutive sequence some of the peptide bonds are split with difficulty. In the sequence shown below, which is present in ACTH (adrenocorticotropic hormone), only the Lys–Lys bond is split to an appreciable extent.

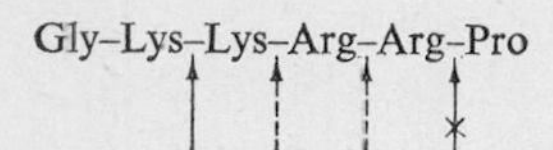

Two others indicated by broken arrows underwent only very limited hydrolysis, whereas Arg–Pro remained intact.[206] Such retardation was particularly evident with AECys peptides. Plapp *et al.*[218] reported that the AECys–Lys bond in the sequence Lys–Asp–Arg–AECys–Lys in a reduced aminoethylated ribonuclease was not split, and AECys–Arg and AECys–Lys bonds in the same protein were split to limited extents. These results may be explained by faster hydrolysis of lysine or arginine peptide bonds leading to either free carboxyl or α-amino group in the

TABLE 15

Kinetic Parameters for Tryptic Hydrolysis of Various Synthetic Substrates†[1]

	Substrate	K_m(app) (mM)	k_{cat} (sec^{-1})	Ref.
i	N^α-Benzoyl-L-argininamide†[2]	2.5	2.8	168
ii	N^α-Tosyl-L-argininamide†[2]	7.4	5.2	168
iii	N^α-Benzoyl-L-lysinamide†[2]	4.6	1.9	168
iv	N^α-Tosyl-L-lysinamide†[2]	16	3.3	168
v	N^α-Benzoyl-N^ϵ-methyl-L-lysinamide†[3]	—	~0.2	243
vi	N^α-Benzoyl-L-arginine ethyl ester	0.0043	14	202
vii	N^α-Tosyl-L-arginine methyl ester	0.013	60	103
viii	N^α-Tosyl-D-arginine methyl ester	0.27	0.74	103
ix	N^α-Tosyl-L-homoarginine methyl ester	0.29	4.1	103
x	N^α-Tosyl-γ-guanidino-L-α-aminobutyric acid methyl ester	0.69	38	249
xi	N^α-Benzoyl-L-lysine methyl ester	0.017	17	104
xii	N^α-Tosyl-L-lysine methyl ester	0.042	67	104
xiii	N^α-Tosyl-L-ornithine methyl ester†[4]	15	3.3	103
xiv	N^α-Acetyl-L-lysine methyl ester†[4]	0.25	52	244
xv	N^α-Acetyl-D-lysine methyl ester†[4]	2.0	0.073	244
xvi	N^α-Benzyloxycarbonyl-L-histidine methyl ester	4.2	6.9	245

xvii	N^{α}-Benzoyl-*S*-aminoethyl-L-cysteinamide†[2]	4.3	0.33	168
xviii	N^{α}-Benzoyl-*S*-aminoethyl-L-cysteine methyl ester	0.094	21	104
xix	N^{α}-Benzoyl-L-α-aminoheptanoic acid methyl ester†[4]	0.010	0.009	244
xx	N^{α}-Benzoyl-L-citrulline methyl ester†[4]	41	0.14	244
xxi	N^{α}-Acetyl-L-tyrosine ethyl ester	42	15	202
xxii	*N*-Acetylglycine ethyl ester†[3]	800	0.028	246
xxiii	*p*-Nitrophenyl acetate†[5]	21	0.013	56
xxiv	N^{α}-Acetyl-L-α-aminoadipamic acid methyl ester	4.5	3.2	244
xxv	N^{α}-Acetyl-*S*-carboxamidomethyl-L-cysteine ethyl ester	40	14.0	244
xxvi	L-Lysine methyl ester†[3]	9.3	14.0	245
xxvii	α-Hydroxy-δ-guanidinovaleric acid methyl ester	0.5	7.8	247
xxviii	ϵ-Aminocaproic acid methyl ester	17	9.3	248
xxix	*p*-Amidinobenzoic acid *p*-nitrophenyl ester	<0.0005	0.04	123
xxx	*p*-Amidinophenylacetic acid methyl ester	0.015	0.028	122
xxxi	*p*-Guanidinobenzoic acid ethyl ester†[6]	0.54	0.0046	122
xxxii	*p*-Guanidinophenylacetic acid ethyl ester	0.011	0.047	122
xxxiii	N^{α}-Benzoyl-L-arginine *p*-nitroanilide†[7]	0.94	0.61	142
xxxiv	L-Lysine *p*-nitroanilide	0.36	0.003	142

†[1] Determinations were made at pH 8.0 and 25 °C in the presence of 2–100 mM $CaCl_2$ unless otherwise indicated.
†[2] 30 °C. †[3] Ca^{2+} was absent. †[4] pH 7.0 in the absence of Ca^{2+}. †[5] pH 7.8 in the absence of Ca^{2+}. †[6] Acylation only.
†[7] 15 °C, available in racemic form.

immediate vicinity of the AECys residues, so preventing tryptic action on this linkage. Incidentally, a lysyl peptide bond seems to be more rapidly hydrolyzed than an arginyl bond,[217] and an AECys–peptide is hydrolyzed by far the slowest.[218] The presence of glutamic acid or aspartic acid residues adjacent to lysine or arginine often has a profound effect on tryptic digestion. The lysyl peptide bonds in some of the following sequences, Glu–Lys–X, Asp–Lys–X and Asp–AECys–X, in cytochrome *c*,[219] α- and β-chains of human hemoglobin,[220,221] and *Clostridium* ferredoxin[222] were found to be resistant to tryptic hydrolysis. The cleavage of the arginine-cysteic acid bond in the performic acid-oxidized ribonuclease was considerably retarded, whereas a cysteic acid-arginyl peptide bond was split at a normal rate.[217] When the cystine residues of ribonuclease were converted to carboxymethyl cysteine (CMCys) (XXV) by reduction followed by carboxymethylation, the introduced carboxyl group was a much weaker acid than the sulfonic acid in cysteic acid, and the tryptic cleavage of Arg–CMCys was not appreciably retarded by CMCys.

$$HOOC\text{–}CH_2\text{–}S\text{–}CH_2\text{–}CH(NH_2)\text{–}COOH$$

(XXV)

The effects of the adjacent acidic amino acid residues on the tryptic hydrolysis may be regarded as resulting from their ability to neutralize the charges on the lysine or arginine residues. The choice of the method of disulfide bond cleavage clearly has an important effect on the subsequent tryptic hydrolysis of a peptide. If a cystine residue should be adjacent to a lysine or arginine residue, there is always some danger that cleavage of the disulfide bond by any of the three methods, namely, performic acid oxidation, reduction followed by carboxymethylation, or reduction followed by aminoethylation, may introduce an effect undesirable for obtaining predictable results from tryptic digestion.

Chemical modification of the ϵ-amino group of lysine or the guanidino group of arginine so as to eliminate the positive charge of these groups leads to loss of susceptibility to tryptic digestion. Similarly polypeptidylation with *N*-carboxy-α-amino acid anhydride,[223] trifluoroacetylation with ethyl thiotrifluoroacetate,[224] carbamylation,[225,226] dinitrophenylation,[227] and reversible maleylation[228] of the ϵ-amino group also made the lysine peptide bond resistant to tryptic digestion. Moreover, guanidination of the same group to a homoarginine residue with *O*-methylisourea (XXVI),[229]

$$R\text{-}NH_2 + CH_3O\text{-}C(=NH)NH_2 \longrightarrow R\text{-}NH\text{-}C(=NH)NH_2 + CH_3OH$$
(XXVI)

or its amidination with methyl acetimidate (XXVII),[230]

$$RNH_2 + CH_3\text{-}C(=NH)OCH_3 \longrightarrow R\text{-}NH\text{-}C(=NH)CH_3 + CH_3OH$$
(XXVII)

also made the lysine bond resistant. In this case the positive charge on the ϵ-amino group was not eliminated but merely shifted by less than 1.5 Å. The extension of peptide analysis to proteins of larger molecular size has required limitation of the number of peptides obtained by tryptic digestion. Blocking the ϵ-amino groups limits the trypsin action to arginine peptide bonds, and so reduced considerably the number of peptides produced. Maleylation has been particularly useful since not only can the maleyl group be removed under mild conditions but the process also increases the solubility of proteins to be digested and of the peptides produced.[228] An attempt to modify the guanidino group of arginine for a similar purpose was made by Itano and Gotlieb.[231]

4.2. Synthetic Substrates for the Study of Specificity

Ever since the original discovery by Bergmann *et al.*[232] and Schwert *et al.*[233] that synthetic amides and esters of lysine and arginine were hydrolyzed by trypsin, many synthetic substrates of small molecular weight have been prepared and used in studies of trypsin catalysis in place of proteins. A precise structure and a simple and accurate method of activity assay offered by the synthetic ester substrate[233] has contributed greatly to our understanding of the mechanism of trypsin catalysis, in spite of the fact that the rates of hydrolysis of small amides or peptides are lower than those of larger oligomeric peptides,[214,216,234,235] and larger quantities of the enzyme are consequently required for their assay. Kinetic parameters, k_{cat} and K_m(app), derived from studies of the steady state kinetics of hydrolysis of some representative synthetic substrates are listed in Table 15. Although trypsin is considered to be much more strict in its specificity with respect to the amino acid side chain compared with chymotrypsin or other proteases, it is surprising that D-isomers of lysine and arginine esters (viii and xv) are hydrolyzed, though at rates much less than those of the L-isomers (vii and xiv). However, in an anilide substrate, the D-isomer is not hydrolyzed as reported by Erlanger *et al.*[142] The distance between the positive electrical charge on the side chain and the bond to be hydrolyzed in the substrate molecule has an important effect on both K_m(app) and k_{cat}. However, such an effect is not of an all-or-none type as was once believed. Thus a homoarginine ester (ix) is hydrolyzed at a rate 7% of that of the corresponding arginine substrate (vii) and a shorter homolog (x) is hydrolyzed at a rate as high as 50% of that of (vii). Similarly an ornithine ester (xiii) is susceptible to tryptic hydrolysis. It may be noted, however, that these ester substrate homologs of lysine and arginine have very high K_m(app) values so making attack by the enzyme even more difficult. When homoarginine is present in peptide form as a result of guanidination of lysine, the homoarginyl peptide bond is not cleaved by tryptic digestion.[229] Such discrepancy between the behavior of an ester and a peptide may be explained by the widely recognized observation that most ester bonds are considerably more labile than a peptide bond, and that K_m(app) for a peptide substrate is usually higher than that for an ester substrate. The critical effect of the distance between the positive electrical charge and the bond to be hydrolyzed is reflected even in substrates containing non-natural amino acids. In AE-cysteine (XXIV) and *O*-aminoethyl-L-serine (AE-serine) (XXVIII) the distance is almost identical to that of lysine.

$$NH_2\text{–}CH_2\text{–}CH_2\text{–}O\text{–}CH_2\text{–}CH(NH_2)\text{–}COOH$$

(XXVIII)

An amide substrate (xvii)[168] and an ester substrate (xviii)[104,236] of AE-cysteine as well as its peptides[237–40] are good substrates of trypsin. Similarly an ester of AE-serine is also a good substrate.[236] Furthermore, in dipeptide esters such as β-alanyl-glycine ethyl ester and *N*-benzoyl-*O*-glycyl-serine ethyl ester the distance in question is not much different from that of lysine, and they have been found to undergo tryptic esterolysis.[241] As purified enzymes have become available in large quantity in recent years, they have been treated as common chemical reagents, and a large quantity of an enzyme and a high concentration of substrate have been frequently used. Such studies revealed that even the positive electrical charge of a substrate is not an absolute requirement for trypsin catalysis, provided that sufficient substrate is added to a concentration high enough to saturate the active site of the enzyme. Under such conditions esters of the neutral amino acid derivatives (xix–xxv) are attacked by trypsin. Substrates (xxvii) and (xxviii) indicate that the α-amino group is not the absolute requirement for a substrate. Hofstee[242] found that even *n*-fatty acid esters of *m*-hydroxybenzoic acid were hydrolyzed by trypsin. Comparison between (xxvi) and (xi), (xii), or (xiv) shows that the acylated α-amino group makes a better substrate. Thus the data in Table 15 indicate that the pattern of specificity of trypsin catalysis with synthetic ester substrates is much broader than that observed with peptides. Particularly striking are the facts that an L-conformation and the presence of a positive electrical charge are not the absolute requirement. However, when trypsin is used for peptide cleavage, its specificity becomes highly stringent, probably because the concentration of the peptide or protein commonly used is not sufficiently high to compensate for a normally high K_m(app) value for the peptide substrate and also because k_{cat} for the peptide bonds of basic amino acids are considerably higher than those for nonspecific amino acid residues.

Esters of *p*-guanidino- and *p*-amidinobenzoic acids were also found to be substrates by Mares-Guia *et al.*[121,122] As discussed in the previous section, it is interesting to note that *p*-guanidinobenzoic acid ethyl ester (xxxi) and *p*-nitrophenyl ester acylate the enzyme but deacylation is extremely slow[121,122] whereas (xxix), (xxx), and (xxxii) undergo continuous tryptic esterolysis.[122,123] The final compounds (xxxiii and xxxiv) are specific chromogenic substrates, which produce a yellow-colored product, *p*-nitroaniline.[142] Also N^{α}-benzoyl-D,L-arginine β-naphthylamide is another chromogenic substrate.[250] The *p*-nitroanilide substrate (xxxiii) is available in racemic form and the D-isomer is a competitive inhibitor (for its K_i, see Table 16). Recently a pure L-isomer has been synthesized by Nishi and Noguchi through a very elegant synthetic procedure.[251]

4.3. Competitive Inhibition

Since trypsin is primarily specific for esters, amides and peptides of lysine and arginine, it can be expected that the specificity determining site of the enzyme has an ability to recognize and bind molecules whose structures resemble at least in part the side chains of lysine and arginine. Inagami and York[252,253] found competitive inhibition of trypsin-catalyzed hydrolysis of BAEE by a series of alkylamines and alkylguanidines. A large number of alkylamines and their derivatives were studied by Mix *et al.*[254] Mares-Guia and Shaw[256,257] demonstrated that alkyl and aromatic amidines and guanidines were potent inhibitors

TABLE 16

Inhibition Constant K_i and Standard Free Energy, ΔG^0, of the Formation of Enzyme-inhibitor Complex for Competitive Inhibitors of Trypsin

Inhibitor	Structural formula	K_i (mM)	ΔG^0 (kcal/mole)	Ref.
Amine				
Methyl†[1]		260	−0.80	252
Ethyl†[1]		62	−1.6	253
n-Propyl†[1]		8.7	−2.8	253
n-Butyl†[1]		1.7	−3.8	253
n-Hexyl†[1]		12	−2.6	253
Benzyl†[2]		0.4	−4.6	253
N,*N*-Dimethylbenzyl†[2]		8.7	−2.8	253
Tyramine†[2]	(XXIX)	3.3	−3.4	253
Tryptamine†[2]	(XXX)	2.4	−3.6	253
3-Dimethylamino-methyl indole†[2]	(XXXI)	280	−0.75	253
Serotonin†[3]	(XXXII)	0.6	−4.1	255
Guanidine†[4]		9.06	−2.7	256
Methyl†[4]	(XXXIII)	4.40	−3.1	256
Methyl†[5]		7.0	−2.7	252
Ethyl†[5]		1.4	−3.7	252
n-Propyl†[5]		0.53	−4.3	252
n-Butyl†[5]		1.3	−3.8	252
Cyclohexyl†[5]		4.4	−3.2	252
Benzyl†[5]		7.4	−2.9	252
Phenyl		0.0725	−5.5	149
Amidine				
Formamidine†[4]	(XXXIV)	60.0	−1.61	256
Acetamidine†[4]	(XXXV)	25.5	−2.10	247
Cyclohexylcarbox-amidine†[4]	(XXXVI)	0.427	−4.4	257
Phenylacetamidine†[4]	(XXXVII)	15.1	−2.4	257
Benzamidine†[4]	(XXXVIII)	0.0166	−6.3	256
p-Aminobenzamidine†[4]		0.00825	−6.7	257
β-Naphthamidine†[4]		0.0146	−6.4	256
m-Toluamidine†[4]		0.0227	−6.1	256

TABLE 16—*Continued*

Inhibitor	Structural formula	K_i (mM)	ΔG^0 (kcal/mole)	Ref.
Others				
Proflavin†[6]	(XXI)	0.034		145
Thionine†[7]	(XXXIX)	0.012		258
p-Aminomethyl benzoic acid†[8]		0.29	−4.8	259
p-Guanidinobenzoic acid		0.55	−4.5	254
4-Guanidinobutyric acid†[3]		1.19	−4.0	255
p-Hydroxyphenyl-pyruvic acid†[3]		1.39	−3.9	255
Indole		15†[1]	−2.5	253
Benzene		1000†[1]		253
Urea		600†[4]		256

†[1] Inhibition of BAEE hydrolysis in 0.1 M KCl, pH 6.6, 25°C.
†[2] Inhibition of BAEE hydrolysis in 0.05 M KCl, pH 6.6, 25°C.
†[3] Inhibition of trypsinogen activation in 0.1 M $CaCl_2$, pH 7.7, 23°C.
†[4] Inhibition of D,L-BANA hydrolysis in 0.1 M Tris, pH 8.15, 15°C.
†[5] Inhibition of BAEE hydrolysis in 0.2 M KCl, pH 8, 25°C.
†[6] Binding measured by an optical method in 0.2 M phosphate buffer pH 7.6, 4°C.
†[7] Similar to †[6], pH 7, 4°C.
†[8] Inhibition of BANA, pH 7.6, 30°C.

of trypsin. Inhibition studies of this type have provided us with important information about the nature of the forces responsible for the binding of inhibitor to the specificity determining site and have led to new understanding of the structure and function of the site.

It is well known that the inhibition constant, K_i, is identical with the dissociation constant of the enzyme-inhibitor complex as long as enzyme and inhibitor are bound reversibly. This relationship is independent of the complexity of K_m(app) or of whether the catalysis involves two steps or more. Therefore, it is possible to obtain the standard free energy change of the enzyme-inhibitor complex formation, ΔG^0, directly from K_i. The two parameters for some inhibitors are listed in Table 16. It is clear that a positively charged group alone, such as is present in

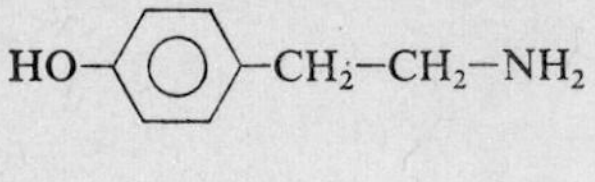

(XXIX) 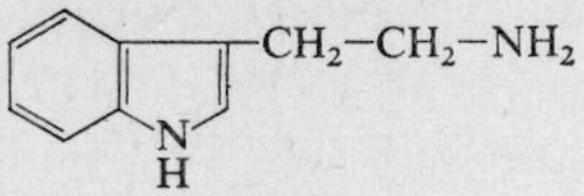 (XXX)

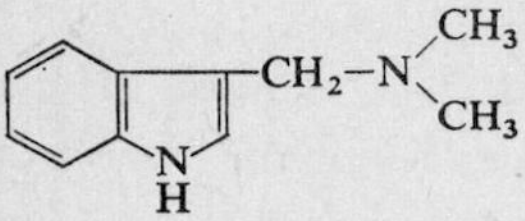

(XXXI) 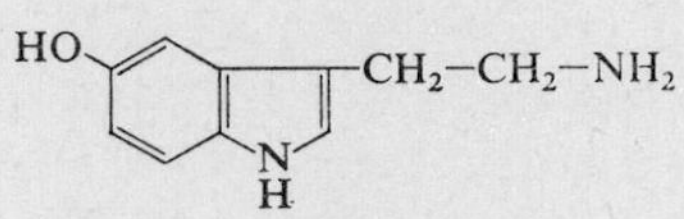 (XXXII)

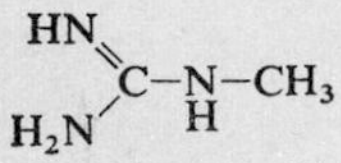

(XXXIII) 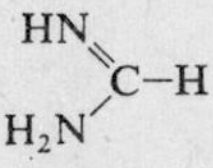 (XXXIV) 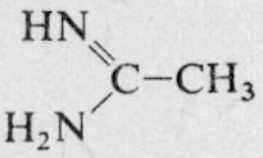 (XXXV)

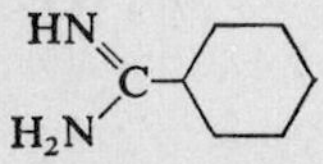

(XXXVI) 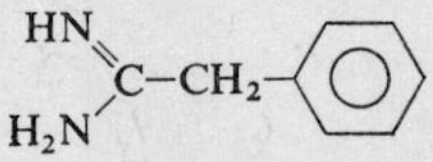 (XXXVII) 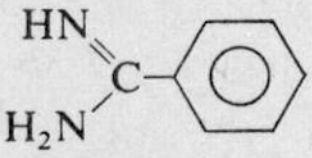 (XXXVIII)

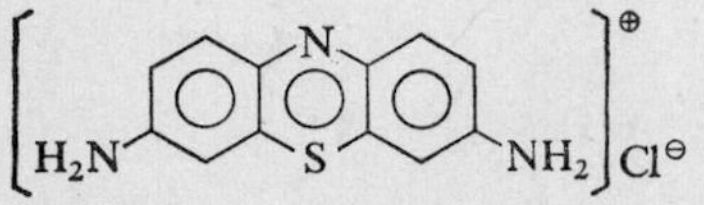

(XXXIX) 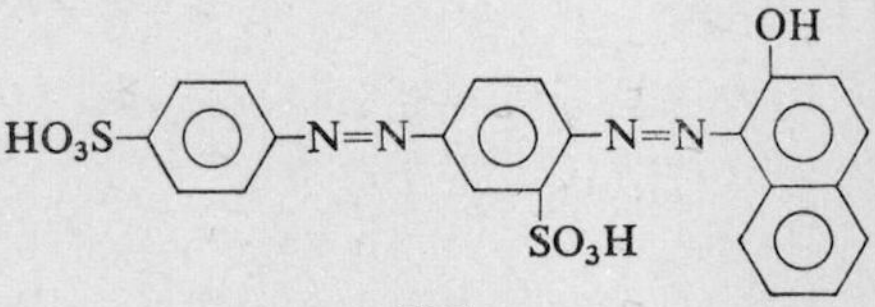 (XC)

TABLE 17

Standard Enthalpy Change, ΔH^0, of Inhibitor Binding to Trypsin

	ΔH^0 (kcal/mole)	
	Alkylamine†[1]	Alkylguanidine†[2]
Methyl	−13	−4.4
Ethyl	−11	−4.8
n-Propyl	−10	−4.5
n-Butyl	−11	−4.5
Benzyl	−10	−4.8
Cyclohexyl	—	−4.3
n-Hexyl	−10	—

†[1] 0.1 M KCl, pH 6.6 and 25 °C (ref. 253).

†[2] 0.2 M KCl, pH 8.0 and 25 °C (ref. 252).

protonated guanidine or formamidine is not sufficient to make a strong inhibitor. The data for the series of alkylamines, guanidines, and amidines reveal that in addition to a positively charged base a hydrocarbon structure of a certain optimum size is required for strong binding of an inhibitor. For example in the amine series as each methylene unit is added to methylamine, ΔG^0 is decreased by 0.8–1.2 kcal/mole. A similar but somewhat less uniform decrease in the ΔG^0 of binding is noted in the guanidine series. Such contribution to the binding strength from the hydrocarbon moiety of the inhibitor is more conspicuous in some of the amidines and guanidines with aromatic hydrocarbon structures, and it is considered to be due to hydrophobic interaction between an apolar area of the specificity determining site and the hydrocarbon structure of the inhibitor.[252,253,256,257]

Quantitatively a -0.8 to -1.2 kcal/mole of ΔG^0 of binding per methylene unit agrees with -0.7 kcal/mole per methylene unit obtained for ΔG^0 of transfer of a methylene unit in the aliphatic side chain of amino acids from aqueous solutions into an organic solvent.[261,262] More convincing support for the involvement of the hydrophobic force comes from a study of temperature dependence of the ΔG^0. From the slope of a plot of log K_i *vs.* $1/T$, where T is the absolute temperature, the standard enthalpy change of binding, ΔH^0, was obtained for alkylamines and alkylguanidines as listed in Table 17. It is clear that all alkylamines have practically identical ΔH^0 of approximately -11 kcal/mole. Similarly ΔH^0 for alkylguanidines are all approximately equal to -4.5 kcal/mole. It is therefore possible to conclude that the enthalpy contribution is due to the interaction, both coulombic and hydrogen-bonding, of the ammonium or guanidinium moiety of the inhibitor with the negatively charged group in the specificity site of the enzyme. Then the lowering of ΔG^0 contributed by the hydrocarbon moiety of an inhibitor should be exclusively due to an increase in the standard entropy change of binding, ΔS^0. Such a pattern agrees with that feature of the hydrophobic force which, as discussed by Kauzmann,[261] derives primarily from a favorable change in entropy. Thus it is most likely that the specificity site of trypsin consists of two areas, a negatively charged site and a hydrophobic area.

There is a limit to the size of the hydrocarbon structure. In the alkylamines, *n*-butylamine is most strongly bound. The binding strength decreases with increasing chain length to *n*-hexylamine.[253] A similar phenomenon occurs in the alkylguanidines.[252] The evidence for optimal molecular size of the hydrophobic part of an inhibitor equally suggests that the size of the hydrophobic area of the enzyme is also limited. Comparison between phenyl- and cyclohexylguanidine and also between cyclohexylcarboxamidine (XXXVI) and benzamidine (XXXVIII) shows that binding of the benzene ring is more favored than that of the cyclohexane ring by 1.8 kcal/mole.[235,236] Precise reasons for such a preference are not known but it might be thought to indicate that the hydrophobic area of the specificity site resembles that of chymotrypsin which preferentially binds aromatic amino acid side chains. Here it must be pointed out that an uncharged aromatic compound alone, such as benzene, is not very strongly bound.

Berhard and Lee[144] and Glazer[145,258] have found that dyes such as proflavin (XXI) and thionine (XXXIX) are strongly bound to the active site of trypsin and that the binding is accompanied by a change in the absorption spectra of these dyes. Such shifts have found use in various kinetic and binding studies.[145,146,258] An interesting feature of the binding of these dyes is the fact that thionine binds specifically to the active site of trypsin but not to the active site of chymotrypsin,[258] whereas another dye, Biebrich Scarlet (XC), is specifically bound to chymotrypsin.[260] A variety of compounds have been shown to inhibit the activation of trypsinogen and also the blood clotting process. Geratz reported that ω-amino acids,[263] ω-guanidino acids,[263] α-keto acids,[264] *p*-aminobenzamidine,[265] as well as serotonin,[255] inhibited such reactions. Kezdy[266] found that *N*-acetyl-L-3,5-dibromotyrosine also inhibited the trypsinogen activation. Hummel[267] reported that quaternary ammonium compounds of large molecular size such as the tetra-*n*-butylammonium and cetyl pyridinium ions were inhibitory to trypsin whereas smaller homologs such as tetramethyl-, tetraethyl-, and tetrapropylammonium ions activated the trypsin-catalyzed hydrolysis of TAME. Systematic studies by Mix, Trettin and Gulzow[124,254,268] encompassed a wide variety of alkylamines,[254] alkylguanidines,[268] benzylamine derivatives,[254] ω-amino aliphatic acids,[254] ω-guanidino aliphatic acids,[268] *p*-aminomethylbenzoic acid,[268] *p*-amidino- and *p*-guanidinobenzoic acids and their derivatives.[124] Landmann and Markwardt[259,269] and their co-workers also reported inhibition by *p*-aminomethyl-, *p*-amidino-, and *p*-guanidinobenzoic acids and their derivatives. Particularly interesting are the findings by both of these groups of German investigators[124,259] that benzyl esters and benzylamides[270] of the above mentioned benzoic acid derivatives are very potent inhibitors with K_i values ranging from 10^{-6} to 10^{-7} M. For example, Mix *et al.*[124] reported the K_i for 4-guanidinobenzoic acid 4′-nitrobenzyl ester to be 3.3×10^{-7} M, and Landmann *et al.*[259] obtained a K_i for 4-amidinobenzoic acid benzyl ester of 2.3×10^{-6} M. There is a possibility that part of such strong interaction is due to the irreversible benzoylation of trypsin similar to that observed by Mares-Guia, Shaw and Cohen.[59,121,122] These strongly bound inhibitors together with benzamidine and *p*-aminobenzamidine used by Mares-Guia and Shaw[257] are of practical importance. For a highly specific but irreversible inhibition of trypsin, use of TLCK[66] may be more practical.

4.4. Site of Inhibitor Binding or Specificity Determining Site

4.4.1. Anionic Site for the Specific Binding

The results outlined in the previous section clearly indicate that the inhibitor binding site consists of a negatively charged group or groups and a hydrophobic area, and attempts to identify functional groups involved in the site have recently been intensified. Stewart and Dobson[271] studied the pH-dependent change of K_m(app) with BAEE as substrate. The plot of $-\log K_m$(app) *vs.* pH according to Dixon[172] gave the plot shown in Fig. 12, which indicates that ionization of a group with a pK_a of 4 is responsible for the increase in K_m(app) in the acidic pH region. Thus it was proposed that a carboxyl group with a pK_a of 4 would be located in the ionic binding site. However, the slope of the plot below pH 4 was higher than 1 but lower than 2 indicating that more than the simple ionization of one carboxyl group was involved in this change in K_m(app). In fact, D'Albis[272] had observed a conformational change of trypsin accompanying the ionization of a group with pK_a 3.7. Studies of changes in absorbance and optical rotation in acid solution led Lazdunski and DeLaage[273] to conclude that trypsin underwent a discrete conformational change as the pH changed from 3.0 to 4.5. They were able to correlate such a change with the unmasking and ionization of two carboxyl groups. Bechet and his coworkers[274-6] studied the effects of inhibitor binding on the conforma-

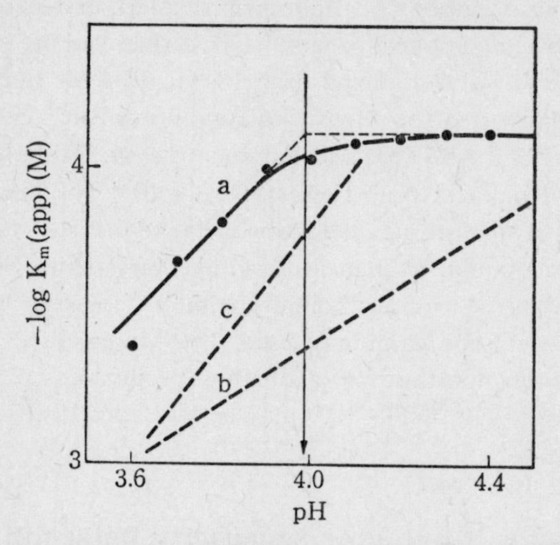

Fig. 12. The pH dependence of $-\log K_m$(app) for the trypsin-catalyzed hydrolysis of BAEE at acid pH. (a) observed, (b) theoretical slope of 1, (c) theoretical slope of 2. Arrow indicates the pK_a value of the ionizing group which causes the pH dependence. Conditions: 0.1 M KCl at 25°C.

tion of trypsin over a wide range of pH. Changes in optical rotation and ultraviolet absorbance were used as indices of the conformational change. As shown by the solid line in Fig. 13, the levorotation of trypsin was increased in both acid and alkaline solutions indicating that the conformational change occurred outside the neutral pH region. The presence of a competitive inhibitor such as *n*-butylamine, benzylamine and benzamidine prevented the increase in levorotation until the pH was shifted to an extreme acid or alkaline value, as shown by the broken line curve. The binding of the competitive inhibitor apparently tends to preserve the native conformation obtainable at neutral pH. Benzamidine, which is very strongly bound, could maintain such a native conformation even below pH 2.0. When the competitive inhibitor was bound to the enzyme in acid solution, proton was liberated. Titration curves for such release of proton as a function of pH indicated that an ionizing group, most probably a carboxyl group, of pK_a 4.5–4.7 was directly involved in the ionic binding with the inhibitor. Furthermore, the affinity of the inhibitor binding was found to be controlled by ionization of a second carboxyl group of pK_a 3.7, which was thought to affect the inhibitor binding by changing the conformation of trypsin.

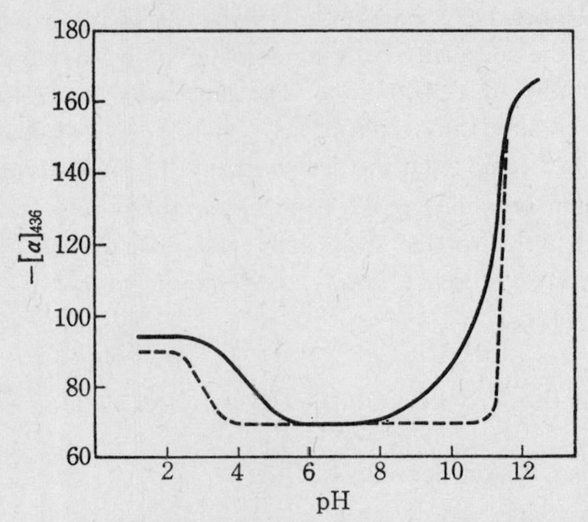

Fig. 13. Effect of pH on the specific rotation of trypsin at 436 nm. The unbroken curve is for trypsin alone. The broken curve was obtained in the presence of 0.4 to 0.6 M *n*-butylamine.

Eyl and Inagami[10] identified a carboxyl group at the inhibitor binding site by a chemical method which involved modification of protein carboxyl groups as developed by Koshland and his co-workers.[277,278] The modification was carried out in two stages. Firstly, the carboxyl group(s) in the specificity site was made inaccessible to modifying reagents by a large excess of a competitive inhibitor, benzamidine. The remaining carboxyl groups were activated by a water-soluble carbodiimide (EDC) and then converted to a peptide containing glycinamide, the latter being used as a nucleophilic acceptor of the activated carboxyl groups as illustrated below:

$$R_1-COOH \; + \; R_2-N=C=N-R_3 \longrightarrow R_1-\overset{\displaystyle O}{\overset{\|}{C}}-O-\underset{\displaystyle NH-R_3}{C}=N-R_2$$

$$\xrightarrow{NH_2-CH_2-\overset{O}{\overset{\|}{C}}NH_2} R_1-\overset{\displaystyle O}{\overset{\|}{C}}NH-CH_2-\overset{\displaystyle O}{\overset{\|}{C}}-NH_2 \; + \; R_2-NH-\overset{\displaystyle O}{\overset{\|}{C}}-NH-R_3$$

After removal of benzamidine by dialysis, the acylation reaction was repeated using ^{14}C-labeled glycinamide in order to label the carboxyl group(s) present in the specificity determining site, which would have been protected initially by benzamidine. The label was found to be bound with Asp-177, originally reported as Asn-177[7,8] Confirmation of Asp instead of Asn came from the determination of the *N*-terminal residue by the Edman degradation of a peptide containing residues 177 through 192, which were isolated from native as well as from DFP-inhibited trypsin[10] (see also ref. 11 and 12). As shown below,

$NH{-}CH_2{-}CONH_2$ (attached through $O{=}C$ to Asp-177)

Trypsin –*Cys*–*Ala*–*Gly*–Trp–Leu–Glu–Gly–Gly–Lys–Asp–*Ser*–*Cys*–
(170 at *Ala*; 177 at Asp)

Chymotrypsin –*Cys*–*Ala*–*Gly*–Ala–Ser–Gly–Val–Ser– –*Ser*–*Cys*–
(184 at *Ala*; 190 at *Ser*)

this residue occurs in a sequence which is not homologous with the corresponding sequence of chymotrypsin. Another acidic residue, Glu-173, is present in this region, and it will be interesting to investigate its function in the substrate binding and/or the conformational change in connection with the findings made by Bechet and D'Albis and discussed above.[274,276] The trypsin sequence in this non-homologous part is longer than that of chymotrypsin by two residues. Addition of the extra amino acid residues is interesting since it may represent an insertion of extra amino acid residues in the course of evolution, in order to create a new enzyme, by adding an anionic binding site to the hydrophobic part of the binding site common to both chymotrypsin and trypsin.

4.4.2. Other Groups

Reversible denaturation of trypsin similar to that occurring in acid solution was also observed in alkaline solution by Bechet and his co-workers,[274,276] and is more pronounced than the acid denaturation as shown in Fig. 13. When the denaturation was followed by measurements of levorotation, it was found to parallel the ionization of a group with a pK_a of 10.1. Furthermore, decrease in the affinity for a competitive inhibitor at alkaline pH was also controlled by the ionization of this group. The inhibitor binding reversed the denaturation. Spectroscopic titrations indicated a decrease of one titrable phenolic hydroxyl group of tyrosine with a pK_a of 10 due to addition of butylamine. These observations all seemed to suggest involvement of a tyrosine residue in the specificity site of trypsin and that its ionization was intimately associated with the conformation and the activity of the enzyme. It is known that ten tyrosine residues are present in the trypsin molecule. Four to six of them are known to be on the outside of the enzyme molecule.[279-84] Chemical modification of the exposed tyrosine residues affected the enzyme activity, but did not cause a complete inhibition of the all-or-none type. For example, Kenner *et al.*[283] found that a rapid partial decrease in the esterase activity occurred after two residues of the tyrosine had been modified by nitration with tetranitromethane. However, further modification of six residues over a prolonged reaction period caused only 30% loss of activity. Trenholm *et al.*[284] acetylated the tyrosine hydroxyl group with *N*-acetylimidazole; a maximum of three residues could be acetylated. The modification caused an approximately two-fold increase in the amidase activity and also increased the extent of substrate activation during the hydrolysis of ester substrates such as TAME and BAEE.[180] However, the degree of acetylation was unaffected by the presence of a competitive inhibitor or a substrate during the modification reaction. These results seem to indicate that the tyrosine residue or residues may not occupy a central position in the active site of trypsin but might possibly be located in the periphery of the site.

Chevallier *et al.*[37] reported observations which seemed to suggest that the ionizing group of pK_a 10, essential for catalysis and maintenance of the native conformation, was the α-amino group of the *N*-terminal isoleucine residue. The ϵ-amino groups of trypsin had previously been selectively acetylated with acetic anhydride at pH 6.8 and 0°C without loss of enzyme activity by Labouesse and Gervais.[285] When 2 M urea was present during the acetylation reaction, the α-amino group was also acetylated and the enzyme activity was then completely lost. On the other hand, *O*-methylation of three exposed tyrosine residues of the fully ϵ-*N*-acetylated trypsin did not affect the dependence of K_m on an ionizing group of pK_a 10.

4.5. Interaction between the Specificity Determining Site and the Catalysis Site, and Significance of the Trypsin Specificity

If the active site of trypsin, or of any enzyme, is rigid, complete by

TABLE 18

AGEE Hydrolysis by Trypsin and α-Chymotrypsin

	Trypsin		α-Chymotrypsin	
	25°C	35°C	25°C	35°C
k_3 (sec^{-1})	0.025	0.054	0.066	0.011
$\Delta H^{\ddagger}$ (kcal/mole)	11		10	
pK of k_3	6.90	6.75	7.03	6.82
K_m (M)	0.79	0.83	0.41	0.46
ΔH^0 (kcal/mole)	0		0	

Determinations were made between pH 6.0 and 7.0. The k_3 values are for pH 8.0, obtained by calculation using the pK of k_3 (ref. 242). The K_m values are independent of pH between pH 6 and 7.

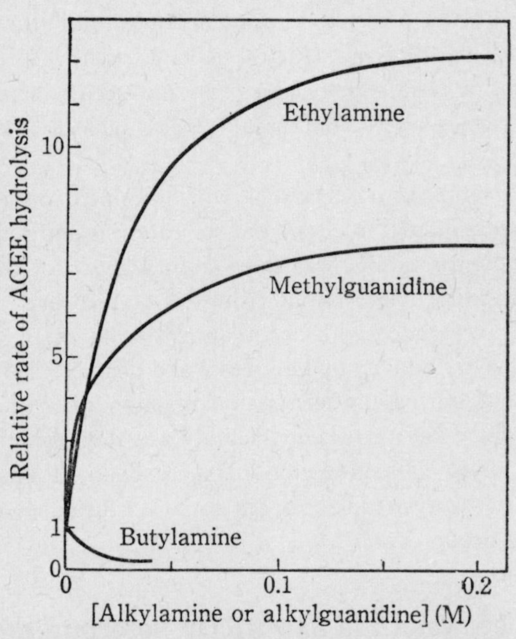

Fig. 14. Activation and inhibition of tryptic hydrolysis of acetylglycine ethyl ester by alkylamines and alkylguanidines. Conditions: pH 6.6 and 25°C.[252,290]

itself and ready for catalysis as proposed by Fisher[286] in his key-lock hypothesis, it is difficult to understand why the rate of hydrolysis of *N*-acetylglycine ethyl ester is lower than that of N^{α}-acetyl-L-lysine methyl ester by as much as 2000-fold (see Table 15) when the former is present at a sufficiently high concentration to saturate the active site. To eliminate such difficulties, Koshland[287,288] proposed a mechanism in which the configuration of the active site is modified to an active form after interaction with a substrate molecule, and coined a term "induced fit" for such a conformational change. The sharp contrast between the efficient catalysis of trypsin on a lysine or arginine substrate and the poor activity towards a glycine substrate indicated that an "induced fit" might take place after binding of the basic side chain structure to the specificity site. In order to see whether such a conformational change might result in the formation of an efficient catalysis site, it was necessary to be able to observe the action of the catalysis sites in isolation. Catalysis of the hydrolysis of *N*-acetylglycine ethyl ester (AGEE), which lacked a side chain and hence did not interact with the specificity site, was chosen as a good model for such an isolated function of the catalysis site. Comparison of the kinetics of trypsin and chymotrypsin catalysis of the hydrolysis of AGEE justified such an approach. As shown in Table 18, the k_3 values for trypsin and chymotrypsin were different only by a factor of two.[246] The difference in the K_m values was also of a similar magnitude. Differences by a factor of 2 indicated a remarkable similarity in view of the large variation, encompassing several orders of magnitude in these constants when different substrates were examined with the same enzyme (see Table 15). A similar closeness in the kinetic parameters was also observed during hydrolysis of NPA, as shown in Table 6.[56,91] We have seen the close similarity between trypsin and chymotrypsin in the amino acid sequence of peptides located in the area of their catalysis sites. We have also discussed the difference in the amino acid sequences of the specificity site. The closeness in the kinetics of the AGEE hydrolysis was therefore interpreted to indicate that catalysis of the AGEE hydrolysis was achieved primarily by the action of the catalysis site alone.

Using this AGEE-trypsin system, Inagami and Murachi[289,290] studied the effect of the side chain binding to the specificity site on the activity of the catalysis site. Alkylamines and alkylguanidines were used as models for the side chains of basic amino acids in a typical substrate

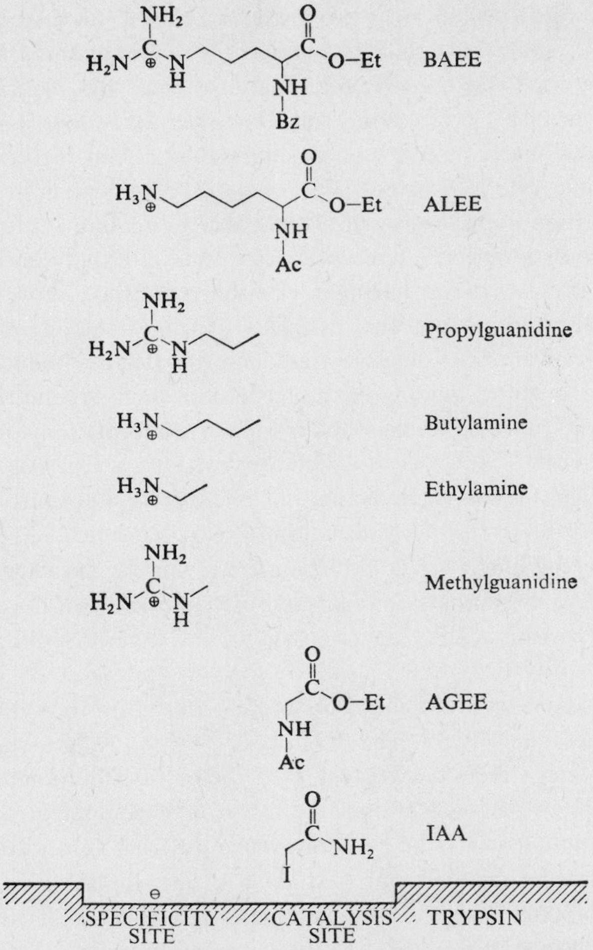

BAEE

ALEE

Propylguanidine

Butylamine

Ethylamine

Methylguanidine

AGEE

IAA

SPECIFICITY SITE

CATALYSIS SITE

TRYPSIN

Fig. 15. Schematic representation of the specificity site and catalysis site of trypsin. Dimensions of the sites are measured by the size of inhibitors and substrates which interact with the enzyme.

molecule. Addition of methyl-, ethyl-, or propylamine or methyl- or ethylguanidine caused a remarkable increase in the rate of catalysis, as shown in Fig. 14. Interestingly, alkylamines or alkylguanidines with longer hydrocarbon chains such as *n*-butylamine or 1-propylguanidine inhibited the AGEE hydrolysis instead of increasing the rate. Fig. 15 illustrates with BAEE and ALEE that the distances between the center of the positive charge on their side chains and the carbonyl carbon atom are approximately identical with those of arginine and lysine. When *n*-butylamine or 1-propylguanidine and AGEE are arranged in a series, the corresponding distance may be only slightly greater because of the overlap of C–H bonds at the junction of the two molecules. This slight difference seems to have caused the inhibition. The dissociation constants for the alkylamines and alkylguanidines, obtained by analysis of the results from the activation experiments, with AGEE as substrate, and K_i for these amines and guanidines, obtained from inhibition experiments using BAEE as substrate, agreed very well. These facts indicated that the alkylamine or alkylguanidine bound at the specificity site caused the activation. In other words, activation by ethylamine can be considered as mimicking the action of the side chain binding of the lysine substrate, ALEE. Thus it has become clear that the function of the specificity site is not limited to the recognition and binding of a specific substrate. The binding causes a conformational change of the catalysis site to a catalytically efficient form. Further detailed study of this "side chain activation" revealed that the activation was caused by an increase in k_3 and not due to stronger binding or decrease in K_m, and also that the increase in k_3 was the result of an increased entropy of activation and the heat of activation was not affected.

Inagami and Hatano[73] found that increase in the catalytic rate of the AGEE hydrolysis due to the side chain activation was accompanied by an increase in the chemical reactivity of an imidazole group of a histidine residue in the catalysis site. The rate of alkylation of the imidazole group by iodoacetamide (IAA) was determined in the presence and absence of methylguanidine. Methylguanidine, at its saturation concentration, increased the rate by a factor of 7. The rate of the AGEE hydrolysis by methylguanidine was also enhanced by a factor of 7. As illustrated in Fig. 15, IAA and AGEE have similar molecular dimensions; therefore, their approach to the catalysis site will not be sterically hindered by a methylguanidine molecule bound at the specificity site. A close relationship between the increase in catalytic efficiency and the increased chemical reactivity in the nucleophilic attack may be taken as an indication that the latter is a mechanism through which the side chain activation works.

The relationship between the side chain activation and the substrate activation, discussed in section 3.5, is not quite clear. Erlanger and Castleman[291] observed that alkylamines with relatively large hydrocarbon chains, such as *n*-hexyl-, *n*-heptylamine, 2-aminopentane and 2-aminoheptane, activated the AGEE hydrolysis to a considerable extent. It is possible that these large amines are bound at the second binding site and activation similar to the substrate activation occurs. Whatever the explanation for this observation may be, there is little doubt that the side chain activation was caused by binding of the alkylamines or guanidines at the primary binding site, as indicated by a very rigid requirement for the activator size and by a good agreement in the measured dissociation constant for the activator-enzyme complex, determined independently by the activation and the inhibition experiments.

Acknowledgments

This work was supported by research grants from the NIH (GM-14703) and NSF (GB-27583). The author is greatly indebted to able assistance by Mrs. C. Blythe in preparing the manuscript.

References

1. L. W. Cunningham, *Comprehensive Biochemistry* (ed. M. Florkin and E. H. Stolz), vol. 16, p. 85, Elsevier, 1965.
2. M. L. Bender and F. J. Kezdy, *Ann. Rev. Biochem.*, **34**, 49 (1965).
3. P. Desnuelle, *The Enzymes* (ed. P. D. Boyer *et al.*), vol. 4, part A, p. 119, Academic Press, 1960.
4. K. A. Walsh, *Methods Enzymol.*, **19**, 41 (1970).
5. B. Keil, *The Enzymes* (ed. P. D. Boyer), vol. 3, p. 250, Academic Press, 1971.
6. K. A. Walsh, D. L. Kauffmann, K. S. V. Sampath Kumar and H. Neurath, *Proc. Natl. Acad. Sci. U.S.*, **51**, 301 (1964).
7. K. A. Walsh and H. Neurath, *ibid.*, **52**, 889 (1964).
8. O. Mikeš, V. Tomášek, V. Holeyšovský and F. Šorm, *Collection Czech. Chem. Commun.*, **32**, 655 (1967).
9. T. Hofmann, *Biochemistry*, **3**, 356 (1964).
10. A. W. Eyl and T. Inagami, *Biochem. Biophys. Res. Commun.*, **38**, 149 (1970); *J. Biol. Chem*, **246**, 738 (1971).
11. K. A. Walsh, see footnote, p. 4710 of ref. 43.
12. B. S. Hartley, see footnote, p. 4710 of ref. 43.

13. H. Neurath, K. A. Walsh and W. P. Winter, *Science*, **158**, 1638 (1967).
14. B. S. Hartley, J. R. Brown, D. L. Kauffman and L. B. Smillie, *Nature*, **207**, 1157 (1965).
15. B. S. Hartley, *ibid.*, **201**, 1284 (1964).
16. D. M. Blow, J. J. Birktoft and B. S. Hartley, *ibid.*, **221**, 337 (1967).
17. D. L. Kauffman, *J. Mol. Biol.*, **12**, 929 (1965).
18. J. R. Brown and B. S. Hartley, *Biochem. J.*, **89**, 59P (1963).
19. B. Keil, Z. Prusík and F. Šorm, *Biochim. Biophys. Acta*, **78**, 559 (1963).
20. M. Rovery, M. Charles, A. Grey, A. Guidoni and P. Desnuelle, *Bull. Soc. Chim. Biol.*, **42**, 1235 (1960).
21. M. Charles, M. Rovery, A. Guidoni and P. Desnuelle, *Biochim. Biophys. Acta*, **69**, 115 (1963).
22. J. Travais and I. E. Liener, *Arch. Biochem. Biophys.*, **97**, 218 (1962).
23. L. P. Chao and I. E. Liener, *Biochim. Biophys. Acta*, **96**, 508 (1965).
24. P. J. Van Melle, S. H. Lewis, E. S. Samsa and R. J. Westfall, *Enzymologia*, **26**, 113 (1963).
25. J. Travais and I. E. Liener, *J. Biol. Chem.*, **240**, 1967 (1965).
26. R. A. Smith and I. E. Liener, *ibid.*, **242**, 4033 (1967).
27. S. Bricteux-Gregoire, R. Schyns and M. Florkin, *Biochim. Biophys. Acta*, **127**, 277 (1966).
28. R. Schyns, S. Bricteux-Gregoire and M. Florkin *ibid.*, **175**, 97 (1969).
29. F. F. Buck, A. J. Vithayathil, M. Bier and F. F. Nord, *Arch. Biochem. Biophys.*, **97**, 417 (1962).
30. A. J. Vithayathil, F. F. Buck, M. Bier and F. F. Nord, *ibid.*, **92**, 532 (1961).
31. F. F. Buck, M. Bier and F. F. Nord, *ibid.*, **98**, 528 (1962).
32. P. J. Keller and B. J. Allan, *J. Biol. Chem.*, **242**, 281 (1967).
33. J. Travais and R. C. Roberts, *Biochemistry*, **8**, 2884 (1969).
34. E. W. Davie and H. Neurath, *J. Biol. Chem.*, **212**, 515 (1955).
35. S. Bricteux-Gregoire, R. Schyns and M. Florkin, *Arch. Intern. Physiol. Biochim.*, **76**, 751 (1968).
36. S. T. Scrimger and T. Hofmann, *J. Biol. Chem.*, **242**, 2528 (1967).
37. J. Chevallier, J. Yon and J. Labouesse, *Biochim. Biophys. Acta*, **181**, 73 (1969).
38. B. Labouesse, H. L. Oppenheimer and G. P. Hess, *Biochem. Biophys. Res. Commun.*, **14**, 318 (1964).
39. H. L. Oppenheimer, B. Labouesse and G. P. Hess, *J. Biol. Chem.*, **241**, 2720 (1966).
40. D. D. Schroeder and E. Shaw, *ibid.*, **243**, 2943 (1968).
41. S. Maroux, M. Rovery and P. Desnuelle, *Biochim. Biophys. Acta*, **140**, 377 (1967).
42. S. Maroux and P. Desnuelle, *ibid.*, **181**, 59 (1969).
43. R. L. Smith and E. Shaw, *J. Biol. Chem.*, **244**, 4704 (1969).
44. E. F. Jansen, M. D. F. Nutting, R. Jang and A. K. Balls, *ibid.*, **179**, 189 (1949).

45. A. K. Balls and E. F. Jansen, *Advan. Enzymol.*, **13**, 321 (1952).
46. G. H. Dixon, D. L. Kauffman and H. Neurath, *J. Biol. Chem.*, **233**. 1373 (1958).
47. R. A. Oosterbaan, P. Kunst, J. van Rotterdam and J. A. Cohen, *Biochim. Biophys. Acta*, **27**, 556 (1958).
48. N. K. Schaffer, R. P. Lang, L. Simet and R. W. Drisco, *J. Biol. Chem.*, **230**, 185 (1958).
49. J. A. Gladner and K. Laki, *J. Am. Chem. Soc.*, **80**, 1263 (1958).
50. M. A. Naughton, F. Sanger, B. S. Hartley and D. C. Shaw, *Biochem, J.*, **77**, 149 (1960).
51. J. R. Brown, D. L. Kauffman and B. S. Hartley, *ibid.*, **103**, 497 (1967).
52. F. Sanger and D. C. Shaw, *Nature,* **187**, 872 (1960).
53. E. L. Smith, F. S. Markland, C. B. Kasper, R. J. DeLange, M. Landon and W. H. Evans, *J. Biol. Chem.*, **241**, 5974 (1966).
54. F. Sanger, *Proc. Chem. Soc.*, **1963**, 76.
55. R. A. Oosterbaan, M. van Adrichem and J. A. Cohen, *Biochim. Biophys. Acta*, **63**, 204 (1962).
56. J. A. Stewart and J. Ouellet, *Can. J. Chem.*, **37**, 751 (1959).
57. M. L. Bender and E. T. Kaiser, *J. Am. Chem. Soc.*, **84**, 2556 (1962).
58. M. Mares-Guia and E. Shaw, *J. Biol. Chem.*, **242**, 5782 (1967).
59. T. Chase and E. Shaw, *Biochem. Biophys. Res. Commun.*, **29**, 508 (1967).
60. B. F. Erlanger and W. Cohen, *J. Am. Chem. Soc.*, **85**, 348 (1963).
61. D. E. Fahrney and A. N. Gold, *ibid.*, **85**, 997 (1963).
62. H. Gutfreund, *Trans. Faraday Soc.*, **51**, 441 (1951).
63. T. Inagami and J. M. Sturtevant, *Biochim. Biophys. Acta*, **38**, 64 (1960).
64. B. R. Hammond and H. Gutfreund, *Biochem. J.*, **61**, 187 (1955).
65. L. W. Cunningham and C. S. Brown, *J. Biol. Chem.*, **221**, 287 (1956).
66. E. Shaw, M. Mares-Guia and W. Cohen, *Biochemistry*, **4**, 2219 (1965).
67. E. S. Severin and V. Tomášek, *Biochem. Biophys. Res. Commun.*, **20**, 496 (1965).
68. E. Shaw and S. Springhorn, *ibid.*, **27**, 391 (1967).
69. V. Tomášek, E. S. Severin and F. Šorm, *ibid.*, **20**, 545 (1965).
70. P. H. Petra, W. Cohen and E. Shaw, *ibid.*, **21**, 612 (1965).
71. J. G. Beeley and H. Neurath, *Biochemistry*, **7**, 1239 (1968).
72. T. Inagami, *J. Biol. Chem.*, **240**, 3453 (1965).
73. T. Inagami and H. Hatano, *ibid.*, **244**, 1176 (1969).
74. C. Schoellmann and E. Shaw, *Biochemistry*, **2**, 252 (1963).
75. Y. Nakagawa and M. L. Bender, *ibid.*, **9**, 259 (1970).
76. B. W. Mathews, P. B. Sigler, R. Henderson and D. M. Blow, *Nature*, **214**, 652 (1967).
77. W. B. Lawson, M. D. Leafer Jr., A. Tewels and G. J. S. Rao, *Z. Physiol. Chem.*, **349**, 251 (1968).
78. R. M. Stroud, L. M. Kay and R. E. Dickerson, *Cold Spring Harbor Symp. Quant. Biol.*, **36** (1971).

79. L. B. Smillie, A. Furka, N. Nagabushan, K. J. Stevenson and C. O. Parkes, *Nature*, **218**, 343 (1968).
80. F. Šorm, V. Holeyšovský, O. Mikeš and V. Tomášek, *Collection Czech. Chem. Commun.*, **30**, 2105 (1965).
81. B. S. Hartley and B. A. Kilby, *Biochem. J.*, **56**, 288 (1954).
82. T. C. Bruice and S. Benkovic, *Bio-organic Mechanisms*, vol. 1, chap. 2, Benjamin, 1966.
83. A. K. Balls and F. L. Aldrich, *Proc. Natl. Acad. Sci. U.S.*, **41**, 190 (1955).
84. A. K. Balls and H. N. Wood, *J. Biol. Chem.*, **219**, 245 (1956).
85. A. K. Balls and C. E. McDonald, *ibid.*, **221**, 993 (1956).
86. G. H. Dixon and H. Neurath, *ibid.*, **225**, 1049 (1957).
87. G. H. Dixon, W. J. Dreyer and H. Neurath, *J. Am. Chem. Soc.*, **78**, 4810 (1956).
88. R. A. Oosterbaan, M. van Adrichem and J. A. Cohen, *Biochim. Biophys. Acta*, **63**, 204 (1962).
89. H. Gutfreund and J. M. Sturtevant, *Proc. Natl. Acad. Sci. U.S.*, **42**, 719 (1956).
90. I. B. Wilson and E. Gabib, *J. Am. Chem. Soc.*, **78**, 202 (1956).
91. H. Gutfreund and J. M. Sturtevant, *Biochem. J.*, **63**, 656 (1956).
92. G. S. Eadie, *J. Biol. Chem.*, **146**, 85 (1942).
93. H. Lineweaver and D. Burk, *J. Am. Chem. Soc.*, **56**, 658 (1934).
94. T. Spencer and J. M. Sturtevant, *ibid.*, **81**, 1874 (1959).
95. L. Faller and J. M. Sturtevant, *J. Biol. Chem.*, **241**, 4825 (1966).
96. H. Gutfreund and B. R. Hammond, *Biochem. J.*, **73**, 526 (1959).
97. F. J. Kezdy, G. E. Clement and M. L. Bender, *J. Am. Chem. Soc.*, **86**, 3690 (1964).
98. B. Zerner and M. L. Bender, *ibid.*, **85**, 356 (1963).
99. G. E. Clement and M. L. Bender, *Biochemistry*, **2**, 836 (1963).
100. M. L. Bender, G. E. Clement, F. J. Kezdy and H. d'A. Heck, *J. Am. Chem. Soc.*, **86**, 3680 (1964).
101. M. L. Bender and W. A. Glasson, *ibid.*, **82**, 3336 (1960).
102. B. Zerner, R. P. M. Bond and M. L. Bender, *ibid.*, **86**, 3678 (1964).
103. N. J. Baines, J. B. Baird and D. T. Elmore, *Biochem. J.*, **90**, 470 (1964).
104. D. T. Elmore, D. V. Roberts and J. J. Smyth, *ibid.*, **102**, 728 (1967).
105. M. L. Bender and F. J. Kezdy, *J. Am. Chem. Soc.*, **87**, 4954 (1965).
106. A. K. Balls, C. E. McDonald and A. S. Brecher, *Proc. Intern. Symp. Enzyme Chem., Tokyo-Kyoto*, p. 392, 1958.
107. M. L. Bender, G. R. Schonbaum and B. Zerner, *J. Am. Chem. Soc.*, **84**, 2562 (1962).
108. G. R. Schonbaum, B. Zerner and M. L. Bender, *J. Biol. Chem.*, **236**, 2930 (1961).
109. M. L. Bender and B. Zerner, *J. Am. Chem. Soc.*, **84**, 2250 (1964).
110. M. L. Bender, G. R. Schonbaum and B. Zerner, *ibid.*, **84**, 2540 (1964).
111. M. L. Bender and E. T. Kaiser, *ibid.*, **84**, 2556 (1964).

112. L. W. Cunningham, *J. Biol. Chem.*, **207**, 443 (1954).
113. W. Cohen and B. F. Erlanger, *J. Am. Chem. Soc.*, **82**, 3928 (1960).
114. W. Cohen and B. F. Erlanger, *Biochemistry*, **1**, 686 (1962).
115. D. K. Myers and A. Kemp Jr., *Nature*, **173**, 33 (1954).
116. V. Massey, W. F. Harrington and B. S. Hartley, *Discussions Faraday Soc.*, **20**, 24 (1955).
117. A. M. Gold and D. Fahrney, *Biochemistry*, **3**, 783 (1964).
118. B. F. Erlanger, A. G. Cooper and W. Cohen, *ibid.*, **5**, 190 (1966).
119. B. F. Erlanger, H. Castleman and A. G. Cooper, *J. Am. Chem. Soc.*, **85**, 1872 (1963).
120. D. T. Elmore and J. J. Smyth, *Biochem. J.*, **107**, 103 (1968).
121. M. Mares-Guia and E. Shaw, *J. Biol. Chem.*, **242**, 5782 (1967).
122. M. Mares-Guia, E. Shaw and W. Cohen, *ibid.*, **242**, 5777 (1967).
123. K. Tanizawa, S. Ishii and Y. Kanaoka, *Biochem. Biophys. Res. Commun.*, **32**, 893 (1968).
124. H. Mix, H. J. Trettin and M. Gulzow, *Z. Physiol. Chem.*, **349**, 1237 (1968).
125. M. L. Bender and G. A. Hamilton, *J. Am. Chem. Soc.*, **84**, 2570 (1962).
126. S. A. Bernhard, W. C. Coles and J. F. Nowell, *ibid.*, **82**, 3043 (1960).
127. R. M. Epand and I. B. Wilson, *J. Biol. Chem.*, **238**, 1718 (1963).
128. T. Inagami and J. M. Sturtevant, *Biochem. Biophys. Res. Commun.*, **14**, 69 (1964).
129. M. Caplow and W. P. Jencks, *J. Biol. Chem.*, **238**, 1907 (1963).
130. M. Caplow and W. P. Jencks, *ibid.*, **239**, 1640 (1964).
131. R. M. Epand and I. B. Wilson, *ibid.*, **238**, 3138 (1963).
132. T. Inagami, *J. Biochem.* (*Tokyo*), **66**, 277 (1969).
133. M. L. Bender, G. E. Clement, C. R. Gunter and F. J. Kezdy, *J. Biol. Chem.*, **238**, 3143 (1963).
134. A. N. Glazer, *ibid.*, **240**, 135 (1965).
135. D. B. Sprinson and F. Rittenberg, *Nature*, **167**, 484 (1951).
136. D. G. Doherty and F. Vaslow, *J. Am. Chem. Soc.*, **74**, 931 (1952).
137. F. Vaslow, *Compt. Rend. Trav. Lab. Carlsberg, Ser. Chim.*, **30**, 45 (1956).
138. M. L. Bender and K. C. Kemp, *J. Am. Chem. Soc.*, **79**, 116 (1957).
139. G. Gutfreund, *Discussions Faraday Soc.*, **20**, 167 (1955).
140. G. Johannin and J. Yon, *Biochem. Biophys. Res. Commun.*, **25**, 320 (1966).
141. B. J. Haverback, B. Dyce and H. A. Edmondson, *Am. J. Med.*, **29**, 424 (1960).
142. B. F. Erlanger, N. Kokowsky and W. Cohen, *Arch. Biochem. Biophys.*, **95**, 271 (1961).
143. H. Tuppy, U. Weisbauer and E. Winterberger, *Z. Physiol. Chem.*, **329**, 278 (1962).
144. S. A. Bernhard and B. F. Lee, *Abstr. 6th Intern. Congr. Biochem., New York*, 1964.

145. A. N. Glazer, *Proc. Natl. Acad. Sci. U.S.*, **54**, 171 (1965).
146. S. A. Bernhard and H. Gutfreund, *ibid.*, **53**, 1238 (1965).
147. T. E. Barman and H. Gutfreund, *Rapid Mixing and Sampling Techniques* (ed. B. Chance, R. H. Eisenhardt and K. K. Lonbert-Holm), p. 339, Academic Press, 1964.
148. T. E. Barman and H. Gutfreund, *Proc. Natl. Acad. Sci. U.S.*, **53**, 1243 (1965).
149. T. E. Barman and H. Gutfreund, *Biochem. J.*, **101**, 411 (1965).
150. W. F. Sager and P. C. Parks, *J. Am. Chem. Soc.*, **85**, 2679 (1963).
151. W. F. Sager and P. C. Parks, *Proc. Natl. Acad. Sci. U.S.*, **52**, 408 (1964).
152. L. P. Hammett, *Physical Organic Chemistry*, 1st ed., p. 269, McGraw-Hill, 1940.
153. T. Inagami, S. S. York and A. Patchornik, *J. Am. Chem. Soc.*, **87**, 126 (1965).
154. T. Inagami, A. Patchornik and S. S. York, *J. Biochem. (Tokyo)*, **65**, 809 (1969).
155. M. Eigen and L. DeMaeyer, *Proc. Roy. Soc. (London)*, **A247**, 505 (1958).
156. J. H. Wang and L. Parker, *Proc. Natl. Acad. Sci. U.S.*, **58**, 2451 (1967).
157. J. H. Wang, *Science*, **161**, 328 (1968).
158. L. Parker and J. H. Wang, *J. Biol. Chem.*, **243**, 3729 (1968).
159. M. Caplow, *J. Am. Chem. Soc.*, **91**, 3639 (1969).
160. L. W. Cunningham, *Science*, **125**, 1145 (1957).
161. H. Neurath, *Advan. Protein Chem.*, **12**, 320 (1957).
162. T. C. Bruice, *Proc. Natl. Acad. Sci. U.S.*, **47**, 1924 (1961).
163. M. L. Bender, *J. Am. Chem. Soc.*, **84**, 2882 (1962).
164. C. H. Johnson and J. R. Knowles, *Biochem. J.*, **101**, 56 (1966).
165. R. Henderson, *ibid.*, **124**, 13 (1971).
166. J. Chevallier and J. Yon, *Biochim. Biophys. Acta*, **122**, 116 (1966).
167. G. W. Schwert and Y. Takenaka, *ibid.*, **16**, 570 (1955).
168. S. S. Wang and F. H. Carpenter, *J. Biol. Chem.*, **243**, 3702 (1968).
169. A. Himoe and G. P. Hess, *Biochem. Biophys. Res. Commun.*, **23**, 234 (1966).
170. A. Himoe, P. C. Parks and G. P. Hess, *J. Biol. Chem.*, **242**, 919 (1967).
171. P. B. Sigler, D. M. Blow, B. W. Mathews and R. Henderson, *J. Mol. Biol.*, **35**, 143 (1968).
172. M. Dixon, *Biochem. J.*, **55**, 161 (1953).
173. J.-J. Bechet, *J. Chim. Phys.*, **61**, 584 (1964).
174. J.-J. Bechet, *ibid.*, **62**, 1095 (1965).
175. M. L. Bender, M. J. Gibian and D. J. Wheelan, *Proc. Natl. Acad. Sci. U.S.*, **56**, 833 (1966).
176. M. Caplow and W. P. Jencks, *Biochemistry*, **1**, 883 (1962).
177. C. G. Trowbridge, A. Krehbiel and M. Laskowski Jr., *ibid.*, **2**, 843 (1963).

178. E. F. Curragh and D. T. Elmore, *Biochem. J.*, **93**, 163 (1964).
179. J.-J. Bechet and J. Yon, *Biochim. Biophys. Acta*, **89**, 117 (1964).
180. H. L. Trenholm, W. E. Spomer and J. F. Wootton, *J. Am. Chem. Soc.*, **88**, 4281 (1966).
181. S. M. Howard and J. W. Mehl, *Biochim. Biophys. Acta*, **105**, 594 (1965).
182. N. Nakata and S. Ishii, *Biochem. Biophys. Res. Commun.*, **41**, 393 (1970).
183. J. F. Wootton and G. P. Hess, *J. Am. Chem. Soc.*, **84**, 440 (1962).
184. B. Labouesse, B. H. Havesteen and G. P. Hess, *Proc. Natl. Acad. Sci. U.S.*, **48**, 2137 (1962).
185. H. L. Oppenheimer, J. Mercouroff and G. P. Hess, *Biochim. Biophys. Acta*, **71**, 78 (1963).
186. B. H. Havesteen and G. P. Hess, *J. Am. Chem. Soc.*, **85**, 791 (1963).
187. A. Himoe, K. G. Brand and G. P. Hess, *J. Biol. Chem.*, **242**, 3963 (1967).
188. K. Imahori, A. Yoshida and H. Hashizume, *Biochim. Biophys. Acta*, **45**, 380 (1960).
189. B. H. Havesteen, B. Labouesse and G. P. Hess, *J. Am. Chem. Soc.*, **85**, 796 (1963).
190. A. Y. Moon, J. Mercouroff and G. P. Hess, *J. Biol. Chem.*, **240**, 717 (1965).
191. A. Y. Moon, J. M. Sturtevant and G. P. Hess, *ibid.*, **240**, 4204 (1965).
192. H. Wiener and D. E. Koshland Jr., *J. Mol. Biol.*, **12**, 881 (1965).
193. R. Biltonen, R. Lumry, V. Madison and H. Parker, *Proc. Natl. Acad. Sci. U.S.*, **54**, 1018 (1965).
194. B. F. Erlanger, H. Castleman and A. G. Cooper, *J. Am. Chem. Soc.*, **85**, 1972 (1963).
195. T. R. Hopkins and J. D. Spikes, *Biochem. Biophys. Res. Commun.*, **30**, 540 (1968).
196. T. R. Hopkins and J. D. Spikes, *ibid.*, **28**, 480 (1967).
197. A. Kotaki, F. Usuki and K. Satake, *Seikagaku* (Japanese), **33**, 475 (1961).
198. A. Kotaki, *J. Biochem. (Tokyo)*, **51**, 301 (1962).
199. J. D. Young and F. H. Carpenter, *J. Biol. Chem.*, **236**, 743 (1961).
200. F. H. Carpenter and W. E. Baum, *ibid.*, **237**, 409 (1962).
201. H. Matsubara, R. M. Sasaki and R. K. Chain, *Proc. Natl. Acad. Sci. U.S.*, **57**, 441 (1967).
202. T. Inagami and J. M. Sturtevant, *J. Biol. Chem.*, **235**, 1019 (1960).
203. R. D. Cole and J. M. Kinkade Jr., *ibid.*, **236**, 2446 (1961).
204. V. Kostka and F. H. Carpenter, *ibid.*, **239**, 1799 (1964).
205. C. H. W. Hirs, S. Moore and W. H. Stein, *ibid.*, **216**, 623 (1956).
206. R. G. Shepherd, S. D. Willson, K. S. Howard, P. H. Bell, D. S. Davies, S. B. Davis, E. A. Eigner and N. E. Shakespeare, *J. Am. Chem. Soc.*, **78**, 5067 (1956).
207. C. H. Li, J. S. Dixon and D. Chung, *ibid.*, **80**, 2587 (1958).

208. C. H. Li, I. I. Geschwind, R. D. Cole, I. D. Raake, J. I. Harris and J. S. Dixon, *Nature*, **176**, 687 (1955).
209. B. Kassell and M. Laskowski Sr., *Biochem. Biophys. Res. Commun.*, **17**, 792 (1964).
210. J. Chauvet, G. Nouvel and R. Acher, *Biochim. Biophys. Acta*, **92**, 200 (1964).
211. T. Ando, S. Ishii and M. Yamasaki, *ibid.*, **34**, 600 (1959).
212. S. Ishii, M. Yamasaki and T. Ando, *J. Biochem.* (*Tokyo*), **61**, 687 (1967).
213. H. Hayashida and M. Yamasaki, *Abstr. 7th Intern. Congr. Biochem., Tokyo*, F-53, 1967.
214. S. G. Waley and J. Watson, *Biochem. J.*, **55**, 328 (1953).
215. Y. Levin, A. Berger and E. Katchalski, *ibid.*, **63**, 308 (1956).
216. M. M. Nachlas, R. E. Plapinger and A. M. Seligman, *Arch. Biochem. Biophys.*, **108**, 266 (1964).
217. A. M. Crestfield, S. Moore and W. H. Stein, *J. Biol. Chem.*, **238**, 622 (1963).
218. B. V. Plapp, M. A. Raftery and R. D. Cole, *ibid.*, **242**, 265 (1967).
219. H. Matsubara and E. L. Smith, *ibid.*, **238**, 2732 (1963).
220. W. Konigsberg and R. J. Hill, *ibid.*, **237**, 2547 (1962).
221. J. Goldstein, W. Konigsberg and R. J. Hill, *ibid.*, **238**, 2016 (1963).
222. M. Tanaka, T. Nakashima, A. M. Benson, H. Mower and K. T. Yasunobu, *Biochemistry*, **5**, 1666 (1966).
223. C. B. Anfinsen, M. Sela and H. Tritch, *Arch. Biochem. Biophys.*, **65**, 156 (1956).
224. R. F. Goldberger and C. B. Anfinsen, *Biochemistry*, **1**, 401 (1962).
225. R. K. Redfield and C. B. Anfinsen, *J. Biol. Chem.*, **221**, 385 (1956).
226. G. R. Stark, *Methods Enzymol.*, **11**, 590 (1967).
227. S. Korman and H. T. Clarke, *J. Biol. Chem.*, **221**, 133 (1956).
228. P. J. G. Butler, J. I. Harris, B. S. Hartley and R. Leberman, *Biochem. J.*, **112**, 679 (1969).
229. G. S. Shields, R. L. Hill and E. L. Smith, *J. Biol. Chem.*, **234**, 1747 (1959).
230. M. J. Hunter and M. L. Ludwig, *J. Am. Chem. Soc.*, **84**, 3491 (1962).
231. H. A. Itano and A. J. Gottlieb, *Biochem. Biophys. Res. Commun.*, **12**, 405 (1963).
232. M. Bergmann, J. S. Fruton and H. Pollock, *J. Biol. Chem.*, **127**, 643 (1937).
233. G. W. Schwert, H. Neurath, S. Kaufman and J. E. Snoke, *ibid.*, **172**, 221 (1948).
234. T. Yamamoto and N. Izumiya, *Arch. Biochem. Biophys.*, **120**, 497 (1967).
235. N. Izumiya and H. Uchio, *J. Biochem.* (*Tokyo*), **46**, 645 (1959).
236. G. I. Tessier, R. J. F. Nivard and M. Gruber, *Biochim. Biophys. Acta*, **89**, 303 (1964).

237. H. Lindley, *Australian J. Chem.*, **12**, 296 (1959).
238. H. Lindley, *Nature*, **178**, 647 (1956).
239. F. Tietze, J. A. Gladner and J. E. Folk, *Biochim. Biophys. Acta*, **26**, 659 (1957).
240. M. A. Raftery and R. D. Cole, *J. Biol. Chem.*, **241**, 3457 (1966).
241. Y. Shalitin, *Bull. Res. Council Israel*, **A10**, 34 (1961).
242. B. H. Hofstee, *Biochim. Biophys. Acta*, **24**, 211 (1957).
243. L. Benoiton and J. Deneault, *ibid.*, **113**, 613 (1965).
244. B. M. Sanborn and G. E. Hein, *Biochemistry*, **7**, 3616 (1968).
245. M. Gorecki and Y. Shalitin, *Biochem. Biophys. Res. Commun.*, **29**, 189 (1967).
246. T. Inagami and H. Mitsuda, *J. Biol. Chem.*, **239**, 1388 (1964).
247. N. M. Green and H. Neurath, *The Proteins* (ed. H. Neurath and K. Bailey), vol. 2B, p. 1057, Academic Press, 1954.
248. M. Ebata and K. Morita, *J. Biochem.* (*Tokyo*), **46**, 407 (1959).
249. J. B. Baird, E. F. Curragh and D. T. Elmore, *Biochem. J.*, **90**, 470 (1964).
250. A. Riedel and E. Wunsch, *Z. Physiol. Chem.*, **316**, 61 (1959).
251. N. Nishi and J. Noguchi, *Proc. 7th Symp. Peptide Chem.*, p. 1, 1970.
252. T. Inagami and S. S. York, *Biochemistry*, **7**, 4045 (1968).
253. T. Inagami, *J. Biol. Chem.*, **239**, 787 (1964).
254. H. Mix, H. J. Trettin and M. Gulzow, *Z. Physiol. Chem.*, **343**, 52 (1965).
255. J. D. Geratz, *Experientia*, **21**, 699 (1965).
256. M. Mares-Guia, *Arch. Biochem. Biophys.*, **127**, 317 (1968).
257. M. Mares-Guia and E. Shaw, *J. Biol. Chem.*, **240**, 1579 (1950).
258. A. N. Glazer, *ibid.*, **242**, 3326 (1967).
259. H. Landmann, F. Markwardt, H.-G. Kazmirowski and P. Neuland, *Z. Physiol. Chem.*, **348**, 745 (1967).
260. A. N. Glazer, *J. Biol. Chem.*, **242**, 4528 (1967).
261. W. Kauzmann, *Advan. Protein Chem.*, **14**, 1 (1959).
262. C. Tanford, *J. Am. Chem. Soc.*, **84**, 4240 (1962).
263. J. D. Geratz, *Arch. Biochem. Biophys.*, **102**, 327 (1963).
264. J. D. Geratz, *ibid.*, **110**, 150 (1965).
265. J. D. Geratz, *Experientia*, **22**, 73 (1966).
266. F. J. Kezdy, *Biochim. Biophys. Acta*, **132**, 197 (1967).
267. R. C. W. Hummel, *Can. J. Biochem. Physiol.*, **39**, 1193 (1961).
268. H. J. Trettin and H. Mix, *Z. Physiol. Chem.*, **340**, 24 (1965).
269. F. Markwardt, H. Landmann and A. Hoffmann, *ibid.*, **340**, 174 (1965).
270. M. Gulzow, H. Mix and H. J. Trettin, *ibid.*, **348**, 285 (1967).
271. J. A. Stewart and J. E. Dobson, *Biochemistry*, **4**, 1086 (1965).
272. A. D'Albis, *Compt. Rend.*, **259**, 1779 (1964).
273. M. Lazdunski and M. DeLaage, *Biochim. Biophys. Acta*, **140**, 417 (1967).
274. A. D'Albis and J.-J. Bechet, *ibid.*, **140**, 435 (1967).

275. J.-J. Bechet, M. C. Gardinnet and J. Yon, *ibid.*, **122**, 101 (1966).
276. J.-J. Bechet and A. D'Albis, *ibid.*, **178**, 561 (1969).
277. D. G. Hoare and D. E. Koshland Jr., *J. Biol. Chem.*, **242**, 2447 (1967).
278. K. L. Carraway and D. E. Koshland Jr., *Biochim. Biophys. Acta*, **160**, 272 (1968).
279. Y. Inada, A. Kamata, A. Matsushima and K. Shibata, *ibid.*, **81**, 323 (1964).
280. M. DeLaage and M. Lazdunski, *ibid.*, **105**, 523 (1965).
281. J. F. Riordan, W. E. C. Wacker and B. L. Vallee, *Nature*, **208**, 1209 (1965).
282. Y. Hachimori, A. Matsushima, M. Suzuki and Y. Inada, *Biochim. Biophys. Acta*, **124**, 395 (1966).
283. R. A. Kenner, K. A. Walsh and H. Neurath, *ibid.*, **33**, 353 (1968).
284. L. Trenholm, W. E. Spomer and J. F. Wootton, *Biochemistry*, **8**, 1741 (1969).
285. J. Labouesse and M. Gervais, *Eur. J. Biochem.*, **2**, 215 (1967).
286. E. Fisher, *Chem. Ber.*, **27**, 2985 (1894).
287. D. E. Koshland Jr., *Proc. Natl. Acad. Sci. U.S.*, **44**, 98 (1958).
288. D. E. Koshland Jr., *J. Cellular Comp. Physiol.*, **54**, 235 (1959).
289. T. Inagami and T. Murachi, *J. Biol. Chem.*, **238**, 1905 (1963).
290. T. Inagami and T. Murachi, *ibid.*, **239**, 1395 (1964).
291. B. F. Erlanger and H. Castleman, *Biochim. Biophys. Acta*, **85**, 507 (1964).

Lysozyme

Kozo HAMAGUCHI*[1] and Katsuya HAYASHI*[2]
*[1]*Department of Biology, Faculty of Science, Osaka University*
Machikaneyama, Toyonaka-shi, Osaka-fu, Japan
*[2]*Department of Agricultural Chemistry, Faculty of Agriculture, Kyushu University*
Hakozaki, Fukuoka-shi, Fukuoka-ken, Japan

---

1. Introduction

About half a century has passed since Fleming discovered lysozyme (EC 3.2.1.17, mucopeptide *N*-acetylmuramylhydrolase) in 1922. In the early stages, it was studied by bacteriologists because of its lytic activity against bacteria. However, following the successful crystallization of lysozyme from hen egg-white by Abraham and Robinson in 1937,[1] the enzyme began to attract the keen attention of protein chemists. There was considerable difficulty attached to studying the structure-function relationship of lysozyme, due to the obscurity of its enzymatic activity. In 1957, Berger and Weiser[2] observed the β-*N*-acetylglucosaminidase activity of lysozyme. Moreover, soluble oligosaccharide substrates were prepared from a hydrolyzate of bacterial cell-walls by Salton and Ghuysen.[3] In 1963, Jolles and his co-workers[4] in France and Canfield[5] in the United States independently determined the amino acid sequence of hen egg-white lysozyme; the first report of its X-ray crystallographic analysis, by the Phillips group in England, appeared in 1965. More detailed information on the three dimensional conformation of the molecule and on the substrate binding mode followed in 1967 from the work of Blake *et al.*[6,7]

Lysozyme is the first enzyme in which it is possible to understand the enzymatic activity on the basis of the three dimensional fine structure of the enzyme molecule. Moreover, much progress has been made in physicochemical and enzymatic studies of lysozyme in solution. In the present review, the conformation of the polypeptide chain and side chains and the stability of hen egg-white lysozyme in solution are described first, followed later by a detailed analysis of the mechanism of enzyme action.

2. Primary Structure

The amino acid compositions of certain avian lysozymes are shown in Table 1.[5,8-14] The amino acid sequence of hen egg-white lysozyme as determined by Canfield is given in Fig. 1.

TABLE 1

Amino Acid Compositions of Avian Lysozymes

	Hen†[1]	Duck				Turkey		Goose		Quail		Pheasant†[7]
		DL–1†[2]	DL–2†[3]	II†[3]	III†[3]	†2	†8	†4	†5	Japanese†[6]	American†[7]	
Asp	21	19	19	19	18–19	20	20	16–17	13–14	20	21	20
Thr	7	7	7	7	7	7	7	10–11	8–9	7	7	7
Ser	10	11	10	10	10–11	10	10	8	6–7	10	10	10
Glu	5	5	5	5	5	3	3	10–11	10	5	5	3
Pro	2	2	2	2	2	2	2	3	2–3	2	2	2
Gly	12	12	12	12	12	13	13	14–15	14	11	12	14
Ala	12	11	11	11	11	13	13	10±1	10	12	12	11
Val	6	7	7	7	7	5	4	6–7	7	7	7	7
Cys/2	8	8	8	8	8	8	8	4	3–4	8	8	8
Met	2	2	2	2	2	2	2	2	2	2	2	3
Ile	6	6	6	6	6	6	6	7–8	9	6	5	6
Leu	8	8	8	8	8	9	8	5–6	4–5	8	8	8
Tyr	3	5	5	5	5	4	5	5–6	6	4	3	4
Phe	3	1	1	1	1	2	2	2	2	2	3	2
Trp	6	—	—	6	6	—	6	2	2–3	—	6	—
Lys	6	5	5	6	6		7	10	11	7	7	8
His	1	0	0	0	0		2	3	3–4	2	1	2
Arg	11	14	14	12–13	13–14		10	7±1	6–7	10	10	9
NH_3	16	16	15									

†[1] Ref. 5. †[2] Ref. 8. †[3] Ref. 9. †[4] Ref. 10. †[5] Ref. 11. †[6] Ref. 12. †[7] Ref. 13. †[8] Ref. 14.

10
Hen Lys–Val–Phe–Gly–Arg–Cys–Glu–Leu–Ala–Ala–Ala–Met–Lys–Arg–
Duck II Lys–Val–Tyr–Ser –Arg–Cys–Glu–Leu–Ala–Ala–Ala–Met–Lys–Arg–
Duck III Lys–Val–Tyr–Gln–Arg–Cys–Glu–Leu–Ala–Ala–Ala–Met–Lys–Arg–
Quail Lys–Val–Tyr–Gly–Arg–Cys–Glu–Leu–Ala–Ala–Ala–Met–Lys–Arg–
Turkey Lys–Val–Tyr–Gly–Arg–Cys–Gln–Leu–Ala–Ala–Ala–Met–Lys–Arg–

20 30
His –Gly–Leu–Asp–Asn–Tyr–Arg–Gly–Tyr–Ser–Leu–Gly–Asn–Trp–Val–Cys–Ala–
Leu–Gly–Leu–Asp–Asn–Tyr–Arg–Gly–Tyr–Ser–Leu–Gly–Asn–Trp–Val–Cys–Ala–
Leu–Gly–Leu–Asp–Asn–Tyr–Arg–Gly–Tyr–Ser–Leu–Gly–Asn–Trp–Val–Cys–Ala–
His –Gly–Leu–Asp–Lys –Tyr–Gln–Gly–Tyr–Ser–Leu–Gly (Asx,Trp) (Val, Cys, Ala,
Leu–Gly–Leu–Asp–Asn–Tyr–Arg–Gly–Tyr–Ser–Leu–Gly–Asn–Trp–Val–Cys–Ala–

40
Ala–Lys–Phe–Glu–Ser–Asn–Phe–Asn–Thr–Gln–Ala–Thr–Asn–Arg–Asn–Thr–
Ala–Asn–Tyr–Glu–Ser–Ser –Phe–Asn–Thr–Gln–Ala–Thr–Asn–Arg–Asn–Thr–
Ala–Asn–Tyr–Glu–Ser–Gly–Phe–Asn (Gln, Ala, Thr, Thr) Asn–Arg–Asp (Thr,
Ala) Lys–Phe–Glu(Ser, Asx, Phe, Asx, Thr, Glx, Ala, Thr, Asx) Arg (Asx, Thr,
Ala–Lys–Phe–Gln–Ser–Asn–Phe–Asn–Thr–His –Ala –Thr–Asn–Arg–Asn–Thr–

50
Asp–Gly–Ser–Thr–Asp–Tyr–
Asp–Gly–Ser–Thr–Asp–Tyr–
Asx, Gly, Ser, Thr–Asx–Tyr–
Asx, Gly, Ser, Thr, Asx, Tyr,
Asp–Gly–Ser–Thr–Asp–Tyr–

60
Gly–Ile–Leu–Gln–Ile–Asn–Ser–Arg–Trp–Trp–Cys–Asn–Asp–Gly–Arg–Thr–
Gly–Ile–Leu–Glu–Ile–Asn–Ser–Arg–Trp–Trp–Cys–Asp–Asn–Gly–Lys–Thr–

Gly, Ile, Leu) (Glx, Ile, Asx, Ser) Arg–Trp–Trp (Cys, Asx, Asx, Gly) Arg–Thr–
Gly–Ile–Leu–Gln–Ile–Asn–Ser–Arg–Trp–Trp–Cys–Asn–Asp–Gly–Arg–Thr–

70 80
Pro–Glv–Ser–Arg–Asn–Leu–Cys–Asn–Ile–Pro–Cys–Ser–Ala–Leu–Leu–Ser–
Pro–Gly–Ser–Lys–Asn–Ala –Cys–Gly –Ile–Pro–Cys–Ser–Val –Leu–Leu–Arg–

Pro–Gly–Ser–Arg(Asx, Leu (Cys, Asx, Ile, Pro, Cys, Ser, Ala, Leu, Leu) (Ser,
Pro–Gly–Ser–Lys–Asn–Leu–Cys–Asn–Ile–Pro–Cys–Ser–Ala–Leu–Leu–Ser–

90 100
Ser, Asp–Ile–Thr–Ala–Ser –Val–Asn–Cys–Ala–Lys–Lys–Ile–Val–Ser–Asp–Gly–
Ser–Asp–Ile–Thr–Glu–Ala–Val–Arg–Cys–Ala–Lys–Arg–Ile–Val–Ser–Asp–Gly–

Ser, Asx, Ile, Thr, Ala, Ser, Val, Asx) Cys, Ala) Lys–Lys–Ile–Val–Ser–Asp–Val–
Ser–Asp–Ile–Thr–Ala–Ser –Val–Asn–Cys–Ala –Lys–Lys–Ile–Ala–Ser–Gly–Gly–

110
Asp–Gly–Met–Asn–Ala–Trp–Val–Ala–Trp–Arg–Asn–Arg–Cys–Lys–Gly–Thr–
Asp–Gly–Met–Asn–Ala–Trp–Val–Ala–Trp–Arg–Asn–Arg–Cys–Arg–Gly–Thr–

His –Gly–Met–Asn–Ala–Trp–Val–Ala–Trp–Arg–Asn–Arg–Cys–Lys–Gly–Thr–
Asp–Gly–Met–Asn–Ala–Trp–Val–Ala–Trp–Arg–Asn–Arg–Cys–Lys–Gly–Thr–

120
Asp–Val–Gln–Ala–Trp–Ile–Arg–Gly–Cys–Arg–Leu
Asp–Val–Ser –Lys–Trp–Ile–Arg–Gly–Cys–Arg–Leu

Asp–Val–Asn–Ala–Trp–Ile–Arg–Gly–Cys–Arg–Leu
Asp–Val–His– Ala–Trp–Ile–Arg–Gly–Cys–Arg–Leu

Fig. 1. The amino acid sequences of hen,[5] duck,[15] quail[12] and turkey[14] egg-white lysozyme.

Certain discrepancies exist between the sequence reported by Canfield and that reported by Jolles, namely at residues 40, 41, 42, 46, 48, 58, 65, 66, 92, and 93. The crystallographic analysis of Blake *et al.*[6] gave results for residues 40, 41, 42, 58, 59, 92, and 93 that are in agreement with the Canfield sequence. The discrepancies at residues 46, 48, 65, and 66 are a difference between Asp or Asn; from the electron density maps it could not be determined whether these residues are amide or free acid.

The amino acid sequences of duck, Japanese quail and turkey egg-white lysozymes were determined by Hermann and Jolles,[15] Kaneda *et al.*[12] and LaRue and Speck Jr.,[14] respectively (Fig. 1). Amino acid sequences for human urine and milk lysozymes have been determined by Canfield[16] and Jolles and Jolles.[17] Comparative studies of sequences for lysozymes of different origins are interesting from the viewpoint of structure-function relationships. For instance, the Asp-101 of hen lysozyme, which is known to be implicated at the substrate binding site, is replaced by Gly in turkey lysozyme. The Trp-62 of hen lysozyme, which also plays an important role in substrate binding, is replaced by Tyr in human lysozyme.

3. Conformation of the Polypeptide Backbone

The structure of lysozyme* was determined by the Phillips group (see illustration in ref. 18). The regions of helix and the features of this helix are summarized in Table 2.[6]

TABLE 2

The Parameters of Helical Regions in Lysozyme[6]

Helix	Residue	Unit rise (Å)	Unit rotation (deg)	Units per turn
A	5–15	1.51	100.2	3.59
B	24–34	1.48_5	98.3	3.66
C	80–85	1.66	106.6	3.38
D	88–96	1.45_5	96.7	3.72

* Unless otherwise stated, the name "lysozyme" is used from now on to indicate hen egg-white lysozyme.

As shown in the table, the parameters for the helical regions of 5–15, 24–34, and 88–96 are close to those of an α-helix. However, some distortion is observed, which gives an intermediate conformation between α-helix and 3.0_{10}-helix. The short helix 80–85 is very close to a 3.0_{10}-helix, while in the region of 119–122 there is only a one turn 3.0_{10}-helix.

The lysozyme molecule contains antiparallel β-structure in the region of 41–54, as illustrated in Fig. 2. The optical rotatory dispersion (ORD)

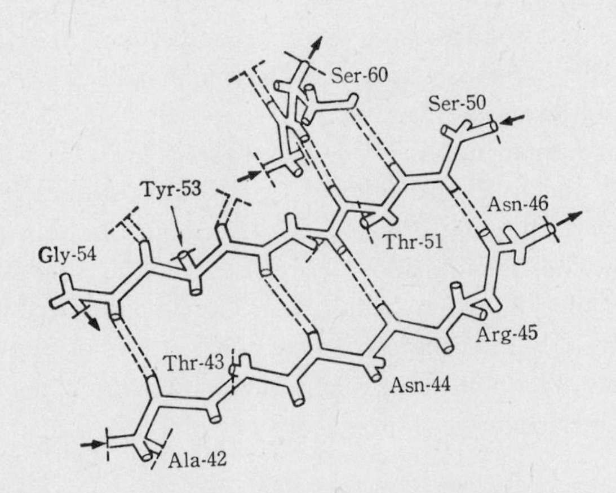

Fig. 2. Antiparallel pleated sheet in the lysozyme molecule.[6]

curve[19] and circular dichroic (CD) spectrum of lysozyme[20] are shown in Fig. 3. The optical rotatory properties of lysozyme are summarized in Table 3.[21–8]

The helical content of lysozyme estimated from the values of the Moffitt parameter, b_0, mean residue rotation at 233 nm ($[m']_{233}$) and mean residue ellipticity at 222 nm ($[\theta]_{222}$) is found to be about 25%. This is in fair agreement with the helical content obtained from X-ray analysis.

The infrared absorption spectrum of lysozyme in D_2O[29] is given in Fig.4: it shows the presence of two absorption maxima at 1650 and 1630 cm^{-1}. The band at 1650 cm^{-1} is due to either helix or random coil, while the band at 1630 cm^{-1} is due to the presence of β-form.[30] Miyazawa *et al.*[31] measured the amide V band of the film infrared absorption spectrum and found absorption maxima at 690, 650, and 600 cm^{-1}. These maxima are assigned to extended β-, disordered, and helical form, respectively. The presence of β-structure in the lysozyme molecule has

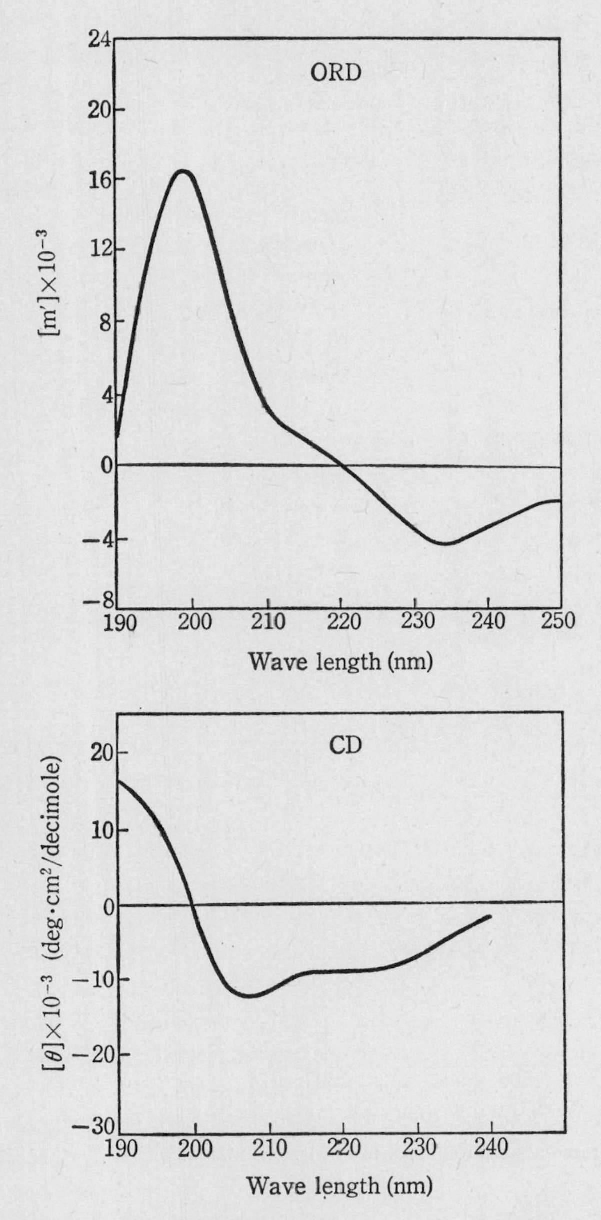

Fig. 3. ORD curve[19] and CD spectrum[20] of lysozyme.

TABLE 3

Optical Rotatory Properties of Lysozyme

Specific rotation at 546 nm ($[\alpha]_{546}$)	−58.8°†1
Reduced mean residue rotation ($[m']$)	
233 nm	−4460†2†7, −4500†3, −4300†6
199 nm	15,900†2†7
436 nm	−97.0†4
Rotatory dispersion constant (λ_c)	250†1, 253†2, 251†4
Moffitt parameters	
a_0	−288†2, −272†4, −275†5
b_0	−161†1, −157†2, −140†4, −145†5
Mean residue ellipticity ($[\theta]$)†8	
220 nm	$-9.40 \pm 0.9 \times 10^{-3}$
208–209 nm	$-11.6 \pm 0.8_6 \times 10^{-3}$
191 nm	$20.8 \pm 2.3 \times 10^{-3}$
186 nm	$16.6 \pm 3._8 \times 10^{-3}$

†1 Ref. 21. †2 Ref. 22. †3 Ref. 23. †4 Ref. 24. †5 Ref. 25. †6 Ref. 26.
†7 Ref. 27. †8 Ref. 28.

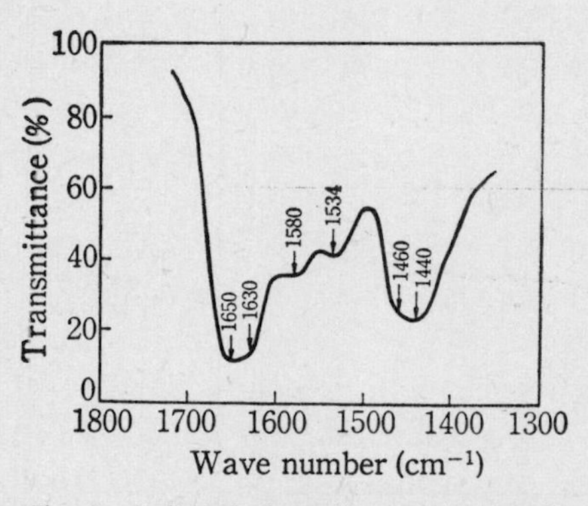

Fig. 4. Infrared absorption spectrum of lysozyme in D_2O.[29]

thus been demonstrated by infrared absorption spectroscopy. It was formerly predicted by Hamaguchi and Imahori[32] on the basis of the optical rotatory properties in various organic solvent-water mixtures.

4. Side Chain States

The states of the side chains of the lysozyme molecule have been investigated in detail by physicochemical methods such as titration and ultraviolet absorption spectroscopy, and by chemical modification.

4.1. Ionizable Groups

4.1.1. Acid-Base Titration

The lysozyme molecule has eleven carboxyl, one imidazole (histidyl), one α-amino, six ε-amino (lysyl), three phenolic (tyrosyl) and eleven guanidyl (arginyl) groups, as ionizable groups. The states of these ionizable groups can be examined by acid-base titration, spectrophotometric titration and chemical modification.

Acid-base titration of lysozyme was first performed by Tanford and Wagner,[33] and Donovan *et al.*,[34,35] who showed that the carboxyl content of lysozyme differs from one preparation to another. Fig. 5 gives the titration curve of lysozyme in 0.2 M KCl at 25°C obtained by Sakakibara and Hamaguchi in 1968.[36] The titration curve was reversible between pH 2 and 11: at pH's below 6, eight groups were titrated.

Fig. 6 shows the titration curve of lysozyme at pH's below 6 in a mixed solvent of 5 M guanidine hydrochloride and 1.2 M urea. In this solvent eleven groups were titrated.

From the titration curve in Fig. 5, the number of protons bound in going from the isoionic point (pH 11.1)[37] to $\bar{r}=0$ was found to be sixteen. However, this figure does not agree with the sum of the basic nitrogen groups (eleven), indicating that three groups remain negatively ionized even at pH's below 2. Donovan *et al.*[34,35] found from proton uptake experiments that three extra groups were masked as carboxylate ions in the native molecule.

The titration data for each type of ionizable group can be analyzed by plotting them according to the equation of Linderstrøm-Lang (see for instance ref.38).

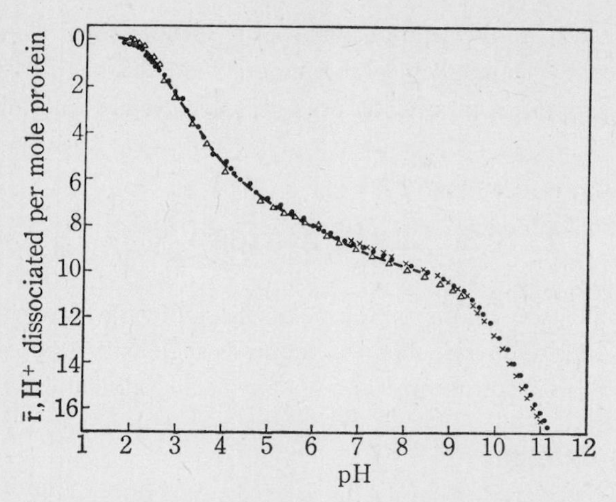

Fig. 5. Titration curve of lysozyme in 0.2 M KCl at 25 °C.[36] ●, Initial titration from the pH value attained after dialysis; △, back titration after exposure to pH 1.8; ×, back titration after exposure to pH 11.1. (The solid line between pH 1.8 and 8.5 represents values calculated from Eq. (1) using the parameters in Table 4.)

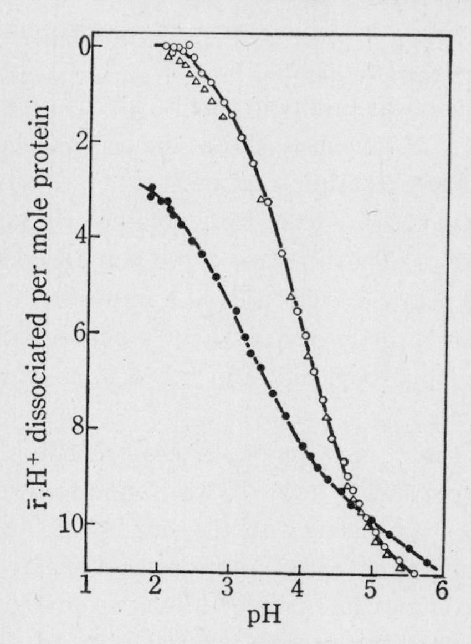

Fig. 6. Titration curve of lysozyme in a mixed solvent of 5 M guanidine hydrochloride and 1.2 M urea at 25 °C.[36] ○, Initial titration from pH 6.0; △, back titration from pH 2.1; ●, titration curve of lysozyme in 0.2 M KCl. (The solid line for the titration in the presence of a denaturing agent represents values calculated from Eq. (1) using $pK_{int}=4.4$ and $w=0.04$.)

$$\mathrm{pH} - \log \frac{r_i}{n_i - r_i} = (\mathrm{p}K_{\mathrm{int}})_i - 0.868\, w\bar{Z}, \tag{1}$$

where n_i is the number of groups of a given type, r_i is the number of hydrogen ions dissociated per mole of protein at a given pH, $(\mathrm{p}K_{\mathrm{int}})_i$ is the negative logarithm of the intrinsic dissociation constant for the i-th type, and $0.868\, w\bar{Z}$ is a correcting term for electrostatic interactions between protons and a protein molecule of mean net charge $\bar{Z}$.

The value of w can be calculated from Eq. (2), assuming the Linderstrøm-Lang protein model as a sphere with uniform charge distribution on its surface.

$$w = \frac{\varepsilon^2}{2DkT}\left(\frac{1}{b} - \frac{\kappa}{1+\kappa a}\right), \tag{2}$$

where ε is the electronic charge, D the dielectric constant, k the Boltzmann constant, T the absolute temperature, b the radius of the sphere which represents the protein molecule, a the radius of exclusion (taken as $b + 2.5$ Å), and κ the Debye-Hückel parameter.

Data for eight carboxyl groups of lysozyme in 0.2 M KCl were plotted according to Eq. (1) and are shown in Fig. 7. The plot for eleven carboxyl groups in the presence of 5 M guanidine hydrochloride and 1.2 M urea is also given.

The Linderstrøm-Lang plot in the presence of the denaturing agent is linear and values of 4.4 and 0.04 are obtained for $\mathrm{p}K_{\mathrm{int}}$ and w, respectively. Donovan *et al.*[34] obtained a $\mathrm{p}K_{\mathrm{int}}$ value of 4.6 and a w value of

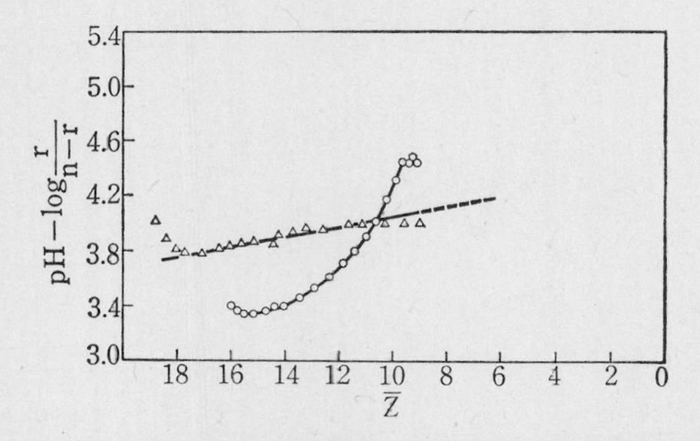

Fig. 7. Plot of the titration data of eight carboxyl groups of lysozyme in 0.2 M KCl (○) and eleven carboxyl groups in a mixed solvent of 5 M guanidine hydrochloride and 1.2 M urea (△) according to Eq. (1).[36]

0.052 for twelve carboxyl groups under the same denaturing conditions. The Linderstrøm-Lang plot for the carboxyl ionization in 0.2 M KCl, on the other hand, was not linear. Tanford and Wagner[33] suggested that the titration curve of the carboxyl groups in the native molecule is a composite of curves with different pK values.

The value of w for lysozyme was calculated to be 0.082 at an ionic strength of 0.2 and 25°C based on a radius of 16 Å.[39] The value of w

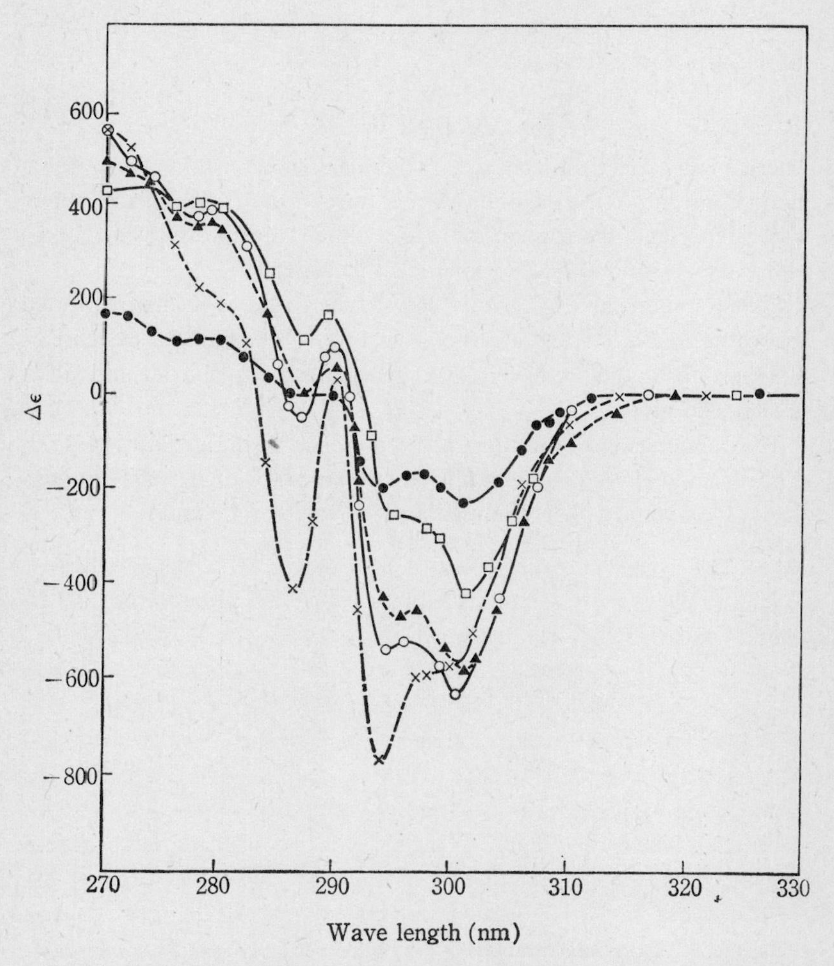

Fig. 8. Acid difference spectra of lysozyme at an ionic strength of 0.16, 28 °C.[40] Reference, lysozyme solution at pH 6.88. ×, pH 0.89; ○, pH 2.89; ▲, pH 3.42; □, pH 5.09; ●, pH 5.94.

may be expected to be unchanged on the acidic side, since there is no evidence for the denaturation of lysozyme until pH 2 (see section 6.2). Nevertheless, difference spectra of lysozyme are obtained on acidification, with a negative peak at 293–294 nm, which is characteristic of the tryptophan residues (Fig. 8).[40]

The difference in the absorptivity at 293–294 nm ($\Delta\varepsilon_{293-294}$) is plotted against pH in Fig. 9.[40] The curves shown in this figure have two

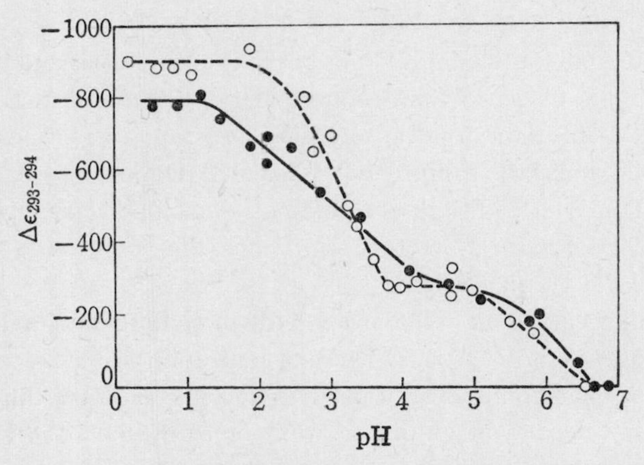

Fig. 9. Dependence on pH of $\Delta\varepsilon$ values for lysozyme at 293–194 nm, 28 °C.[40] ●, Without guanidine hydrochloride; ionic strength, 0.16: ○, 1.6 M guanidine hydrochloride.

stages: the apparent p*K* values are 3.05 and 6.25 in the absence of guanidine hydrochloride, and 3.10 and 5.80 in the presence of 1.60 M guanidine hydrochloride. Donovan *et al.*[41] suggested that carboxyl groups with p*K* values of 3.1 and 6.2 interact strongly with tryptophan residues in the native lysozyme and the ionization of the carboxyl groups affects the ultraviolet absorption spectrum of the tryptophan residues. When lysozyme is denatured by 5 M guanidine hydrochloride, the strong interaction between the carboxyl groups and tryptophan residues is eliminated and no difference spectra are produced on acidification (Fig. 9).

From these facts, Sakakibara and Hamaguchi[36] attempted to construct a theoretical titration curve on the acidic side, assuming a reasonable w value of 0.08 and a p*K* value of 3.5 for one group, 6.0 for another group, and 4.3 for the remaining six groups of the eight titratable groups of the native lysozyme molecule. A value of 4.3 is an expected normal pK_{int} value for carboxyl groups. Furthermore, they assumed $pK_{int} = 6.8$ and $w = 0.07$ for one imidazole group and $pK_{int} = 7.8$ and $w = 0.07$ for one α-amino group. The values of 6.8 and 7.8 are normal pK_{int} values for imidazole and α-amino groups, respectively. It was also assumed that no ion other than hydrogen ions is bound to the protein. The theoretical titration curve constructed from these assumptions fits the experimental data quite well in the pH range of 2.0–8.5, as shown by the solid line in Fig. 5.

At pH's above 8.5 the ε-amino groups of the lysine residues and tyrosine residues begin to ionize. Spectrophotometric titration of the three tyrosine residues of the lysozyme molecule has been carried out by Fromageot and Shneck,[42] Tanford and Wagner,[33] Donovan *et al.*,[35,41] Inada[43] and Tojo *et al.*[44] The results obtained by Tojo *et al.* are shown in Fig. 10 and 11. The titration curve at 295 nm (Fig. 10) runs parallel with the curve at 245 nm (Fig. 11).

These titration curves show the presence of three different stages of ionization of the phenolic groups. The first ionization occurs between pH 8 and 11.3, the second between pH 11.3 and 12.3, and the third at pH> 12.3. Time-dependent ionization was observed above the latter pH, indicating that lysozyme undergoes denaturation under such conditions.

The total changes in molar extinction coefficient at 295 and 245 nm were 7950 and 39,700, respectively, i.e. 2650 at 295 nm and 13,200 at

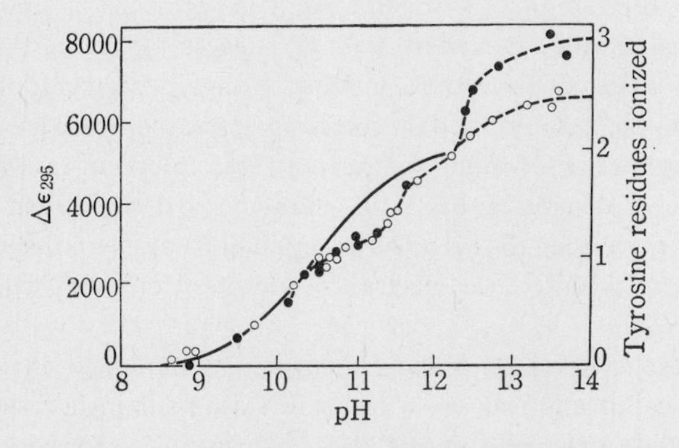

Fig. 10. Spectrophotometric titration of lysozyme at 295 nm, 25 °C; ionic strength, 0.20.[36,44] ○, Initial readings; ●, final readings.

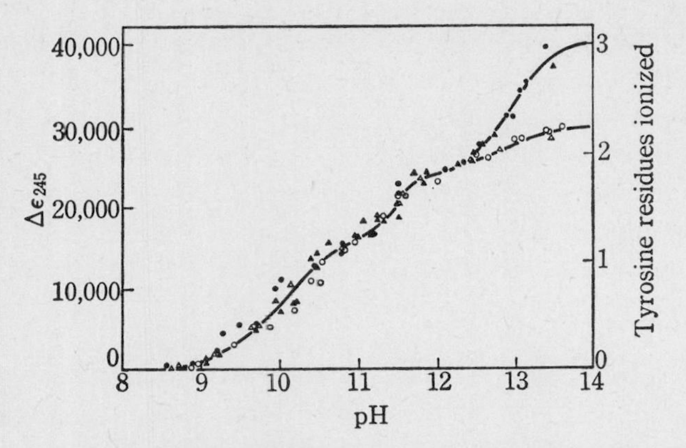

Fig. 11. Spectrophotometric titration of lysozyme at 245 nm, 25 °C; ionic strength, 0.20.[44] ○, Initial readings; ●, final readings.

245 nm per tyrosine residue in the lysozyme molecule. The right ordinates in Fig. 10 and 11 represent the number of residues ionized assuming these values. The titration curves given in these figures show that the three tyrosine residues in the lysozyme molecule have different pK values and that two of the three tyrosine residues ionize reversibly with apparent pK values of 9.95 and 11.6, respectively, and the third ionizes irreversibly with an apparent pK value of 12.6.

The Linderstrøm-Lang plot for the two phenolic groups is shown in Fig. 12. The plot consists of two linear portions with a break at $\bar{Z}=2$. The values of pK_{int} and w obtained from the linear portion above $\bar{Z}=2$ were 10.66 and 0.07, respectively.

The solid line in Fig. 10 shows a theoretical curve constructed by assuming p$K_{int}=10.66$ and $w=0.07$. The theoretical curve thus constructed fits the experimental data of the first ionization stage but deviates considerably from the second stage. Attempts to construct a theoretical curve which satisfactorily fits both the first and second ionization stages by using proper parameters have not been successful. The nature of the phenolic ionization of the lysozyme molecule is as yet unknown.

As will be described in section 4.1.2.E, Hayashi *et al.*[45,46] carried out iodination of the tyrosine residues and found that Tyr-20 and 23 were both iodinated, but at different rates, and that Tyr-53 was nonreactive towards iodination. They also found that the iodination of Tyr-20 affected the extent of the iodination of Tyr-23, and vice versa. It may be

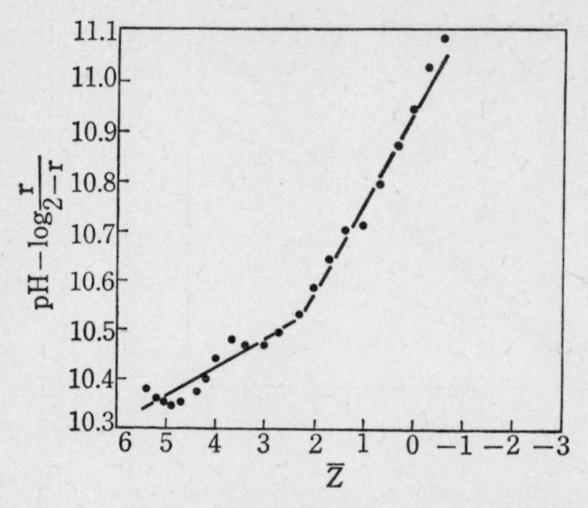

Fig. 12. Plot of the titration data for two phenolic groups of lysozyme in 0.2 M KCl according to Eq. (1).[36]

possible therefore that the ionization of either Tyr-20 or 23 affects the ionization tendency of Tyr-23 or 20, respectively.

The number of protons binding to the ε-amino groups of the lysine residues is computed by subtraction of the spectrophotometric titration data for tyrosyl ionization from the potentiometric titration data shown in Fig. 5. Plots of the data according to the Linderstrøm-Lang equation for the six ε-amino groups give a value of 0.07 for w and 10.66 for pK_{int}.

TABLE 4
Parameters Used in Drawing Calculated Titration Curves of Lysozyme in 0.2 M KCl, at 25.0°C

Group	Number		pK_{Int}	w
	Analysis	Titration		
Carboxyl	11	8	3.5 for 1	0.08
			4.3 for 6	0.08
			6.0 for 1	0.08
Imidazole	1	1	6.8	0.08
δ-Amino	1	1	7.8	0.07
ε-Amino	6	6	10.66	0.07
Phenolic	3	2 (pH<12)		
Guanidyl	11			

Table 4 summarizes the parameters used in calculating the titration curve of lysozyme in 0.2 M KCl.

4.1.2. Other Methods

A. Carboxyl Groups

X-ray crystallographic analysis shows that Glu-35 and Asp-52 play an important role in the enzymatic activity of lysozyme. Glu-35 locates in the nonpolar region at one side of the cleft where the substrate is bound, and Asp-52 locates in the complex network of hydrogen bonds formed by Asn-46, Ser-50 and Asn-59. In such an environment, Asp-52 seems to be ionized, and Glu-35 appears to exist in unionized form at relatively high pH's. Therefore, the carboxyl group with $pK=6$, obtained from titration and ultraviolet absorption spectra, may be assigned to Glu-35.[47] Changes in the ultraviolet fluorescence spectrum with pH accompanying the interaction of chitotriose are considered to be due to an interaction between Glu-35 and Trp-108.[48] Nuclear magnetic resonance studies of the interaction of lysozyme with the β-methylglycoside of *N*-acetylglucosamine confirm the assignment of the carboxyl group with $pK=6$ to Glu-35.[49] It is as yet unknown which group corresponds to the carboxyl group with a p*K* of about 3.5. Lehrer and Fasman[50] suggested that the pH-dependence of the ultraviolet fluorescence spectrum of lysozyme-inhibitor complex between 5.0 and 2.0 is due to the interaction of Asp-101 and Trp-63 or Asp-52 and Trp-63.

Another ionizable group with an apparent p*K* of about 4.5 has also been found to be involved in the binding of chitotriose.[48,50-2] Since trisaccharide is known to interact with Asp-101, while monosaccharide does not, Dahlquist and Raftery[49,53] assigned an apparent p*K* of 4.2 to this group. Furthermore, they suggested that Asp-52 would have a low p*K* of 2. They also demonstrated that the ionization of Asp-52 and Glu-35 affects the binding of the β-methylglycoside of *N*-acetylglucosamine on the basis of a space-filling model of lysozyme. As will be described later, studies on the denaturation of lysozyme also give information on the states of the carboxyl groups.[54,55]

Hoare and Koshland Jr.[56] found that eight of the eleven carboxyl groups were modified by a water-soluble carbodiimide, 1-benzyl-3-(3-dimethylamino-(*N*)-propyl)–carbodiimide metho-*p*-toluenesulfonate, in the presence of glycine methyl ester. Horinishi *et al.*[57] reported that six of the eleven carboxyl groups were modified by treatment with another carbodiimide, 1-ethyl-3-(3-morpholinyl-(4)-propyl)-carbodiimide, in the presence of glycine ethyl ester. Therefore, it is tempting to assume that the six groups with p*K* 4.3 obtained from the titration are of the reactive type.

B. Amino Groups

As described above, acid-base titration has revealed that all the amino groups of lysozyme are titrated with normal p*K* values. However, tests of the reactivity towards β-naphthoquinone-4-sulfonic acid (NQS) have shown that four of the seven amino groups react with NQS at relatively low concentrations, one reacts at higher concentrations, and the remaining two groups have no reaction.[58] Similar results were also obtained for the reactivity towards β-naphthoquinone disulfonic acid.[59] Only three of the seven amino groups react with monochlorotrifluoro-*p*-benzoic acid.[60,61]

Herzig *et al.*[62] investigated the reaction of the bifunctional protein reagents phenol-2,4-disulfonyl chloride and α-naphthol-2,4-disulfonyl chloride with lysozyme. By ultracentrifugal analysis it was shown that no appreciable intermolecular cross-linking occurred. The location of the reacted lysine residues within the sequence of lysozyme was not determined.

Hiremath and Day[63] studied the cross-linking reaction of α,α'-dibromoxylenesulfonic acid with lysozyme and observed the presence of intramolecular cross-links between Lys-33 and Lys-116, and Lys-96 and Lys-97. Marfey *et al.*[64] found from the reaction with 1,5-difluoro-2,4-dinitrobenzene that a cross-linking reaction occurred between Lys-7 and Lys-41.

C. Imidazole Group

Lysozyme contains only one histidine residue. As described in section 4.1.1, the value of pK_{int} of the histidine residue was found to be 6.8 from acid-base titration experiments.[36] This is a normal p*K* value for the imidazole group. The same value was also obtained by Tanford and Wagner.[33] However, Meadows *et al.*[65] obtained a p*K* value of 5.8 from the pH dependence of the imidazole C-2 proton resonance of hen egg-white and human lysozymes. This value was unaffected by the presence of inhibitor.

Diazo-1-*H*-tetrazole is known to react with the histidine residue of lysozyme.[66]

D. Guanidyl Groups

The states of arginine residues cannot be examined by acid-base titration since the guanidyl group has a high p*K* value above 12.

Eleven arginine residues of lysozyme react with glyoxal to the extent of approximately 80%. The reaction is even and partial: there is no selective reactivity of any specific arginine residues.[60,61]

E. Phenolic Groups

As described in section 4.1.1, two of the three tyrosine residues are titrated reversibly and the other ionizes only after alkali denaturation has occurred at pH's above 12.3.[44] Two tyrosine residues react with cyanuric fluoride while the other is nonreactive.[67] From iodination experiments, it was also shown that two tyrosine residues are iodinated: Covelli and Wolff *et al.*[68,69] reported that the one nonreactive tyrosine residue was Tyr-23. However, as mentioned above, the iodination experiments of Hayashi *et al.*[45,46] indicated that the nonreactive tyrosine residue was Tyr-53, and that the two reactive tyrosine residues were iodinated at different rates. They suggested that during the period when either of Tyr-20 or 23 is being iodinated, Tyr-23 or 20 cannot be iodinated owing to steric hindrance by the already formed diiodotyrosine. Recently, Atassi and Habeeb[70] modified the tyrosine residues with tetranitromethane and found that Tyr-20 and 23 were nitrated but that Tyr-53 was non-reactive.

The non-reactivity of Tyr-53 is easily presumed from the three dimensional structure of lysozyme constructed on the basis of the X-ray analysis performed by Blake *et al.*;[6,7,71] the phenolic group is linked to the NH_2 group of Asn-66 by a hydrogen bond.

As shown in Table 1, duck egg-white lysozyme contains five tyrosine residues. The results of the spectrophotometric titration of duck lysozyme are given in Fig. 13.[44]

As can be seen from this figure, there are two stages of ionization; the first corresponds to three tyrosine residues and the second to two tyrosine residues. The titration showed no time dependence below pH 12, but above this pH a time effect was observed. The three tyrosine residues were titrated at a p*K* value of 10.15.

In the sequence of duck egg-white lysozyme, the Phe-3 and 34 of hen egg-white lysozyme are replaced by tyrosine residues. According to the results of X-ray crystallographic analysis,[6,7] Phe-3 and 34 locate on

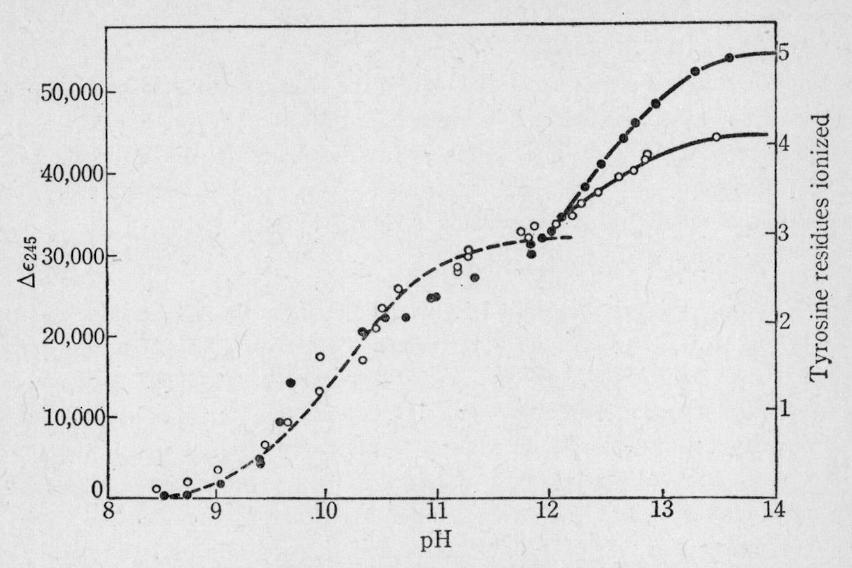

Fig. 13. Spectrophotometric titration of duck egg-white lysozyme at 245 nm, 25 °C; ionic strength, 0.20.[44] ○, Initial readings; ●, final readings.

the surface of the hen lysozyme molecule. Therefore, if the Tyr-3 and 34 of duck lysozyme are assumed to be titrated freely, the total number of normal tyrosine residues in duck lysozyme would be four (Tyr-3, 20, 23, and 34). However, the results of spectrophotometric titration have shown that only three of the five tyrosine residues are titrated freely. This suggests that the states of the tyrosine residues in duck lysozyme are different from those in hen lysozyme.

The change in the CD spectrum of lysozyme with pH is shown in Fig. 14.[72] The CD spectrum of the native lysozyme at pH 7.0 has positive maxima at 294, 289, 283, and 278 nm and a broad negative band close to 260 nm. With increasing pH, the CD band at around 294 nm increased and a positive maximum appeared at 253 nm. At pH 11.2, for example, the CD spectrum had positive maxima at 253, 284, 289, and 294 nm. The appearance of a maximum at 253 nm was ascribed to the ionization of tyrosine residues. At pH 12.7, where a time-dependent conformational change occurs (see section 4.1.1), the CD spectrum in the region of 200–300 nm decreased greatly when the solution was allowed to stand overnight. No change in the entire CD spectrum was observed on acidification to pH 2.0.

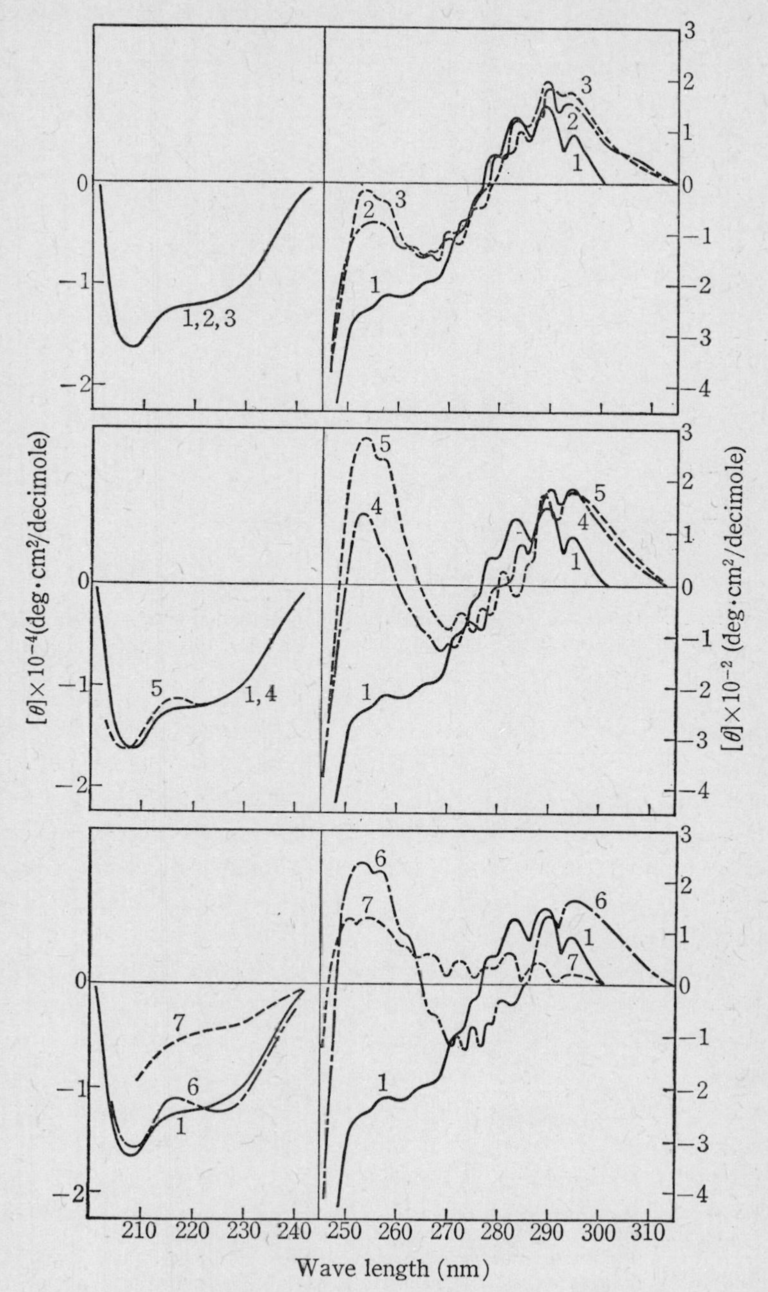

Fig. 14. CD spectra of lysozyme at various pH's; ionic strength, 0.1.[72] 1, pH 7.0; 2, pH 9.8; 3, pH 10.1; 4, pH 11.2; 5, pH 12.1; 6, pH 12.8 after 30 min; 7, pH 12.7 after 20 hr.

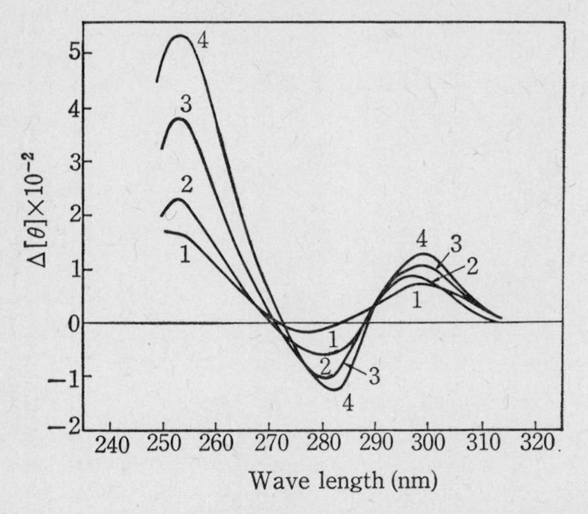

Fig. 15. Difference CD spectra obtained by subtracting the CD spectrum of lysozyme at pH 7.0 from that at pH 9.8 (1), 10.1 (2), 11.2 (3), and 12.1 (4).[72]

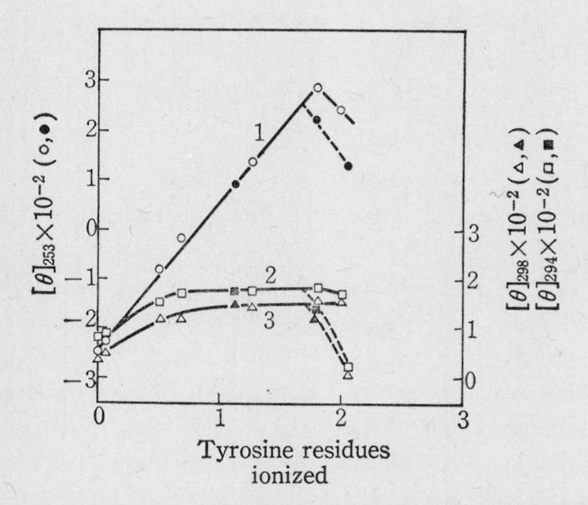

Fig. 16. Values of [θ] at 253 nm (1), 194 nm (2), and 298 nm (3) plotted against the number of tyrosine residues ionized.[72] Open symbols indicate data obtained after 30 min; closed symbols, those after 20 hr.

Fig. 15 shows the difference CD spectra obtained by subtracting the CD spectrum of lysozyme at pH 7.0 from that at various alkaline pH values. The difference CD spectrum has positive maxima at 253 and 298 nm. The difference ultraviolet absorption spectrum of lysozyme at an alkaline pH referred to the protein at neutral pH also has maxima at 245 and 295 nm.[44]

When the values of $[\theta]$ at 253 nm were plotted against the number of tyrosine residues ionized (which is obtained from the spectrophotometric titration curve in Fig. 10), a linear relation was obtained with a slope of $[\theta] = 300$ per ionized tyrosine residue (Fig. 16). When more than two tyrosine residues were ionized, however, this linearity was lost. On the other hand, the values of $[\theta]$ at 294 and 298 nm increased until one tyrosine residue was ionized. On ionization of more than one tyrosine residue there was no further increase in the CD band at 294 nm. Therefore, the changes in the CD bands at 294 and 253 nm with ionization of the tyrosine residues were not parallel.

4.2. Nonpolar Groups

Nonpolar side chains are mostly in the interior of the lysozyme molecule. However, there are a number of nonpolar residues which locate on the surface of the molecule; for example, Val-2, Phe-3, Leu-17, Phe-34, Leu-75, and Trp-123. Trp-62, Trp-63, Trp-108, Ile-98, and Val-109 are on the surface of the cleft where substrates are bound.[6,7]

4.2.1. Tryptophan Residues

Among the nonpolar residues, tryptophan residues are important in relation to enzymatic function and their states have been studied in detail. The lysozyme molecule contains six tryptophan residues: Trp-62, 63, and 108, as well as carboxyl groups, play an important role in the enzyme activity.

Hamaguchi and his co-workers[24,29,73–77] carried out extensive studies on the ultraviolet difference spectra of lysozyme produced by various reagents. Fig. 17 illustrates the difference spectra of tryptophan and a mixture of tryptophan and tyrosine produced by guanidine hydrochloride.[24]

The difference spectra of tryptophan in Fig. 17 have positive peaks at 291 and 284 nm, although there are generally positive peaks at around 275, 285, and 292 nm. It was found that the difference spectra of a 2:1

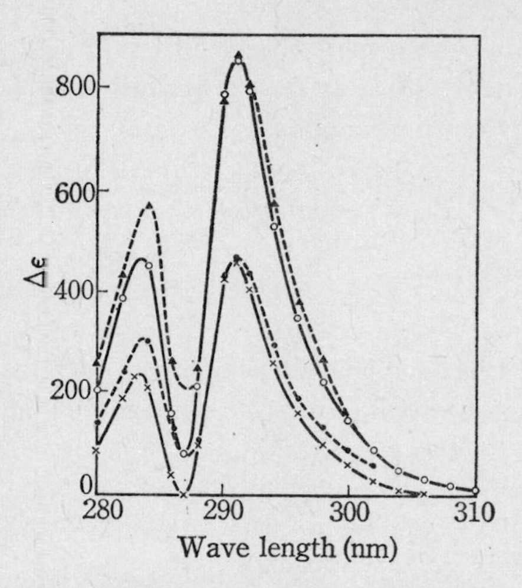

Fig. 17. Difference spectra of tryptophan and a mixture of tryptophan and tyrosine produced by guanidine hydrochloride in 0.01 M EDTA, pH 7.0.[24] Tryptophan in 3.2 M (×) and 6.4 M (○) guanidine hydrochloride; reference, tryptophan in 0.01 M EDTA, pH 7.0. Mixture in 3.2 M (●) and 6.4 M (△) guanidine hydrochloride; reference, mixture in 0.01 M EDTA, pH 7.0.

mixture of tryptophan and tyrosine had positive peaks at 291 and 284 nm. The difference in the absorbancy ($\Delta\varepsilon$) at 291 nm was not affected by the presence of tyrosine. At 284 nm, however, the value of $\Delta\varepsilon$ increased in the presence of tyrosine. Therefore, in the case of proteins containing tryptophan and tyrosine residues, the states of the tryptophan residues can be examined using the 291 nm peak in the difference spectra.

Fig. 18 shows the values of $\Delta\varepsilon$ at 292–293 nm for L-leucyl-L-tryptophan and lysozyme plotted against the concentration of dimethylformamide (DMF).[73] The values of $\Delta\varepsilon$ for L-leucyl-L-tryptophan and lysozyme increase linearly with the concentration of DMF. By comparing the slope of the straight line for lysozyme with the slope for L-leucyl-L-tryptophan, the number of tryptophan residues exposed to the solvent can be estimated. The data shown in the figure give a value of 4.0 for this ratio.

Table 5 summarizes the number of tryptophan residues which are accessible to the solvent by using the solvent perturbation technique. It is clear from this table that the increment of $\Delta\varepsilon$ at 292 nm for lysozyme is about four times greater than the increment for L-leucyl-L-tryptophan,

TABLE 5A

Increment of εΔ at 292 nm per Mole of Reagent for Tryptophan and Lysozyme†

Reagent	Increment of Δε at 292 nm		Ratio	Ref.
	Tryptophan	Lysozyme		
GuCl	138	373	2.7	24,74
Urea	80	300	3.8	
LiCl	48	160	3.3	
LiBr	104	363	3.5	75
NaBr	138	380	2.8	

† Reference: aqueous solution.

TABLE 5B

Increment of Δε at *ca.* 292 nm per 10% Organic Solvent for L-Leucyl-L-tryptophan (or Tryptophan) and Lysozyme†[1]

Solvent	L-Leucyl-L-tryptophan		Lysozyme		Ratio	Ref.
	λ_{max}	Δε/10%	λ_{max}	Δε/10%		
MeOH	290	100	292	†[2]		
EtOH	291	130	292	†[2]		74,77
n-PrOH	291.5	170	292	†[2]		
Iso-PrOH	291.5	170	292	†[2]		
Ethylene glycol monomethyl ether	292	160	292	610	4.0	77
DMF	291–293	192	292–295	763	4.0	73
DMSO	292–294	242	293–295	850	3.5	29
Ethylene glycol	290	148	292	550	3.7	
Polyethylene glycol 300	291	197	292	700	3.6	76
Sucrose	292	81	293	540	6.7	

†[1] Reference: aqueous solution.

†[2] The relation between Δε at 292 nm and the concentration of alcohol is not linear.

except for sucrose and alcohols. (The effect of alcohols on the ultraviolet absorption spectrum of lysozyme will be described in section 6.7). This indicates that four of the six tryptophan residues are located on the surface of the lysozyme molecule and are accessible to the solvent, and that the remaining two tryptophan residues are in the interior of the molecule.

X-ray analysis shows that Trp-62, 63, 108, and 123 exist on the surface of the lysozyme molecule. Thus, the solvent perturbation technique provides a good agreement with the X-ray result.

Fig. 19 gives the difference spectra of lysozyme produced by guanidine hydrochloride.[24] With up to 3.2 M guanidine hydrochloride, the lysozyme molecule is not denatured and the difference spectra have positive peaks at 284 and 292 nm. Above 3.2 M guanidine hydrochloride, the lysozyme molecule is denatured and the difference spectra have negative peaks at 284 and 292 nm and a positive peak at 300 nm.

The values of $\Delta\varepsilon$ at 292 nm are plotted against concentration of guanidine hydrochloride in Fig. 20. The value of $\Delta\varepsilon$ at 292 nm increases linearly up to 3.2 M guanidine hydrochloride and decreases abruptly above this concentration, which itself corresponds to the concentration at which the optical rotatory parameters and viscosity change. This abrupt change in $\Delta\varepsilon$ at 292 nm thus corresponds to the denaturation blue shift

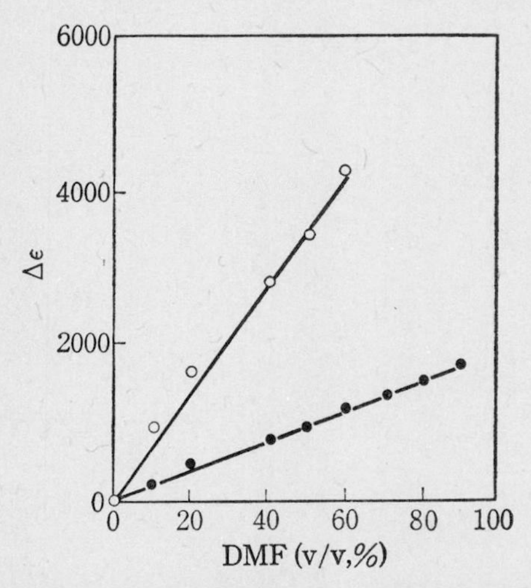

Fig. 18. Dependence of $\Delta\varepsilon$ for lysozyme at 292–295 nm (○) and L-leucyl-L-tryptophan at 292–293 nm (●) on concentration of dimethylformamide (DMF).[73]

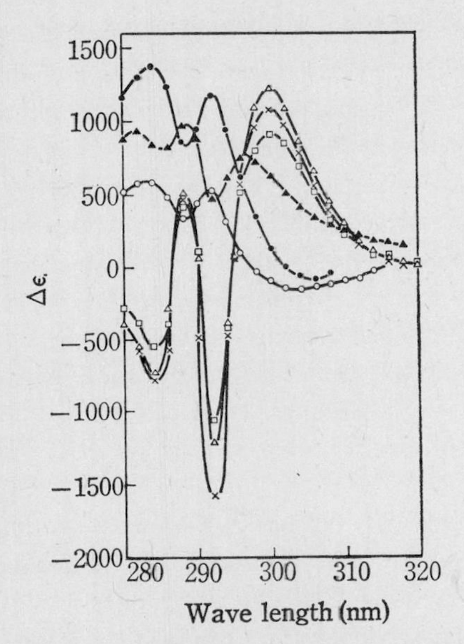

Fig. 19. Final difference spectra of lysozyme as a function of guanidine hydrochloride concentration at 25 °C in 0.01 M EDTA, pH 7.0.[24] ○, 1.3 M; ●, 3.2 M; ▲, 3.8 M; □, 4.5 M; ×, 5.8 M; △, 6.4 M guanidine hydrochloride; reference, lysozyme in 0.01 M EDTA, pH 7.0.

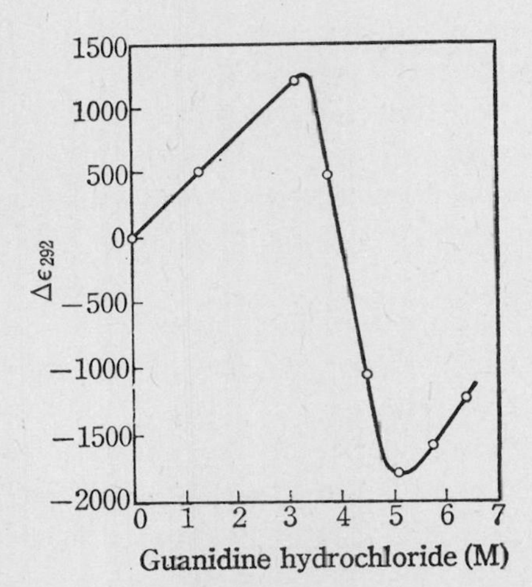

Fig. 20. Dependence of final value of $\Delta\varepsilon$ of lysozyme at 292 nm on the concentration of guanidine hydrochloride at 25 °C (0.01 M EDTA, pH 7.0).[24]

of the tryptophan residues as designated by Bigelow and Geschwind.[78,79] The value of $\Delta\varepsilon$ at 292 nm increases again in $>$ 5 M guanidine hydrochloride. The increment of $\Delta\varepsilon$ per mole of guanidine hydrochloride in this region ($>$5 M) is approximately 1.5 times as great as that below 3.2 M. The increase in the value of $\Delta\varepsilon$ above 5 M guanidine hydrochloride is due to the solvent effect of the completely denatured lysozyme on the tryptophan residues. The magnitude of the blue shift at 292 nm occurring after total denaturation by guanidine hydrochloride was −4700. If, as described above, two tryptophan residues are masked in the interior of the lysozyme molecule, the magnitude of the blue shift for one tryptophan residue is −2350. This same value has also been obtained from the results of acid denaturation of lysozyme in the presence of 3.8 M guanidine hydrochloride.[40]

The magnitude of the blue shift at 292 nm when one tryptophan residue is exposed to the solvent from the interior of the molecule is −1900 to −2800 for α-lactalbumin,[80] −2000 for aldolase,[81] and −2500 for chymotrypsin.[82,83] These values are in good agreement with one another. It thus appears that two tryptophan residues are normally masked in the interior of the lysozyme molecule but are exposed to the solvent by denaturation.

Williams *et al.*[84] and Williams and Laskowski Jr.[85] on the other hand have suggested that six tryptophan residues are partially exposed, and that there are variations in the degree of exposure.

Modification of the tryptophan residues of lysozyme with *N*-bromosuccinimide (NBS) was carried out by Hayashi *et al.*[86,87] It was found that Trp-62 is rapidly oxidized by NBS. This is interesting in view of the role of tryptophan residues in the enzymatic activity, as will be described below in section 8.

Kronman *et al.*[88] showed that the oxidation of lysozyme by NBS at higher concentrations results in destruction of tyrosine and histidine residues in addition to tryptophan residues.

Hachimori *et al.*[89] studied the oxidation of lysozyme in a H_2O_2–dioxane system and found that five out of the six tryptophan residues were oxidized at low concentrations of H_2O_2, and that the remaining one residue was oxidized at higher concentrations.

The reaction of lysozyme with iodine gave a product in which one tryptophan residue was oxidized. This had no enzymatic activity.[90,91] Ozonization of lysozyme in formic acid selectively converted two tryptophan residues (108 and 111) to N'-formylkynurenine residues without loss of lytic activity.[92]

2-Hydroxyl-5-nitro-benzylbromide (Koshland reagent) does not react with the tryptophan residues of lysozyme (less than one). However, in the presence of 30% 2-chloroethanol, all the tryptophan residues react with the reagent.[93]

The CD spectrum of lysozyme in the aromatic absorption region (Fig. 14) involves a contribution from the tryptophan residues. This is shown by the change in the CD spectrum accompanying the interaction of *N*-acetylglucosamine, which is known to interact with tryptophan residues.[72,94,95]

An ultraviolet fluorescence spectrum of lysozyme arises from tryptophan residues. Fluorescence studies have been carried out by Chirchich[96,97] and Lehrer and Fasman.[48,50] The quantum yield of lysozyme is 0.065, which is considerably lower than the quantum yield of tryptophan (0.20); the fluorescence maximum is at 337 nm. The pH dependence of the fluorescence maxima is illustrated in Fig. 21.[50]

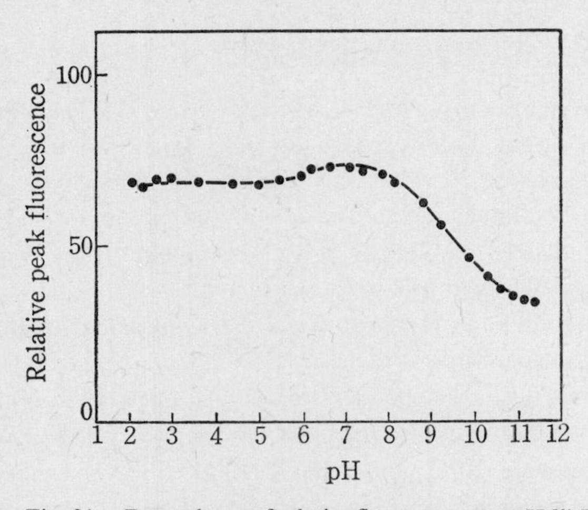

Fig. 21. Dependence of relative fluorescence on pH.[50] Excitation at 280 nm.

4.2.2. Methionine Residues

Jori *et al.*[98] found that photooxidation of lysozyme in 84% acetic acid in the presence of methylene blue modified two methionine residues selectively to sulfoxide. The product had only 5% activity when compared with native lysozyme.

4.2.3. Disulfide Bonds

There are four disulfide bonds (64–80, 76–94, 6–127, and 30–115) in the lysozyme molecule. X-ray analysis has shown that the dihedral angles about the S—S bonds are all within the range of $100 \pm 5°$, and that the conformations of the disulfide bonds 64–80 and 76–94 are mirror images to those of 6–127 and 30–115.[6,7] The disulfide bonds of native lysozyme do not react with mercaptoethanol[29,73] or sodium sulfite.[99–101] However, the four disulfide bonds of native lysozyme are quantitatively reduced by dithiothreitol.[102] Reduction of the disulfide bonds with mercaptoethanol in the presence of various denaturing agents occurs at concentrations lower than those at which the optical rotatory properties begin to change.[103] Beychok[104] has reported that the disulfide bonds contribute to the CD bands of lysozyme in the region of 200 to 300 nm.

5. Size and Shape

The molecular weight of lysozyme is calculated to be 14,307 from its amino acid composition. The molecular shape is an approximate ellipsoid of dimensions 45 × 30 × 30 Å.[6,7] The physicochemical properties of lysozyme are summarized in Table 6. Lysozyme molecules aggregate reversibly at pH's above 5.5.[119–121] The change in the weight average molecular weight with pH is shown in Fig. 22.[120]

Lysozyme in solution at a pH of 5 to 9 is a self-associating system of the monomer-dimer type, while at higher pH's polymers larger than the dimer appear. Dimerization is dependent on the pH, protein concentration, temperature, and ionic strength. At pH 5.0–5.5, protein concentration 0.15%, temperature 20°C, and ionic strength 0.15, lysozyme is monodisperse. The dimerization reaction is related to the ionization of a single group with an apparent p*K* of 6.2.[120] As can be seen from section 4.1.2.A, the group with an apparent p*K* of 6.2 involved in dimerization is very close in p*K* value to Glu-35.

Recently, Sophianopoulos[122] demonstrated that the presence of inhibitors, such as *N*-acetylglucosamine, chitobiose, and chitotriose, greatly reduces the extent of the reversible dimerization of lysozyme molecules. The order of effectiveness of inhibition of dimerization is chitotriose> chitobiose> *N*-acetylglucosamine. This is in accord with the order of the association constants of these saccharides with ly-

TABLE 6

Physicochemical Properties of Lysozyme

Sedimentation constant (*s*)	1.87†[1], 2.1†[2], 1.91†[4], 1.9†[14]
Diffusion constant (*D*)	10.4†[1], 10.2†[2], 11.2†[2], 11.8†[3], 11.2†[4]
Intrinsic viscosity (cc/g)	2.98†[3], 3.00†[4], 3.7†[19], 3.3†[8], 3.2†[9], 3.1†[15], 2.5†[16]
Partial specific volume	0.189 (546 nm)†[17], 0.1955 (436 nm)†[17], 0.195 (436 nm)†[18], 0.188 (578 nm)†[11]
Molecular weight	14,100 (*s*-*D*)†[1], 17,200 (*s*-*D*)†[2],
	14,100 (small-angle X-ray scattering)†[3]
	14,540 (sedimentation equilibrium)†[4]
	14,800†[17], 18,000†[18] (light scattering)
	14,307 (amino acid analysis)†[7]
Size	16 Å†[4], 15.3 Å†[3] (radius)
	28×30×32 Å†[5], 45×30×30 Å†[13]
Isoelectric point	11.35†[2], 11.1†[10], 11.18†[20]
$E_{1\text{cm}}^{1\%}$ at 280 nm	26.9†[19], 26.35†[4]

†[1] Ref. 105. †[2] Ref. 106. †[3] Ref. 107. †[4] Ref. 39. †[5] Ref. 108. †[6] Ref. 29. †[7] Ref. 5. †[8] Ref. 24. †[9] Ref. 109. †[10] Ref. 33. †[11] Ref. 110. †[12] Ref. 111. †[13] Ref. 112. †[14] Ref. 113. †[15] Ref. 114. †[16] Ref. 115. †[17] Ref. 116 †[18] Ref. 117. †[19] Ref. 118. †[20] Ref. 37.

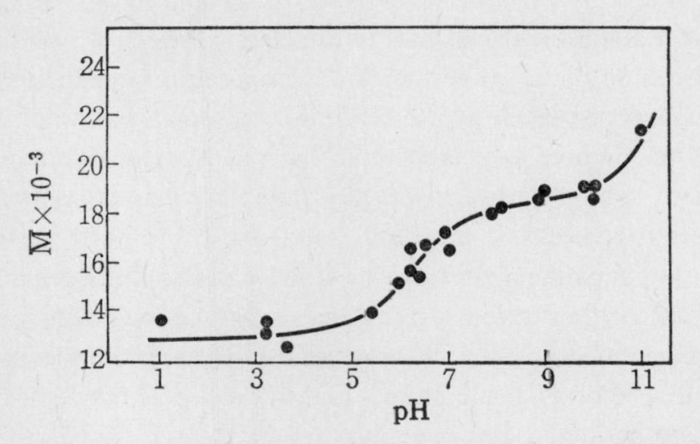

Fig. 22. Change in the apparent weight average molecular weight of lysozyme with pH at 20 °C.[120] Lysozyme concentration, *ca.* 1.4%, in 0.15 M KCl.

sozyme.[51,123] The dissociation constant of the dimer in the absence of inhibitor, at 20°C, an ionic strength of 0.15 and pH 8.0, is $(2.71 \pm 0.24) \times 10^{-3}$ M. Sophianopoulos have also calculated the dissociation constants of lysozyme saccharide complexes indirectly, using only the results from the sedimentation equilibrium. The fact that inhibitor reduces the dimerization of lysozyme implies that the active site is involved in the dimerization of lysozyme.

In contrast to this, Zehavi and Lustig[124] reported that *N*-acetylglucosamine, chitobiose, chitotriose, and *N*-acetylglucosamine (NAG)–*N*-acetylmuramic acid (NAM) did not exhibit any significant effect on the dimerization of lysozyme. They also showed that chitotetrose, chitopentaose and bacterial cell-wall tetrasaccharide (NAG–NAM–NAG–NAM) reduce the apparent weight average molecular weight of lysozyme. These results are thus not in agreement with those obtained by Sophianopoulos; however, the experimental conditions employed in each case were different.

6. Denaturation

6.1. Thermal Denaturation

The effect of temperature on the optical rotatory properties and viscosity of lysozyme was investigated by Hamaguchi and Sakai.[125] The temperature dependence of the Moffitt parameters a_0 and b_0 at various pH values is shown in Fig. 23. The transition temperature became lower as the pH was decreased below 5. The conformational change of lysozyme denatured by heat is smaller than that of lysozyme denatured by guanidine hydrochloride. For instance, the values of a_0 and b_0 were $-344°$ and $-100°$ at pH 5.14 and 86°C, but $-510°$ and $-40°$ in the presence of 6.4 M guanidine hydrochloride at pH 7 and 25°C. In general, an increase in the concentration of the denaturant is more effective in denaturing lysozyme than raising the temperature. Even in the presence of 1.9 M guanidine hydrochloride, the b_0 value at the higher temperature end was found to be $-90°$.

Fig. 24 illustrates the change in the reduced viscosity (η_{sp}/c) with temperature at pH 2.60. The value of η_{sp}/c was 4.7 cc/g above 60°C; the

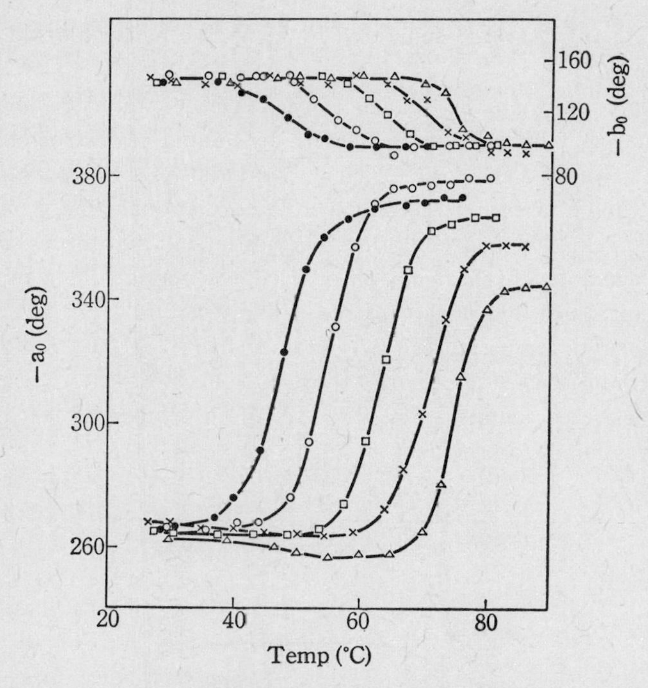

Fig. 23. Temperature dependence of a_0 and b_0 values for lysozyme at various pH values.[125] ●, pH 1.94; ○, pH 3.11; □, pH 3.37; ×, pH 3.97; △, pH 5.14; ionic strength, 0.1.

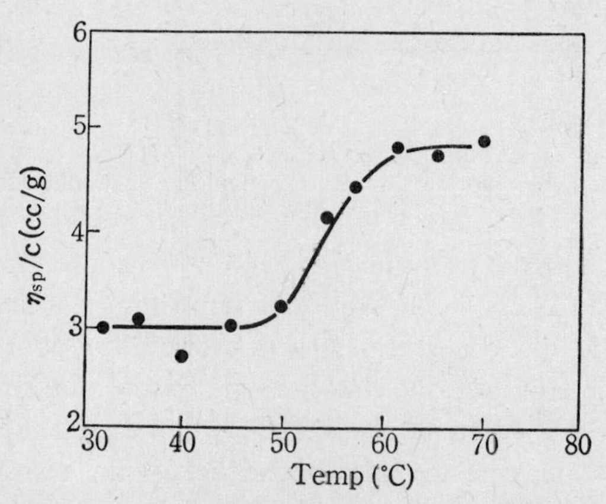

Fig. 24. Reduced viscosity of lysozyme (0.74%) at pH 2.60 as a function of temperature.[125] Ionic strength, 0.1.

value of η_{sp}/c of lysozyme denatured by guanidine hydrochloride was 6.7 cc/g at 25°C.

That the conformational change of lysozyme denatured by heat is smaller that that of lysozyme denatured by guanidine hydrochloride was also shown by Aune *et al.*[126] They found that lysozyme denatured at pH 1.65 and 60.5°C undergoes further conformational change on the addition of guanidine hydrochloride. They estimated that the lysozyme molecule denatured by heat has an ordered polypeptide chain of one fourth to one third that of the ordered native structure.

The transition temperatures (T_m) obtained from Fig. 23 are plotted against pH in Fig. 25.[40,125] At pH's above 6, lysozyme solutions became turbid at elevated temperatures. The transition temperature at pH's above 5 was found to be 75.0°C.

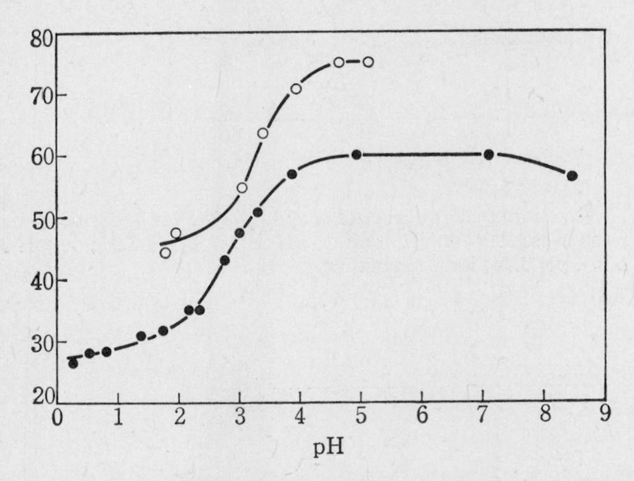

Fig. 25. Transition temperature, T_m, as a function of pH.[40,125] ●, 1.60 M guanidine hydrochloride; ○, without guanidine hydrochloride.

The pH dependence of the transition temperature was also studied by Sophianopoulos and Weiss.[54] The transition temperature at neutral pH was 75°C, which is in good agreement with the result obtained by Hamaguchi and Sakai. However, the change in the transition temperature on the acidic side was different from that shown in Fig. 25. Sophianopoulos and Weiss analyzed the pH dependence of the transition temperature and interpreted the denaturation process assuming that three carboxyl groups of abnormally low pK value (pK_1 = 1.46, pK_2 =

1.66, and $pK_3 = 1.66$) are involved and that these pK values become 3.40 on heat denaturation.

6.2. Effect of pH

Lysozyme is very stable in the low pH region. The optical rotatory properties, viscosity, and solvent perturbation of the ultraviolet spectrum do not change on acidifcation to a pH below 2. On the alkaline side, lysozyme is denatured at pH's above 12. This has been demonstrated by both spectrophotometric titration[43,44] and CD spectroscopy.[72]

6.3. Urea Denaturation

Lysozyme is very resistant to urea. It is denatured only to a small extent even in the presence of 9 or 10 M urea.[24,127]

6.4. Denaturation by Guanidine Hydrochloride

Guanidine hydrochloride is a very strong denaturant of lysozyme. The intrinsic viscosity of lysozyme in 6 M guanidine hydrochloride is shown in Table 7. It is considerably smaller than the viscosity observed when four sulfide bonds are reduced by mercaptoethanol in 6 M guanidine hydrochloride. One of the four disulfide bonds is bridged between residues 6 and 127 so that the chain ends are linked to make a loop; this restricts the expansion of the denatured lysozyme molecule. The three other disulfide bonds also restrict the expansion.[129]

The optical rotatory properties of lysozyme denatured by guanidine

TABLE 7

Intrinsic Viscosity of Lysozyme at 25°C

Solvent	$[\eta]$ (cc/g)
0.2 M NaCl	2.98†1
	2.7†2
6 M GuCl	7.1†1
	6.5†2
6 M GuCl + mercaptoethanol	16.1†1
	17.1†2

†1 Ref. 128. †2 K.C. Aune, in ref. 129.

TABLE 8
Optical Rotatory Properties of Lysozyme in Concentrated Guanidine Hydrochloride

Disulfide bond intact		Disulfide bond reduced		Ref.
a_0	b_0	a_0	b_0	
−494	−22	−449	0	130
−510	−40	−440	−45	103

hydrochloride are shown in Table 8. Hamaguchi and Kurono[24,74] studied the denaturation of lysozyme by guanidine hydrochloride by measurement of its optical rotatory properties, viscosity and ultraviolet absorption spectrum. When guanidine hydrochloride was added to lysozyme solutions, these properties changed with time and attained final values after one or two hours. The variations in the final values of b_0 and the mean residue rotation at 500 nm ($[m']_{500}$) with guanidine hydrochloride concentration are given in Fig. 26. The value of b_0 was constant up to 3.2 M guanidine hydrochloride. Further increase in concentration caused an abrupt decrease in the value of b_0.

The levrorotation at 500 nm decreased by about 10° at up to 3.2 M guanidine hydrochloride, due to the solvent effect on the specific rotation. The reduced viscosity also changed above 3.2 M guanidine hydrochloride. A result similar to the curve given in Fig. 26 has also been obtained by Tanford *et al.*[131] from measurements of a_0, b_0, and $[m']$ at 234 nm.

These authors assumed a two-state transition for the denaturation of lysozyme by guanidine hydrochloride:

$$N \rightleftharpoons U \tag{3}$$

From the change in the value of $[m']_{234}$ with the concentration of guanidine hydrochloride, they found the following relation between the equilibrium constant K and the concentration of guanidine hydrochloride ([GuCl]) at pH 5.0–5.5 and 0.1 M NaCl (Fig. 27):

$$\log K = -9.43 + 15.85 \log ([\mathrm{GuCl}]) \tag{4}$$

The denaturation of lysozyme by guanidine hydrochloride follows a first order reaction.[24,40,74,131] The rate constants obtained from the changes in the values of $[m']_{500}$, η_{sp}/c and $\Delta\varepsilon_{292}$ were 0.063, 0.063, and 0.072 min^{-1}, respectively. The rate constant obtained from viscosity measurements was a fifth order dependent of the guanidine hydrochlo-

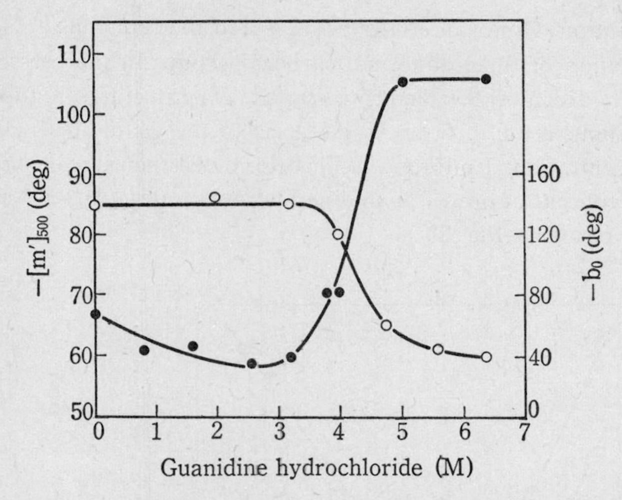

Fig. 26. Dependence of final values of mean residual rotation at 500 nm (●) and parameter b_0 (○) of lysozyme on guanidine hydrochloride concentration at 25 °C in 0.01 M EDTA, pH 7.0.

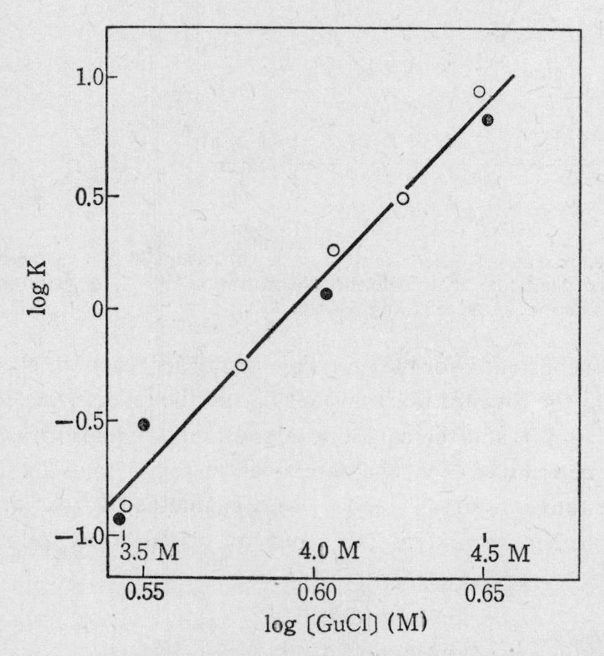

Fig. 27. Dependence of the equilibrium constant on guanidine hydrochloride concentration.[131] The straight line is drawn according to Eq. (4).

ride concentration. Donovan *et al.*[34,35] measured the rate constants of the normalization of three abnormal carboxyl groups in the presence of guanidine hydrochloride from proton uptake experiments, and found that the constants are fifth order with respect to the guanidine hydrochloride concentration. Tanford *et al.*[131] determined the rate constants k at various concentrations of guanidine hydrochloride at pH 5.5 and obtained the results in Fig. 28.

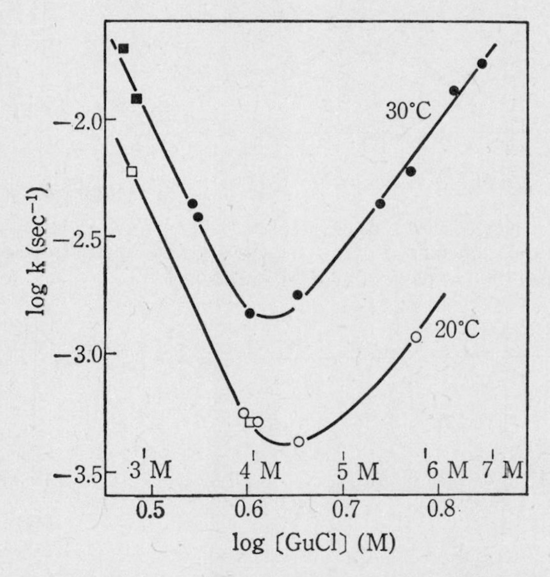

Fig. 28. Apparent first-order rate constants at 20 °C and 30 °C as a function of guanidine hydrochloride concentration.[131] ○, ●, Unfolding reaction; □, ■, refolding reaction.

Since the rate constant k for the reversible first order reaction, $N \rightleftharpoons U$, is the sum of the forward rate constant k_1 and the reverse rate constant k_2 ($k = k_1 + k_2$), and the equilibrium constant K equals k_1/k_2, k_1 and k_2 can be determined from the data given in Fig. 27 and 28. The changes in the rate constants k_1 and k_2 with guanidine hydrochloride concentration are shown in Fig. 29[131] and are expressed as follows:

$$\left.\begin{aligned} k_1 &= \text{constant } [\text{GuCl}]^5 \\ k_2 &= \text{constant } [\text{GuCl}]^{-11} \\ K &= \text{constant } [\text{GuCl}]^{16} \end{aligned}\right\} \qquad (5)$$

From these facts, Tanford *et al.*[131] proposed the following reaction

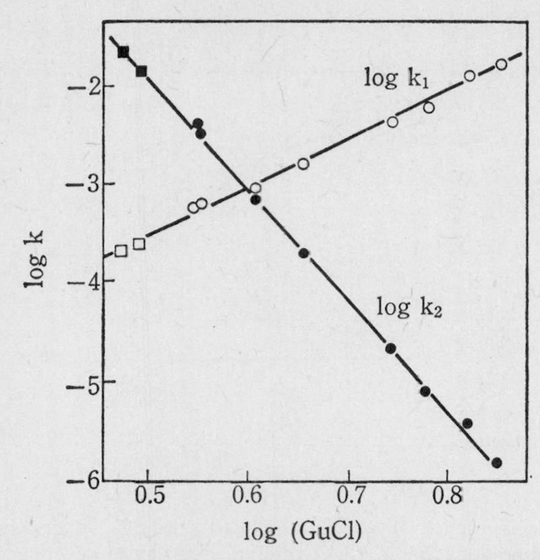

Fig. 29. Rate constants for unfolding (k_1) and refolding (k_2) at 30 °C as a function of guanidine hydrochloride concentration.[131]

scheme for the denaturation of lysozyme by guanidine hydrochloride:

$$N \underset{}{\overset{+5Gu^+}{\rightleftharpoons}} (PGu_5)^* \underset{}{\overset{+11Gu^+}{\rightleftharpoons}} UGu_{16}\,, \tag{6}$$

where P represents the protein in the activated state. They also obtained the thermodynamic parameters shown in Table 9 for the formation of the activated complex.

Ogasahara and Hamaguchi[40] examined the pH dependence of a_0 and b_0 at various concentrations of guanidine hydrochloride (Fig. 30). The equilibrium constant between native and denatured lysozyme in the presence of 3.84 M guanidine hydrochloride was proportional to the 1.7th power of the hydrogen ion concentration. The acid denaturation of lysozyme in the presence of guanidine hydrochloride followed revers-

TABLE 9

Thermodynamic Parameters for the Formation of Activated Complex at 25 °C[131]

	ΔF_u (kcal/mole)	ΔH (kcal/mole)	ΔS_u (e.u.)
From native lysozyme	14	30	+52
From unfolded lysozyme	39	16	−77

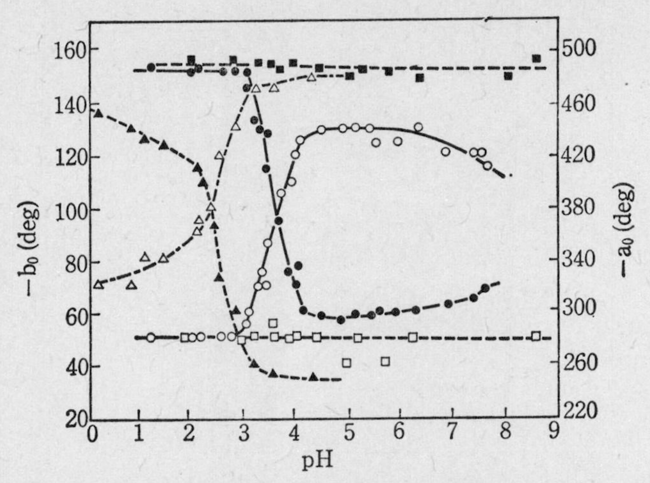

Fig. 30. The pH dependence of a_0 values (closed symbols) and b_0 values (open symbols) at 25 °C.[40] Guanidine hydrochloride concentrations: ▲, △, 2.56 M; ●, ○, 3.84 M; ■, □, 5.06 M.

ible first order reaction kinetics. The apparent rate constant, k, in the presence of 3.84 M guanidine hydrochloride was expressed by the equation

$$k = k_1' \, [\mathrm{H}^+]^{1.2} + k_2' \, [\mathrm{H}^+]^{0.8} \tag{7}$$

The denaturation of lysozyme by guanidine hydrochloride as a function of pH, temperature and the concentration of guanidine hydrochloride was studied by Aune and Tanford,[55,132] and by Tanford and Aune[133] in detail. The results were analyzed by using the two-state transition scheme in Eq. (3). The equilibrium constants at various concentrations of guanidine hydrochloride are plotted against pH in Fig. 31.

The data shown in the figure were interpreted satisfactorily by assuming that four groups are involved in denaturation in the region from acidic to neutral pH. These groups were assumed to be Asp-52 (or 103) (pK = 1.9), Glu-7 (pK = 1.9), Glu-35 (pK = 6.1), and His-15 (pK = 5.8) on the basis of X-ray crystallographic model and saccharide binding data. Aune and Tanford[132] proposed various hypotheses concerning the equilibrium of native and denatured states of lysozyme in guanidine hydrochloride, but did not obtain any definitive conclusion on the mode of action of guanidine hydrochloride. However, thermodynamic analysis led them to the conclusion that the free energy of stabilization of

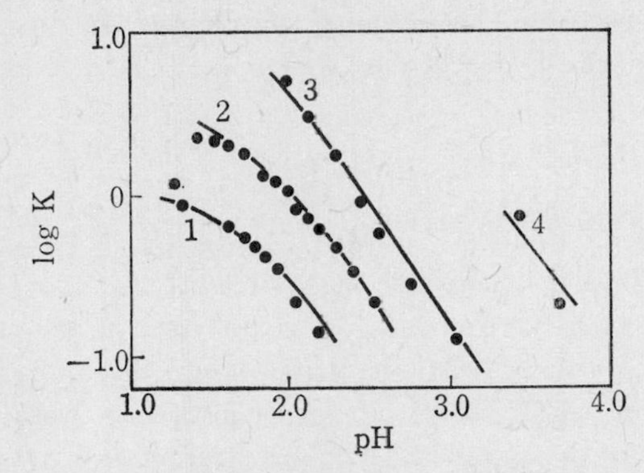

Fig. 31. Equilibrium constants at various guanidine hydrochloride concentrations as a function of pH.[55] Guanidine hydrochloride concentrations: 1.98 M (1), 2.25 M (2), 2.56 M (3), and 3.34 M (4).

TABLE 10

Thermodynamic Parameters in 1 M Guanidine Hydrochloride at 25°C, pH 7.[133]

Reaction	log K	ΔG^0 (kcal/mole)	ΔH (kcal/mole)	ΔS^0 (e.u.)	ΔC_p (cal/deg per mole)
$N \rightleftharpoons U$	−5.8	7.9	22.4	49	1375
$N \rightleftharpoons X$	−5.7	7·8	41.2	112	950
$X \rightleftharpoons D$	−0.1	0.1	−18.8	−63	425

native lysozyme in the absence of denaturant is small (10–20 kcal/mole).

As described above, the denatured form (X) obtained by raising the temperature in the absence of denaturants is different from the form (D) obtained on denaturing by guanidine hydrochloride. Tanford and Aune[133] obtained thermodynamic parameters for the reactions $N \rightleftharpoons X$, $X \rightleftharpoons D$, as well as for $N \rightleftharpoons D$ (Table 10).

The results are consistent with those obtained by Sophianopoulos and Weiss[54] and with the estimate that form X has an ordered chain of one fourth to one third that of the native structure[126] (see section 6.1).

6.5. Denaturation by Formamide

Formamide acts as a denaturant against lysozyme.[32] The values of a_0 and b_0 changed above 60% (15 M) formamide but became constant at 80%. The values of a_0 and b_0 of lysozyme in 80% formamide were

$-350°$ and $-60°$, respectively. These values are larger than those of lysozyme in 6 M guanidine hydrochloride. The value of η_{sp}/c was 9.8 cc/g.

6.6. Effects of Amide Derivatives

The effects of *N*,*N*-dimethylformamide, *N*,*N*-dimethylacetamide, and tetramethylurea were different from those of guanidine hydrochloride and formamide in that a decrease in the value of $-b_0$ accompanied decrease in the value of $-a_0$.[29,32,73] A similar phenomenon was also observed for the effect of dioxane.[24,74] The concentration at which the a_0 and b_0 values began to change was 35%, 50%, and 70% for dimethylformamide, dimethylacetamide, and tetramethylurea, respectively. It is interesting to note that dioxane, dimethylformamide, dimethylaectamide, and tetramethylurea serve only as proton acceptors and not as proton donors. The value of b_0 for lysozyme did not change with up to 11 M acetamide and 7.6 M 1,3-dimethylurea.

6.7. Effects of Alcohols

The effects of methanol, ethanol, *n*-propanol, isopropanol, glycerol, α-monochlorohydrin, ethylene glycol, ethylene glycol monomethyl ether, and 2-chloroethanol on the optical rotatory and ultraviolet spectral properties of lysozyme were studied by Hamaguchi and his co-workers.[24,74,76,77] The effectiveness in denaturing lysozyme increases with chain length and increasing hydrocarbon content. On the other hand, the hydroxyl content of the solvent molecule has no significant effect on the lysozyme molecule. Thus, ethylene glycol and glycerol do not affect the conformation of lysozyme. These observations suggest that the internal fold of the lysozyme molecule is stabilized by hydrophobic interactions. Methanol, ethanol, and propanol increase the helical content of the lysozyme molecule. The effect of ethanol on the CD spectrum of lysozyme is illustrated in Fig. 32.[95] The negative CD bands at 208 and 222 nm increased above 60% ethanol, indicating an increase in the helical content of the lysozyme molecule.

As described in section 4.2.1, the molar absorptivity at 292 nm in the difference spectrum of lysozyme, which originates from tryptophan residues, does not increase linearly with the concentrations of methanol,

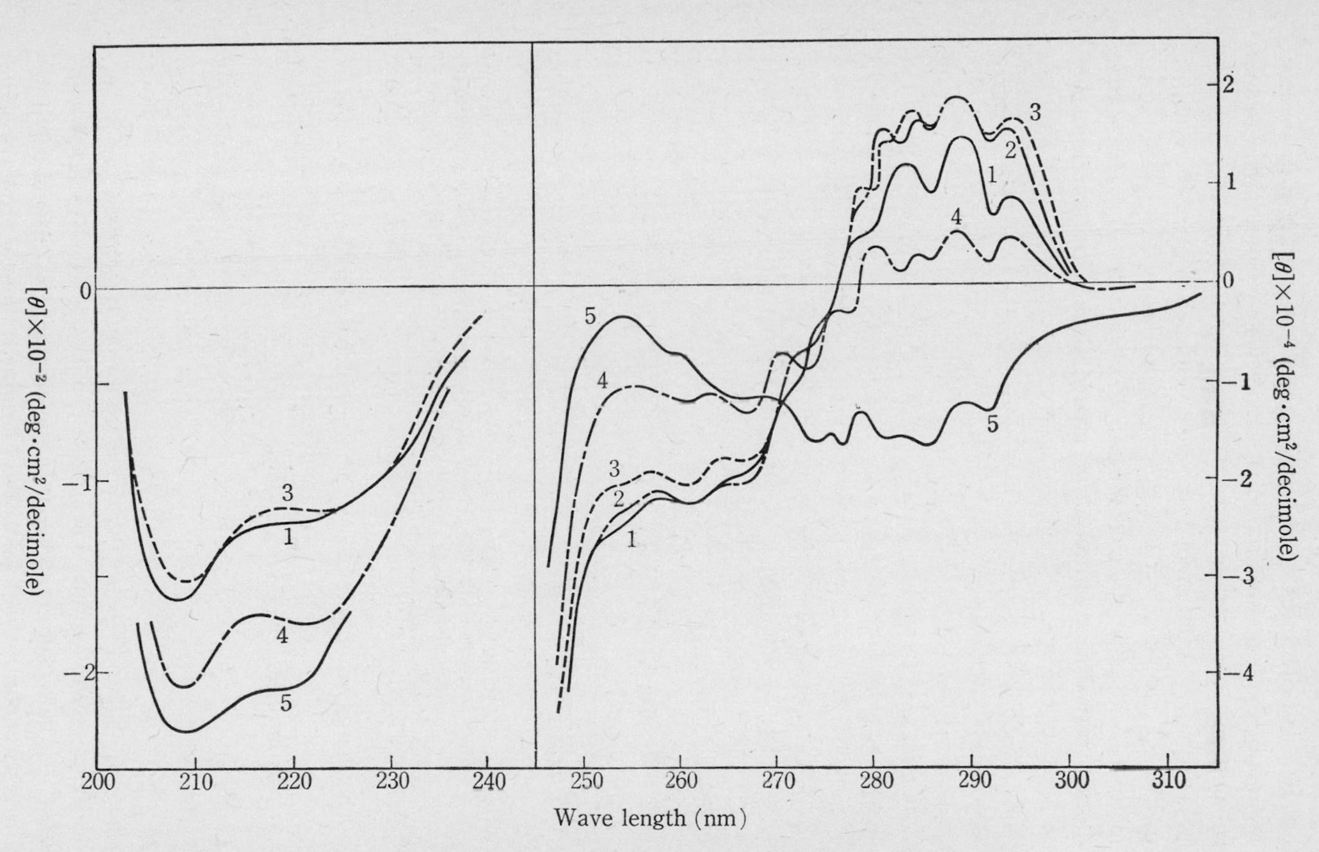

Fig. 32. CD spectra of lysozyme in water (1) and in 20% (2), 60% (3), 80% (4), and 90% (5) ethanol.[95]

ethanol, and propanol.[24,74,77] This suggests that a strong short-range interaction exists between the alcohol molecule and tryptophan residues in addition to a long-range solvent effect. Therefore, the effect of alcohols on the CD spectrum of lysozyme was studied in detail.[95] As shown in Fig. 32, the positive CD bands in the aromatic absorption region increased with ethanol concentration. The values of $[\theta]$ at 294 nm are plotted against the logarithm of the ethanol concentration in Fig. 33.[95]

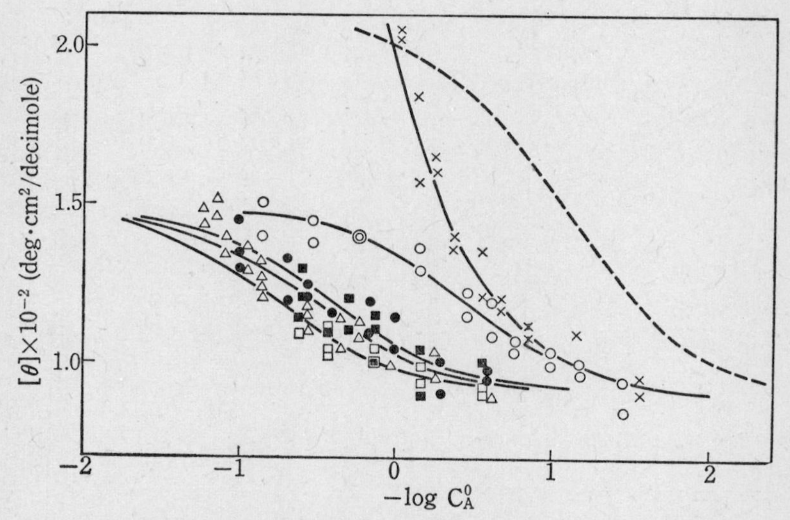

Fig. 33. Changes in values of $[\theta]$ at 294 nm with the concentration of various reagents at 25 °C.[95] ○, Ethanol; ●, methanol; □, isopropanol; ■, *n*-propanol; △, ethylene glycol; ×, sucrose; ---, *N*-acetylglucosamine at pH 7.4.

The change in the ellipticity at 294 nm which is considered to originate from tryptophan residues was analyzed by assuming the following 1:1 interaction of lysozyme and alcohol:

$$E + A \rightleftharpoons EA \tag{8}$$

$$K = \frac{C_{EA}}{C_E \cdot C_A}, \tag{9}$$

where C_E, C_A, and C_{EA} are the concentrations of lysozyme, alcohol, and lysozyme–alcohol complex, respectively, and K is the association constant. Assuming that the change in the ellipticity at 294 nm ($\Delta[\theta]$) for any given molar ratio of alcohol to lysozyme is directly proportional to the fraction of lysozyme which is associated with the alcohol molecule, we obtain the following equation at high molar ratios of alcohol to lysozyme:

$$\frac{1}{\Delta[\theta]} = \frac{1}{\Delta[\theta]_\infty}\left[\frac{1}{K \cdot C_A{}^0} + 1\right], \tag{10}$$

where $\Delta[\theta]$ is the difference between the values of ellipticity for the complete lysozyme-alcohol complex and for lysozyme, and $C_A{}^0$ is the initial concentration of alcohol. Therefore, the value of $\Delta[\theta]_\infty$ and the association constant K are obtained from the plot of the reciprocals of $\Delta[\theta]$ against the reciprocals of the initial concentrations of alcohol. Such a plot gives a straight line. The following equation is also obtained for the interaction of lysozyme and alcohol on the basis of scheme (8), making the same assumption as for Eq. (10).

$$\log\frac{[\theta]-[\theta]_0}{[\theta]_\infty-[\theta]} = \log K + \log C_A{}^0\,, \tag{11}$$

where $[\theta]_0$ is the ellipticity of lysozyme at 294 nm in the absence of alcohol.

As shown in Fig. 34, a plot of $\log([\theta]-[\theta]_0)/([\theta]_\infty-[\theta])$ against log

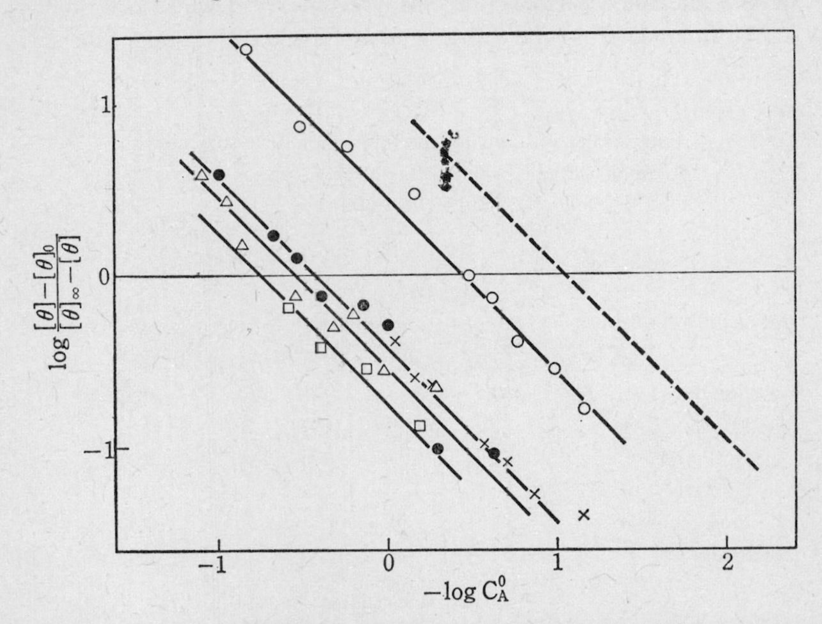

Fig. 34. Plot of $\log([\theta] - [\theta]_0)/([\theta]_{00} - [\theta])$ against the log of the concentration of various reagents.[95] (Symbols as in Fig. 33.)

C_A^0 is linear with a slope of unity. This indicates that one mole of alcohol binds to one mole of lysozyme. The association constants are summarized in Table 11.

It should be pointed out that the association constant for ethanol is greater than that for methanol or propanol. Of course, a number of alcohol molecules must bind to the enzyme molecule in aqueous alcohol solutions. Nevertheless, so far as the interaction is revealed by CD measurements, the results obtained indicated that one mole of alcohol interacts with one mole of lysozyme. This suggests that the change in the state of certain tryptophan residues resulted from interaction with an alcohol molecule. The binding site of alcohol was inferred to be the substrate binding site from experiments to determine the effect of alcohol on lysozyme molecules in which Trp-62 was selectively oxidized by NBS. It appeared that Trp-62 made an appreciable contribution to the enhancement of the CD spectrum resulting from the interaction of ethanol.

Analysis of the variation in molar absorptivity difference at 293 nm with alcohol concentration also led to the conclusion that one mole of alcohol interacted with one mole of lysozyme.[134] The association constants obtained from the 293 nm peak and 305 nm trough are sum-

TABLE 11

Association Constants for the Interaction of Lysozyme with Alcohols Obtained from Ultraviolet Absorption Spectra[134] and CD Spectra[95]

	Native lysozyme		
	UV		CD
	293 nm	305 nm	294 nm
MeOH	0.5	0.3	0.4
EtOH	2.7	0.7	2.8
Iso-PrOH	2.9	1.1	0.3
n-PrOH			0.2
Ethylene glycol			0.3
Sucrose			0.4
	Lysozyme oxidized by NBS		
EtOH	2.1	–	2.8

marized in Table 11. The origin of the 305 nm trough was recently discussed by Ananthanarayanan and Bigelow,[135,136] who considered it to be due to tryptophan residues in the vicinity of charged groups.

6.8. Effect of Dimethyl Sulfoxide

The properties of lysozyme in DMSO-water mixtures were studied by Hamaguchi.[29] No aggregation of the lysozyme molecules occurred in DMSO. The values of $[m']$ at 436 nm, of a_0, and b_0 for lysozyme in DMSO were $-81.3°$, $-238°$, and $-80°$, respectively. The reduced viscosity was 1.1 cc/g. It was found by infrared absorption spectroscopy that there was no indication of the presence of β-structure in the lysozyme molecule in DMSO. Changes in the optical rotatory properties and ultraviolet absorption spectrum occurred above 70% DMSO.

6.9. Effects of Inorganic Salts

Among inorganic salts, LiCl and LiBr have a strong denaturing action on lysozyme.[75] The value of $-b_0$ was constant up to 6 M LiCl and 4.5 M LiBr, and then decreased with further increases in concentration, with an accompanying increase in the levorotation. The values of a_0 and b_0 were $-280°$ and $-65°$ in 8 M LiCl and $-300°$ and $-60°$ in 6.5 M LiBr at pH 7 and 25°C. These values are larger than those of lysozyme in 6 M guanidine hydrochloride. The reduced viscosity was 5.6 cc/g in 8 M LiCl and 5.5 cc/g in 6.2 M LiBr. The denaturation blue shift of lysozyme in LiCl and LiBr was also considerably smaller than that in 6 M guanidine hydrochloride. These facts suggest that denaturation by these salts is not so complete as that by guanidine hydrochloride. Sodium bromide has no denaturing action on lysozyme at concentrations up to 5.6 M at pH 7 and 25°C.

6.10. Effect of Temperature in the Presence of Various Denaturing Agents

The thermal stability of lysozyme in the presence of various reagents was studied by Hamaguchi and Sakai[125] from measurements of the optical rotatory dispersion. The molar effectiveness in increasing the value of $\Delta[m']_{436}$ (or Δa_0) (the difference between the mean residue rotation, or the value of a_0, at the high and low temperature ends) follows the order, tetramethylurea < dimethylurea, dimethylformamide, dimethylacetamide < acetamide < formamide < urea < guanidine hydrochloride. The value of Δb_0 (the difference between the value of b_0 at the high and low temperature ends) decreases with the concentration of methyl-substituted ureas and amides. Furthermore, the ability of these compounds to lower the transition temperature is enhanced by alkyl substitution of the NH_2 groups. The same situation was also found for the thermal transition in aqueous solutions of methanol, ethanol, ethylene glycol, propylene glycol, dimethyl sulfoxide and dioxane. At higher concentrations of dimethylformamide, methanol, ethanol, or ethylene glycol, the value of b_0 did not change with temperature, while the value of $-a_0$ increased steeply within a narrow temperature range. These facts suggest that as more methyl groups are introduced into the solvent molecule, only the internal fold of lysozyme (stabilized by hydrophobic interactions) is disrupted on raising the temperature.

6.11. Surface Denaturation

It was found difficult to spread lysozyme as a monolayer on the surface of water.[99,137-40] When lysozyme had been denatured by urea and 40% propyl alcohol prior to spreading, a complete lysozyme monolayer was obtained on water at pH 10.5. Raising the temperature of the substrate water was also effective in spreading a monolayer. Yamashita and Bull[141] obtained a complete lysozyme monolayer on an aqueous solution of 3.5 M KCl.

When native lysozyme was spread on water at pH 10.5, the surface pressure-area curve shifted to enclose a larger area with time. The increase in area at constant pressure followed first order reaction kinetics.[99]

7. Enzymatic Action

7.1. Substrate

7.1.1. Bacterial Cell-bodies

The cell-bodies of certain Gram-positive bacteria are normally used for assaying the lytic activity of lysozyme. Bacteria known to be most sensitive to lysozyme action are *Micrococcus lysodeikticus* and *Bacillus megatherium*. Gram-negative bacteria show great resistance to lysozyme action. However, the presence of ethylenediamine tetraacetate accelerates the lysis of Gram-negative bacteria by lysozyme.[142] Heating or pre-treatment of cell-bodies of Gram-negative bacteria by an organic solvent also causes an appreciable increase in sensitivity towards the lytic action of lysozyme.

7.1.2. Bacterial Cell-walls

It has long been known that the digestion of cell-walls by lysozyme is due to the catalytic hydrolysis of the polysaccharide moiety of certain cell-wall components.[143] Salton[144] observed that insoluble cell-walls separated from the cell-bodies of sensitive bacteria were digested rapidly and became soluble by the action of lysozyme. Thereafter, much research has been directed towards solving the problem of the mode or pattern of digestion of bacterial cell-walls by lysozyme.

7.1.3. Soluble Substrates

A. Polysaccharide Substrate

Many attempts to follow lysozyme action by using was tersoluble cell-wall polysaccharides of high polymerization degree have been made. If polysaccharides were readily available in suitable amounts, the action of lysozyme could be easily and precisely assayed by viscometry or by the reducing power method. However, in fact, the separation of sufficiently large amounts of cell-wall polysaccharide is a very difficult task.[145]

Berger and Weiser[2] observed that lysozyme possesses β-*N*-acetylglucosaminidase activity, which can cause the hydrolysis of chitin, a polysaccharide of *N*-acetylglucosamine. Nevertheless, in practice, chitin is not well suited as a substrate for lysozyme assay because, being an insoluble micellar substance, it is itself not readily hydrolyzed by lysozyme. Hamaguchi and Funatsu[146] found that a soluble chitin derivative (prepared by reaction with ethylene oxide in concentrated sodium hydroxide solution) was easily hydrolyzed by lysozyme. This soluble derivative contains partly a 6-*O* β-hydroxyethylated *N*-acetylglucosamine residue, and is called glycol chitin (I). A viscometric assay method has been established using this soluble substrate. Another

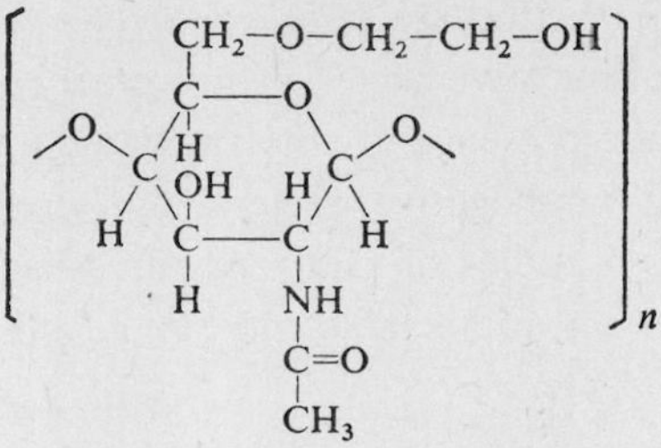

$CH_2-O-CH_2-CH_2-OH$
C—O
O, C, H, C, O
C, OH, H, C
H, C—C, H
H, NH
C=O
CH_3
]$_n$

(I)

related soluble substance useful for the assay of lysozyme is carboxylmethyl chitin (II), which is also a 6-*O* derivative of chitin.[147]

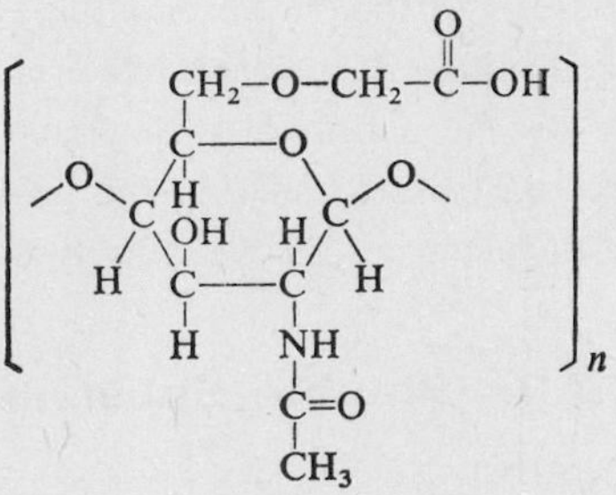

O
CH_2-O-CH_2-C-OH
C—O
O, C, H, C, O
C, OH, H, C
H, C—C, H
H, NH
C=O
CH_3
]$_n$

(II)

B. Oligosaccharide Substrate

Soluble oligisaccharide substrates are prepared from hydrolyzates of bacterial cell-walls or chitin. The oligosaccharide substrate prepared from cell-walls has the formula (NAG–NAM)$_n$ (III). Separation

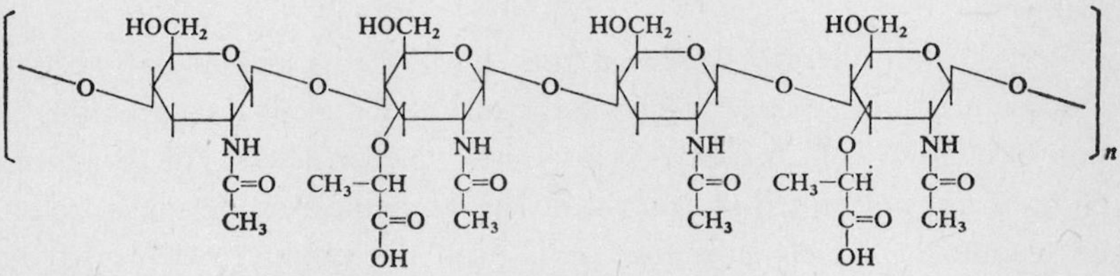

$HOCH_2$ $HOCH_2$ $HOCH_2$ $HOCH_2$
O O O O O O O O O
NH O NH NH O NH
C=O CH_3-CH C=O C=O CH_3-CH C=O
CH_3 C=O CH_3 CH_3 C=O CH_3
OH OH
]$_n$

(III)

methods of cell-wall oligosaccharides have been reported in detail by Jolles,[148] Salton[149] and Leyh-Bouille *et al.*[150] The isolation and purification of oligosaccharides of *N*-acetylglucosamine from partial hydrolyzates of chitin by concentrated hydrochloric acid was studied by Rupley.[47]

C. Synthetic Substrate

Osawa[151] found that *p*-nitrophenyl β-*N*,*N*′-diacetylchitobioside (IV) was slowly hydrolyzed by lysozyme, releasing *p*-nitrophenol. Neither glycoside of *N*-acetylglucosamine nor D-glucose are hydrolyzed by lysozyme. However, Raftery and Rand-Meir[152] reported that *p*-nitrophenyl β-*N*-acetylglucosaminide (V), *p*-nitrophenyl β-glucoside (VI) and *p*-nitrophenyl β-2-deoxyglucoside (VII) are only hydrolyzed by lysozyme in the presence of oligosaccharides of *N*-acetylglucosamine such as tetrasaccharide.

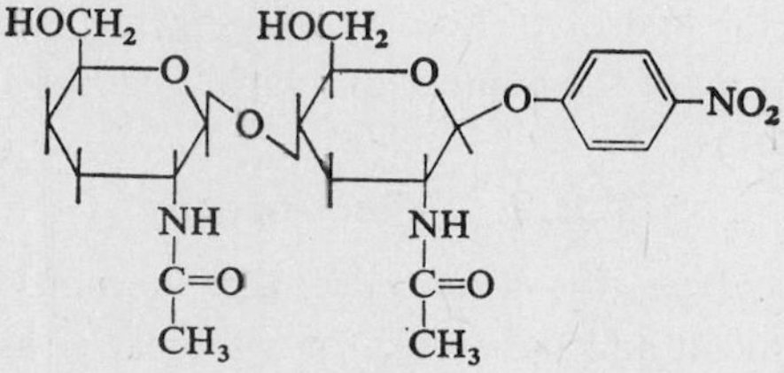

(IV)

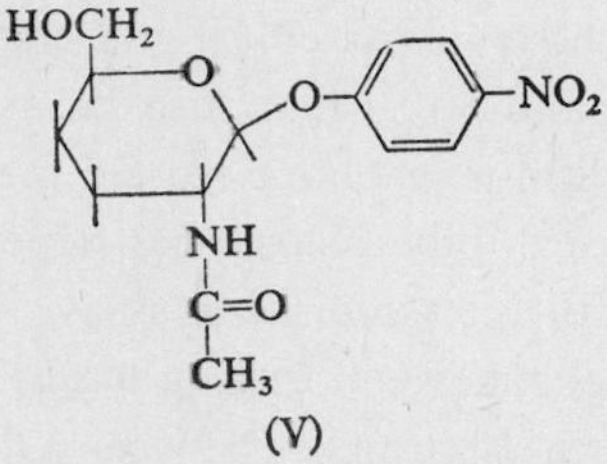

(V)

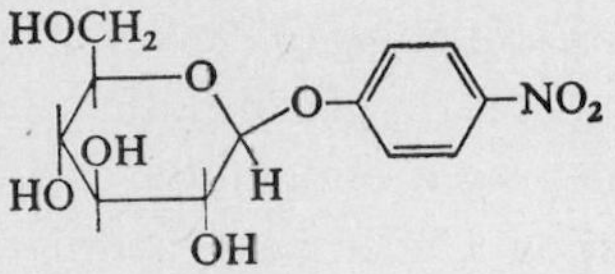

(VI)

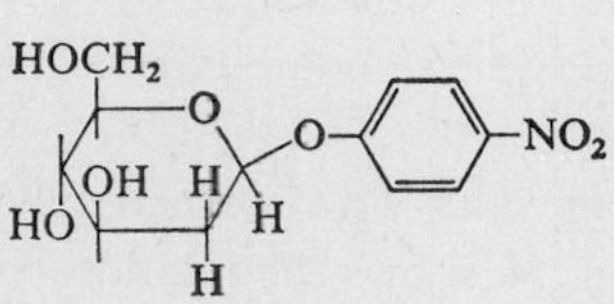

(VII)

7.2. Assay Method

7.2.1. Lytic Method

Dead cell suspensions of *M. lysodeikticus* are widely used for the measurement of the lytic activity of lysozyme. Although the lytic method is the most sensitive among the various assay methods, it is necessary to control the assay conditions carefully in order to obtain reproducible results.[153] The lytic phenomenon is considered to consist of two distinct steps; hydrolysis of cell-wall composing polysaccharides, and dissolution or bursting of the damaged bacterial cell-bodies. When it is required to assay the hydrolytic activity of lysozyme, other methods giving a direct measure of the number of hydrolyzed bonds are preferable.

7.2.2. Viscometry

As mentioned above, the isolation of large amounts of viscous polysaccharide is difficult and therefore, for viscometric assays,[154] soluble chitin derivatives are adopted as substrate. Hamaguchi and Funatsu[146] established a viscometric method for the assay of lysozyme using glycol chitin as substrate. Carboxymethyl chitin and chitosan (degrees of deacetylation between 20 and 30%) may also be used for the assay.[147b] Since carboxymethyl chitin is insoluble in acid media and the viscosity of its aqueous solution exhibits a sharp pH dependence, glycol chitin is at present preferred for the viscometric assay.

An Ostward-type viscometer is normally used. The relative viscosity (η_{rel}) of 0.1% glycol chitin solution lies between 1.4 and 1.6. The enzyme amount which is sufficient to lower an η_{rel} of 1.4 to 1.1–1.2, for 10 ml of substrate solution, is about 50 μg (0.2 ml of 0.25% lysozyme solution is rapidly added to 10 ml of 0.1% glycol chitin solution). The activity of lysozyme was represented by a diminution of η_{rel} at 3 min after the addition of enzyme solution. Under these conditions, no reducing power due to hydrolysis of β-*N*-acetylglucosaminide linkages could be detected. Plots of the extent of diminution of η_{rel} at 3 min against the amount of enzyme added (the standard curve) did not reveal any linear proportionality.

Fig. 35 illustrates the results of a viscometric assay with 0.08% glycol chitin as substrate; Fig. 36 shows the relation between the amount of lysozyme and the viscosity decrease at 3 min.

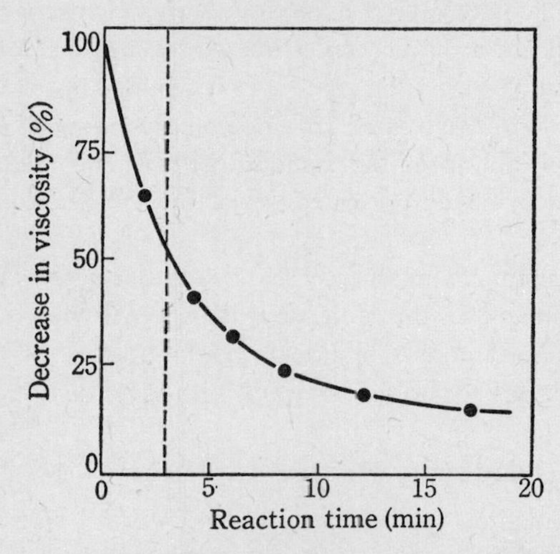

Fig. 35. Viscometric assay of lysozyme activity. Substrate, glycol chitin (0.08%); lysozyme, 0.76 μmole; 0.1 M phosphate buffer at pH 5.6, 30 °C. For representing the activity, the decrease in viscosity at 3 min is used.

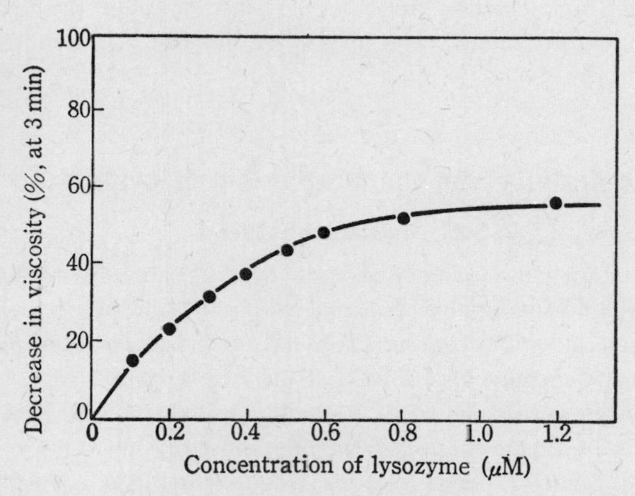

Fig. 36. Relation between amount of lysozyme and viscosity decrease at 3 min. Glycol chitin, 0.08%, 10 ml; lysozyme, 0.2 ml at various concentrations. Conditions of measurement were as for Fig. 35.

Changes in lysozyme activity caused by various forms of treatment, such as chemical modification, have been estimated by measuring the diminution of η_{rel} at 3 min for a certain amount (x) of treated lysozyme, by reading the amount (y) of untreated enzyme corresponding to the measured diminution of η_{rel} from the standard curve. The relative activity of the treated lysozyme is represented by the ratio of the enzyme amounts, $(x)/(y) \times 100$.

Changes in activity accompanying the physical conditions of the solution have also been estimated in the same way. However, since viscometric assay requires large amounts of lysozyme, this method is not suited to direct enzyme assay for crude extracts from tissues or organs.

7.2.3. Reducing Power Method[155]

The reducing power method can generally be used for all types of substrates, without limitation according to their solubility. The method is preferred to all others in that it gives a direct measure of the number of glycoside linkages hydrolyzed. However, high concentrations of enzyme and substrate and a prolonged reaction time are necessary for the reducing power method. Furthermore, the transglycosylation occurring in the bond-cleaving process of the substrate makes the quantitative counting of the number of hydrolyzed linkages of the substrate difficult.

7.3. Lytic Activity and the Mechanism of Lysis

7.3.1. Lytic Activity

The maximum lytic activity of lysozyme occurs at about pH 7; below pH 4, lysis is not observable. Phenomenological studies on the effect of neutral salts at various concentrations have been carried out by many investigators. Generally, lysis is most accelerated by *ca.* 0.1 M of salts with a univalent cation and by *ca.* 0.025 M of salts with a divalent cation.[156] However, these effects vary with the concentration of lysozyme. Wilcox and Daniel[157] found that the lytic activity of lysozyme was linearly proportional to concentration up to 0.02 mg/ml, but that at higher concentrations of the enzyme the activity was somewhat reduced. This reduction in activity could be prevented by the addition of suitable amounts of sodium ion; nevertheless, no proportional relationship between activity and amount of lysozyme at high concentrations was observed under these conditions.

7.3.2. Mechanism of Lysis

Cell-walls of sensitive bacteria which had previously been partially treated with lysozyme were abruptly dissolved by the addition of an alkaline solution. Cell-bodies which had been incubated with lysozyme at low pH's, where no lysis was observed, were instantaneously dissolved by adjusting the pH of the medium to 7.0 or higher. Fig. 37[158] illustrates this pH dependence of the lytic activity. The lysis rate was also accelerated in the presence of proteolytic enzymes such as trypsin.[159]

These findings suggest that the lytic phenomenon may consist of two different steps, where the first results in the damage (partial hydrolysis) of cell-wall polysaccharide and the second results in the bursting of damaged cell-bodies. The former seems to represent the true enzymatic action of lysozyme, while the latter appears to be a physical change probably concurrent with autolysis of the bacterial cell-bodies.

Many investigators have long focused their attention on the mode of action of lysozyme on cell-bodies. Electron-microscopic observations of damaged cell-bodies, and also the finding that isolated cell-walls are

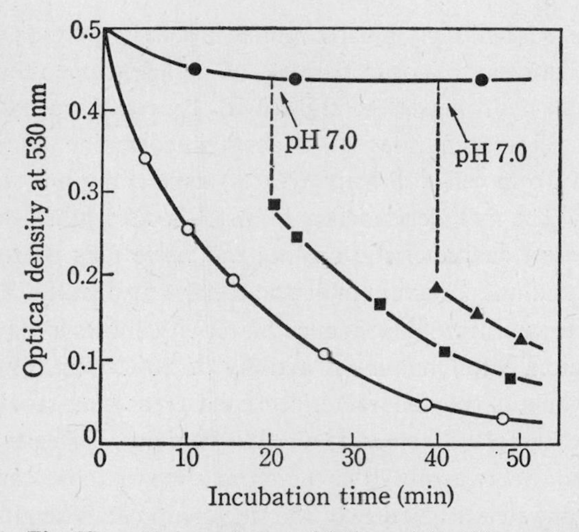

Fig. 37. pH dependence of lytic activity.[158] ○, pH 7.0; ●, pH 4.5; ■, pH changed instaneously to 7.0 after incubation at pH 4.5 for 20 min; ▲, pH changed to 7.0 after incubation at pH 4.5 for 40 min.

capable of being digested by lysozyme, indicate that the action point of lysozyme is the cell-wall components of the bacterial cell.[160]

About 50% of the cell-wall material becomes dialyzable through a semipermeable membrane after complete hydrolysis by lysozyme, but no appreciable amounts of free amino acid and free amino sugar are formed. Berger and Weiser[2] found that lysozyme hydrolyzed the β-1,4-*N*-acetylglucosaminide linkage of the chitin molecule, and concluded that the enzyme lysozyme should be categorized with β-*N*-acetylglucoaminidase. At the same time, they reported that lysozyme showed no ability to hydrolyze phenyl β-*N*-acetylglucosaminide. Salton and Ghuysen[3] reported that the hydrolyzate of bacterial cell-walls contained a disaccharide NAG–NAM with β-1,6-glycoside linkage and a tetrasaccharide, NAG–β-1,6-NAM–β-1,4-NAG–β-1,6-NAM. They asserted, according to their observations, that lysozyme possessed β-1,4-*N*-acetylglucosaminidase activity, but that the disaccharide NAG–β-1,6-NAM was hydrolyzed no further by lysozyme. More recently it was shown that the monosaccharide residues of the disaccharide NAG–NAM isolated from the hydrolyzate are connected by a β-1,4-glucosaminide linkage.[161]

7.4. β-1,4-*N*-Acetylglucosaminidase Activity

In the assay of the β-1,4-*N*-acetylglucosaminidase activity of lysozyme, soluble chitin derivatives, oligosaccharides of *N*-acetylglucosamine (isolated from partial hydrolyzates of chitin with concentrated hydrochloric acid), and oligosaccharides of *N*-acetylglucosamine and *N*-acetylmuramic acid (from cell-wall hydrolyzates) have commonly been used as substrates. The pH dependence of β-1,4-*N*-acetylglucosaminidase activity shows a characteristic profile; the curve falls gently in the alkaline region and has a maximum at about pH 4.5 (Fig. 38).[162,163] The profile for the temperature dependence of activity shows a plateau in the 60–70°C region, with maximum activity at 50°C (Fig. 39).[164] The lysozyme molecule is thus characterized by a great heat-stability, retaining its native structure below 75°C. The activity decrease observable in the region from 50 to 70°C therefore appears to be caused by a change in the fine stereostructure of the side chain catalytic groups rather than by a denaturation of the main structure of the lysozyme molecule.[164] The activity of lysozyme is inhibited by salt solutions at

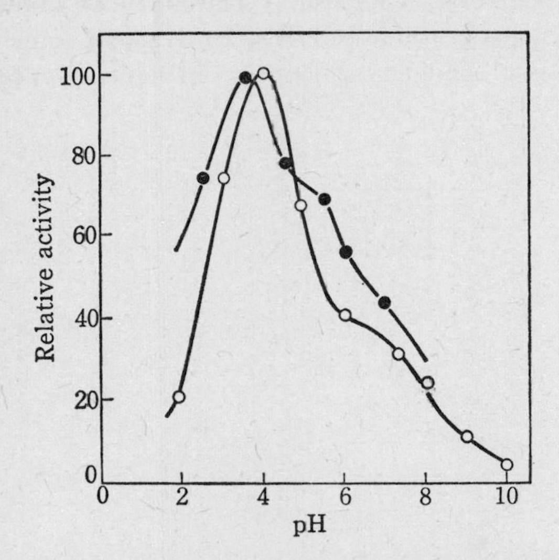

Fig. 38. pH dependence of β-1,4-*N*-acetylglucosaminidase activity.[162,163] ○, Viscometry; ●, reducing power method (incubation time, 120 min).

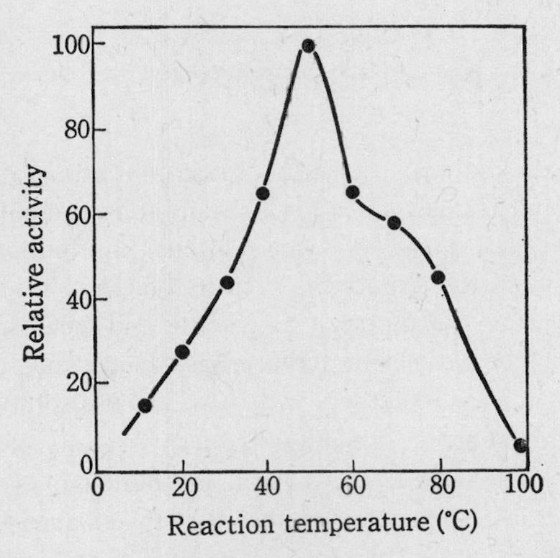

Fig. 39. Temperature dependence of β-1,4-*N*-acetylglucosaminidase activity.[164] Data obtained by the reducing power method (incubation time, 60 min).

high concentrations such as 1.0 M (Fig. 40).[165] The inhibitory action of salts or ions depends upon their ionic radii or their transport numbers. The precise mechanism of inhibitory action however has not yet been fully elucidated.

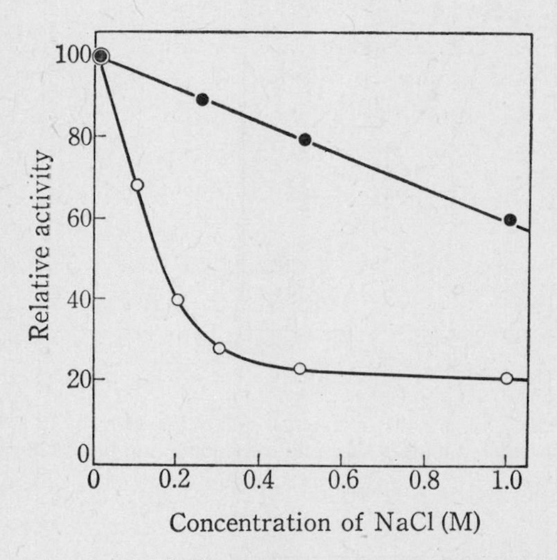

Fig. 40. Inhibition of lysozyme activity by sodium chloride.[165] ○, 0.1 M acetate, pH 4.5; ●, 0.1 M phosphate, pH 8.0. Data obtained by viscometry at 30 °C.

The presence of organic solvents may also cause inhibition of lysozyme activity. Ikeda and Hamaguchi[95] found that aliphatic alcohols bound specifically to the cleft of the lysozyme molecule, inhibiting the activity. On the other hand, the unpublished results of Hayashi *et al.* have also shown that certain sugars, such as glucose and fructose at high concentrations, enhance the enzyme activity. Viscometric measurements indicated a 50% increase in activity with 30% glucose solutions at pH 5.6. Since in such glucose solutions the reduced viscosity of the glycol chitin solution was the same as that in ordinary buffer solutions, it may be that the factor responsible for the activity enhancement resides in the enzyme itself.

Denaturant solutions at moderate concentrations, at which the unfolding of the main chain of the lysozyme molecule does not take place, reversibly inhibited the activity. Inhibition was also observed with anionic polymers and anionic detergents. Such inhibition may be explained merely by an electrostatic interaction between positive charges on the enzyme and negative charges on the additives. In this connection, the inhibitory action of cationic detergents is of greatest interest, and will be discussed in a later section.

7.5. Transglycosylation

Many fragmental findings obtained in experiments on the enzymatic action of lysozyme suggest that the enzyme may possess a transglycosylation activity or synthetic activity. For example, when a large amount of lysozyme is added to complete lysozyme hydrolyzates of glycol chitin, the reducing power in the hydrolyzate decreases gradually over a long period. This has been explained by the new enzymatic synthesis of *N*-acetylglucosaminide linkages, or by transglycosylation by lysozyme. The first clear evidence for the transglycosylation activity of lysozyme was obtained by Kravchenko, Rupley and Sharon using oligosaccharides of *N*-acetylglucosamine as substrate.

Kravchenko[166)] observed that the rate of hydrolysis of *N,N'*-diacetylchitobiose NAG–NAG was accelerated by the presence of *N*-acetylglucosamine, and that for the hydrolysis of *N,N'*-diacetylchitobiose, an induction period of 8 hr was needed to start the hydrolytic reaction. It is easy to assume that *N,N'*-diacetylchitobiose may be subject to some kind of change during the induction period; reaction products may be accumulated during the induction period, in sufficient amounts to start the abrupt reaction. When a trace of tetrasaccharide of *N*-acetylglucosamine is added to the solution of *N,N'*-diacetylchitobiose, the hydrolysis proceeds rapidly without an induction period. This observation strongly indicates that products are accumulated during the induction period; these may be oligosaccharides with a degree of polymerization higher than two, and the hydrolysis of *N,N'*-diacetylchitobiose may therefore proceed mainly through transglycosylation.

8. Structure and Function

8.1. Chemical Modification and Activity

The structure of the active site (binding and catalytic sites) of lysozyme has been clearly demonstrated by various experimental data, of which a great part are derived from X-ray crystallographic studies or chemical studies on the enzyme-substrate (or more precisely, enzyme-competitive inhibitor) complex. Before 1966, when the X-ray crystallographic results were obtained, studies on the active site of lysozyme were focused primarily on the chemical modification of the lysozyme molecule. The present section treats mainly the details of such chemical modification, although much of the material perhaps now carries only an historical interest.

8.1.1. Terminal Amino Acids

Jolles[167] examined the role of the *N*-terminal sequence in the activity. After hydrolysis by bacterial aminopeptidase (1/100, wt/wt) at pH 7.6, 37°C for 6 hr, there was no change in activity. However, hydrolysis for 24 hr under the same conditions caused a considerable decrease in activity. The aminopeptidase digest was first fractionated on a column of Amberlite CG-50 and then purified on a column of Sephadex G-25. The results indicated that the removal of lysine and phenylalanine residues from the *N*-terminal sequence did not affect the activity. The release of the two amino acid residues, leucine and arginine, in the *C*-terminal sequence by carboxypeptidase A and B did not cause a loss in activity.

8.1.2. Disulfide Linkages

The disulfide linkages of protein molecules are generally considered to play an important role in the maintenance of the tertiary structure of the molecule, but not to be included in the active site. This is the case for lysozyme. However, changes in activity accompanying the cleavage of disulfide linkages have been extensively studied in anticipation of gaining useful information about the fine structure at the active locus of the enzyme. Selective cleavage of the disulfide linkage between Cys-6 and Cys-127 did not cause any loss in activity.[167]

Reduction of the lysozyme molecule in concentrated urea solutions, by appropriate reducing agents such as mercaptoethanol, resulted in the complete cleavage of the four disulfide linkages and a complete loss in activity. Removal of urea from the reaction mixture, followed by oxidation with atmospheric oxygen, caused regeneration of the disulfide linkages. Complete regeneration of the same structure as that in the native enzyme, together with complete recovery of activity, was attained by a careful choice of conditions for the oxidation process.[118,168,169] The regeneration of the disulfide linkages and tertiary structure, and the recovery in activity, were followed in detail throughout the oxidation process. After two disulfide linkages had on average been regenerated, measurable activity began to appear, implying that the regeneration of at least two disulfide linkages is necessary to construct the main chain structure indispensable for the exhibition of activity.

8.1.3. Histidine Residue

Many hydrolytic enzymes contain histidine residues as catalytic groups, and the imidazole ring of the histidine residue, as a nucleophile or general acid-base, is well known to be an efficient catalyst for the hydrolysis of esters, amides, and acid anhydrides. In this connection, the single histidine residue of the lysozyme molecule formerly attracted the interest of many investigators. It was reported from several laboratories that photooxidation by visible light, in the presence of a sensitizing dye such as methylene blue, and oxidation by a variety of reagents caused a considerable reduction in enzyme activity. Since the histidine residue is oxidized by such processes, the possible function of the histidine residue as a catalytic group was actively discussed. However, no exact conclusions could be drawn from the results of these experiments, since other amino acid residues can also be oxidized in the oxidation process.

Kravchenko *et al.*[170] and Jolles[167] attempted carboxymethylation of the histidine residue with monoiodoacetic acid. Jolles reported that the histidine residue was not modified by monoiodoacetic acid at pH 5.5, 37°C, while Kravchenko reported that carboxymethylation of the histidine residue did not affect the activity. Recently, the possible role of the histidine residue as a catalytic group has virtually been precluded by the finding that duck egg-white lysozyme, while showing the same order of activity as chicken egg-white lysozyme, lacks the histidine residue.[171]

8.1.4. Arginine Residues

The possibility that arginine residues may act as catalytic groups for hydrolytic reactions, as indicated by experiments with various model systems, seems remote. However, if in fact arginine residues do play some role in the activity of lysozyme, then it can be expected that this function would concern the binding of substrate, since the basicity of the guanidyl group of arginine residues appears too high for the exhibition of catalytic activity as a nucleophile or general acid at neutrality. Lysozyme contains eleven arginine residues: all react easily with glyoxal at alkaline pH,[60,61] and the modification caused does not result in a loss of activity (K.Hayashi *et al.*, unpublished). It is well known that the basicity of the basic amino acid residues of lysozyme is essential to the binding of the enzyme to bacterial cell-bodies at neutral pH in the lytic process. The basicity of the arginine residues is thus vital to the binding of lysozyme and bacterial cell-bodies or cell-walls, and hence for the lytic activity of the enzyme.

8.1.5. Lysine Residues

The ε-amino group of lysine residues may act as a general catalyst for hydrolytic reactions at neutrality, since its pK value lies near pH 10; the relation between the lysine residues and activity of hydrolytic enzymes has thus been investigated by many workers. Fraenkel-Conrat[172] reported that acetylation of the ε-amino group of lysine residues in the lysozyme molecule by acetic anhydride caused a considerable loss in lytic activity. Wauters and Leonis[173] observed that the activity decreased according to the number of introduced dinitrophenyl groups at the ε-amino group of lysine residues. Carboxymethylation of three lysine residues did not affect the activity, but additional carboxymethylation resulted in inactivation, due probably to a conformational change in the modified lysozyme molecule.[174]

Yamasaki *et al.*[159,175] investigated the role of lysine residues by acetylation with acetic anhydride. Lysozyme acetylated to various degrees was fractionated by column chromatography with carboxymethyl cellulose. The lytic activity decreased according to the extent of acetylation. As shown in Fig. 41, the pH-activity profile of the modified enzyme preparations was quite characteristic; the decrease in activity was observed only at the more alkaline region, beyond an optimum pH for the activity of each preparation. The value of this optimum pH decreased linearly as a function of the degree of acetylation. Similar results were obtained in experiments with isolated bacterial cell-walls as substrate. On the other hand, activity determinations by viscometry and the reducing power method, with glycol chitin as the substrate, clearly demonstrated that acetylation of six out of the seven amino groups did not cause any decrease in the activity of lysozyme. Trinitro-

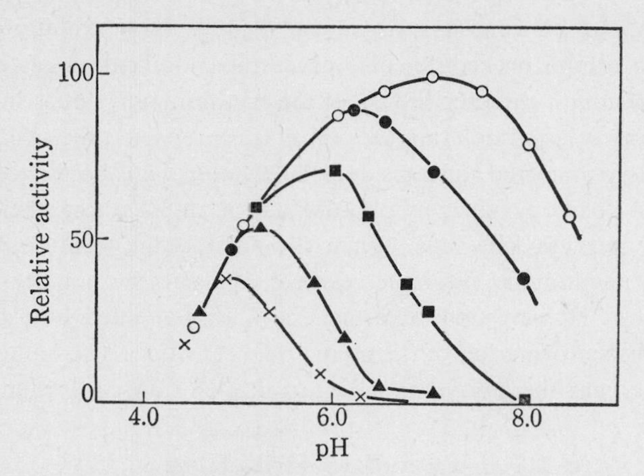

Fig. 41. pH dependence of activity of acetylated lysozyme.[175] ○, Untreated enzyme. Free amino groups per mole of enzyme: 4.0 (●), 2.4 (■), 1.1 (▲), and 0.3 (×).

phenylation of the remaining group brought about a 25% loss in activity. These facts led us to conclude that the amino groups, one α-amino group of *N*-terminal leucine and six ε-amino groups of lysine residues, do not participate in the β-1,4-*N*-acetylglucosaminidase activity of lysozyme. Parsons *et al.*[176] reported that the affinity constant for lysozyme acetylated with trisaccharide of *N*-acetylglucosamine was 4.7×10^{-5} M, slightly larger than that of native lysozyme (6.0×10^{-6} M).

The above conclusion has been corroborated by many other experimental results. Hiremath and Day[63] reported that lysozyme modified by the bifunctional reagent dibromo-*p*-xylenesulfonic acid contained two bridges, one between Lys-96 and Lys-97 and the other between Lys-33 and Lys-116. Their preparation exhibited full activity for the lysis of bacterial cell-bodies. McCoubrey and Smith[177] attempted to convert lysine residues to arginine residues by amidylation of the ε-amino groups with iminoester. Eighty percent amidylation of the lysine residues did not cause any decrease in activity. Geschwind and Lee[178] successfully converted lysine residues to arginine residues with methyl isourea; again, no change in activity was observed. This result was confirmed later by Parsons *et al.*[176]

8.1.6. Methionine Residues

Chemical modification of the methionine residues of lysozyme has been carried out by carboxymethylation[179] and photooxidation.[180] However, since both procedures also cause the modification of other amino acid residues, the precise role of the methionine residues in the enzyme activity was not determined. Jori *et al.*[98] reported the selective photooxidation of methionine residues: the latter were converted to methionine sulfoxide residues by photooxidation in 84% acetic acid in the presence of methylene blue. When two methionine residues were converted to methionine sulfoxide residues, the enzyme activity was completely lost. However, the treatment was accompanied by a concomitant transconformation of the main chain structure. The complete loss of activity may therefore have arisen from such transconformation.

8.1.7. Tyrosine Residues

Lysozyme contains three tyrosine residues, of which Tyr-20 and Tyr-23 locate on the outside of the molecule and are reactive.[45,46] Chemical modification of these two residues does not cause any decrease in activity.[181] The third residue, Tyr-53, is abnormal and unreactive unless the enzyme is denatured. Parsons *et al.*[176] confirmed that *O*-acetylation of the three tyrosine residues with acetylimidazole does not affect the activity.

8.1.8. Tryptophan Residues

In has long been assumed that tryptophan residues play some role in the activity of lysozyme. Fraenkel-Conrat[172] reported that the oxidation of lysozyme by periodate, iodine or peroxide caused a decrease in activity, and that subsequent reduction resulted in a restoration of the activity to some extent. Weil *et al.*[180] observed that in photooxidation the consumption of two moles of molecular oxygen per mole of lysozyme caused a 70% loss in activity. In this reaction, 1.0 mole of histidine residues and 1.2 moles of tryptophan residues were oxidized. With the consumption of six moles of molecular oxygen, one mole of histidine, three moles of tryptophan and half a mole of tyrosine residues were oxidized and the modified lysozyme lost its activity completely.

Bernier and Jolles[182] and Rao and Ramachandran[183] attempted the oxidation of tryptophan residues with *N*-bromosuccinimide (NBS) and confirmed that the oxidation of one or four tryptophan residues caused a 46% or 96% decrease in lytic activity, respectively. Hayashi *et al.*[86,163] found that oxidation of lysozyme with equimolar NBS at pH 4.5 caused a 80% decrease in both the β-1,4-*N*-acetylglucosaminidase activity and the optical density at 293 nm, judged from the difference spectrum arising from the lysozyme-substrate complex. The optical rotatory dispersion data showed that the modified lysozyme retained the same conformation as the native enzyme.[184] Since the difference spectrum of the lysozyme-substrate complex referred to the native lysozyme was considered to originate from a change in the environment around a specific tryptophan residue, it was suggested that this specific tryptophan residue might perhaps be oxidized specifically by equimolar NBS.

The extent of the decrease in activity due to oxidation by equimolar NBS varied according to the conditions adopted. The amount of lysozyme-substrate complex measured by difference spectrophotometry also varied with the conditions or the state of the solution. Oxidation by equimolar NBS under conditions that permitted the ready formation of lysozyme-substrate complex caused a considerable inactivation of the enzyme. Thus, the ability of the specific tryptophan residue to participate in the complex formation was exactly proportional to its selective oxidizability by NBS. This may mean that a specific tryptophan residue can be selectively oxidized when it is in the active state for binding of the substrate. Oxidation in 5 M urea solutions, in which lysozyme hardly binds with the substrate, caused an appreciably lower inactivation of the enzyme. This suggests that a considerable amount of tryptophan residues other than the specific tryptophan residue is oxidized in the urea solution. On the basis of the above facts, attempts were made by chemical techniques to determine the position of the specific tryptophan residue which was selectively oxidizable by NBS at pH 4.5, and which participated in the formation of the lysozyme-substrate complex. NBS-oxidized lysozyme exhibited characteristic absorption bands in the 250–260 nm region, which seemed to be due to oxidation products of the oxindole nucleus. The peptide containing the oxidized tryptophan residue was isolated by following the characteristic absorption bands. Lysozyme oxidized by equimolar NBS at pH 4.5 was reduced with mercaptoethanol and carboxymethylated with monoiodoacetic acid (RCM–lysozyme). The RCM–lysozyme was then subjected to tryptic digestion, followed by titration at pH 8.0. Trace amounts of insoluble material in the tryptic digest were removed by centrifugation and the supernatant was lyophilized. The lyophilized material was dissolved in 0.2 M acetic acid and subjected to gel-filtration on a column of Sephadex

G-25. Three fractions were separated. The fraction eluted last exhibited enhanced absorption close to 250 nm, suggesting that it perhaps included a peptide containing the oxidized tryptophan residue. After further gel-filtration on the column of Sephadex G-25, the amino acid composition of the peptide was analyzed. The results showed that the peptide contained Gly(1), Arg(1), CM-Cys(1), Asp(2), Trp(1) and oxidized-Trp(1). The content of the tryptophan and oxidized tryptophan residues was determined by spectrophotometry, and the peptide identified as a derivative of Canfield's T–9 peptide, Trp_{62}–Trp–Asn–Asp–Gly–Arg_{67}. Next, the position of the oxidized tryptophan residue was determined. After removing the *N*-terminal amino acid by the Edman degradation method, characterization of the remaining peptide was carried out. This peptide exhibited a typical absorption band of the amino acid tryptophan. The split PTH-amino acid was identified as PTH-oxidized tryptophan from the *Rf*-value by paper chromatography and from its absorption bands. This fact clearly indicated that the selectively oxidized tryptophan residue was Trp-62. Contrary to the particular reactivity of Trp-62, the adjacent Trp-63 was not oxidized even by 20 moles of NBS per mole of lysozyme. This may mean that the nucleus of Trp-63 is protected from contact with the reagents in the medium by that of Trp-62.

Hartdegen and Rupley[90,185] oxidized the lysozyme molecule by 0.5 mole equivalent of I_3^- (containing radioactive iodine, ^{131}I) at pH 5.5, and separated with a column of Bio-Rex 70 a modified lysozyme fraction which was inactive and contained only one oxidized tryptophan residue. Oxidation of this tryptophan residue was inhibited by the presence of *N*-acetylglucosamine or its oligosaccharides, indicating that the tryptophan residue sensitive to iodine oxidation is located in the binding site for the substrate. Modified and inactive lysozyme thus separated was digested by trypsin after carboxymethylation. The tryptic digest was fractionated on a column of Dowex-1. Analysis of the peptide containing the oxidized tryptophan residue demonstrated that the selectively oxidizable residue was Trp-108. The same result was obtained by X-ray crystallographic studies.[91] It is an interesting fact that the bulky iodine selectively oxidizes only Trp-108. In this connection, Rupley and Imoto[186] discussed the selective oxidizability of Trp-108 in regard to the effect of the nearby γ-carboxyl group of Glu-35.

Previero *et al.*[92] carried out ozone oxidation of lysozyme. The gas was slowly blown into a lysozyme solution in anhydrous formic acid at low temperatures. Oxidation by ozone resulted in the formation of *N*′-formylkynurenine, and inactivation followed suddenly after the *N*′-formylkynurenine produced had reached a certain level. Two preparations, one retaining complete activity and the other completely without activity, were prepared and the positions of the *N*′-formylkynurenine residues determined by the usual tryptic digestion method. In the active preparation, Trp-108 and Trp-111 were selectively oxidized, while in the inactive preparation residues other than Trp-108 and Trp-111 were also partially oxidized. As reported by Rupley *et al.*, the oxidation of Trp-108 by iodine to oxindole causes inactivation of the enzyme. The inactivation is thus considered to be due to a distortion of the steric position of the side chain of Glu-35 by oxindole derived from Trp-108. In fact *N*′-formylkynurenine may have no interaction with the side chain of Glu-35, and so be unrelated to the distortion of the steric position.

8.1.9. Carboxyl Groups

It has long been known that esterification of the carboxyl groups of lysozyme causes inactivation.[172] However, although a variety of chemical modifications of lysozyme have been carried out in connection with its activity, clear positive evidence concerning the catalytic or binding groups was not accumulated until very recently, except in the case of the tryptophan residues. Therefore, the possible role of carboxyl groups in the action of lysozyme has only gradually come to light. Chemical modification of the carboxyl groups was attempted by several investigators, but the difficulty attached to experiments employing aqueous solutions, and the large number of carboxyl groups in the lysozyme molecule, restricted progress in this field. Water-soluble carbodiimide (condensing reagent) was first synthesized by Sheehan and Hess in the 1950's,[187] and subsequently application of the carbodiimide method for the chemical modification of the carboxyl groups of protein molecules has frequently been attempted. Condensing reagents enable nucleophiles to attack the carbonyl carbon of carboxylic acids, resulting in the formation of substituted carbonyl compounds such as esters and amides.

$$\mathrm{R{-}N{=}C{=}N{-}R'} + \mathrm{R''{-}\overset{\displaystyle O}{\overset{\|}{C}}{-}OH} \longrightarrow \mathrm{R{-}NH{-}\underset{\displaystyle O{-}C(=O){-}R''}{C}{=}N{-}R'} \tag{12}$$

$$\longrightarrow \mathrm{R{-}NH{-}\overset{\displaystyle O}{\overset{\|}{C}}{-}\overset{\displaystyle R''{-}C{=}O}{\overset{|}{N}}{-}R'} \xrightarrow{+B} \mathrm{R''{-}\overset{\displaystyle O}{\overset{\|}{C}}{-}B} + \mathrm{R{-}NH{-}\overset{\displaystyle O}{\overset{\|}{C}}{-}NH{-}R'}$$

Hoare and Koshland[56,188] first adopted this method for the modification of the carboxyl groups of lysozyme. They employed 1-benzyl-3-(3-dimethylpropyl) carbodiimide, 1-methyl-3-(3-dimethylaminopropyl) carbodiimide and 1-cyclohexyl-3-(2-morpholinoethyl) carbodiimide as condensing reagents, and carried out the coupling of the carboxyl groups of lysozyme with nucleophilic amino acid esters. Eight of the eleven carboxyl groups were easily condensed with amino acid esters. Horinishi *et al.*[57] attempted the condensation of the carboxyl groups with glycine methyl ester through the action of the condensing reagent, 1-ethyl-3-(3-morpholinopropyl) carbodiimide. They found that six carboxyl groups were modified in aqueous solutions. The activity was unchanged until two carboxyl groups had on average been modified. Modification of three carboxyl groups caused a 60% reduction in the original activity and modification of six groups resulted in complete inactivation. Hayashi *et al.*[189] carried out condensation of ^{14}C-glycine methyl ester using 1-cyclohexyl-3-(2-morpholinoethyl) carbodiimide by choosing conditions appropriate for the step-wise reaction. Preparations with, on average, one or two modified carboxyl groups were isolated with a column of Bio-Rex 70. The fraction with one modified carboxyl group retained about 70% of the original activity. In order to determine the position(s) of the modified groups, the fraction was subjected to digestion by trypsin. Radioactivity was distributed widely over several peptide fractions, and it was therefore concluded that several different carboxyl groups were in fact modified at almost the same rate, and that selective modification of a specific group was not possible by this method. Recently, Lin and Koshland[190] carried out the condensation of carboxyl groups by the carbodiimide method using aminomethanesulfonic acid, and reported that all carboxyl groups except Glu-35 could be modified under normal conditions. The presence of the substrate prevented the modification of Asp-52. When all of the carboxyl groups except Glu-35 and Asp-52 had been modified, the remaining activity of the modified lysozyme was about 50% of the original value, suggesting that only Glu-35 and Asp-52 are responsible to the enzyme activity, as evidenced by the results of X-ray crystallographic studies.

Ethyl-esterification of lysozyme carboxyl groups was carried out by Parsons and Raftery[191] and Parsons *et al.*,[176] using triethyloxonium fluoroborate. A single ethyl ester derivative of lysozyme, which was inactive but still retained the capacity for binding *N,N'*-diacetylchitobiose, was isolated. The ^{14}C-labeled preparation was digested in chymotrypsin after oxidation with performic acid and a peptide containing ^{14}C was isolated by column chromatography. The sequence of this peptide was determined by digestion with carboxypeptidase, and the results showed that Asp-52 had been selectively esterified in the single ethyl ester derivative. It thus became clear that Asp-52 is essential for normal lysozyme activity.

8.2. Binding of the Substrate

8.2.1. Determination of the Binding Site by X-ray Analysis

Blake *et al.*[7] elucidated the structure of crystals of lysozyme complexes with oligosaccharides of *N*-acetylglucosamine, and described both the substrate binding sites and an outline mechanism for the hydrolysis of the substrate. Substrate bound to the cleft of the lysozyme molecule, and the complete binding site could accommodate six *N*-acetylglucosamine residues. The subsites for the sugar residues were denoted A,B,C,D,E, and F, respectively (counting from the residue at the non-reducing terminal). Attempts to crystallize complexes with oligosaccharides of *N*-acetylglucosamine in which the subsite D was occupied by a sugar residue were unsuccessful. Consequently, it was not possible to determine the substrate binding at sites including subsites D,E and F. Complexes with mono-, di- and trisaccharides of *N*-acetylglucosamine and their derivatives were however subjected to X-ray analysis. Since all of these oligosaccharides occupied the region of the site which contained the subsites A,B and C, the mode of binding on the lysozyme molecule was determined only at the subsites A,B and C. The mode of binding of sugar residues at subsites D,E and F was determined by fitting model structures to these subsites.

If the chair conformiaton of the sugar residue of an oligosaccharide was placed at subsite D, C-6 of the sugar residue approached too closely to >C=O of the peptide bond of Asp-52, and the 2-acetamide group of the sugar residue at subsite C approached too closely to the indole nucleus of Trp-108. In order to avoid such restrictions, it was necessary to change the conformation of the sugar residue at subsite D to the half-chair conformation, in which C-1, C-2, O-5 and C-5 lay in the same plane. With this distorted sugar residue at subsite D a hydrogen bond was readily formed between 6-OH and >C=O of the peptide bond of Asp-52. Furthermore, it was then possible for sugar residues to locate at subsites E and F without difficulty. The modes of binding of the substrate thus determined are illustrated in Fig. 42.

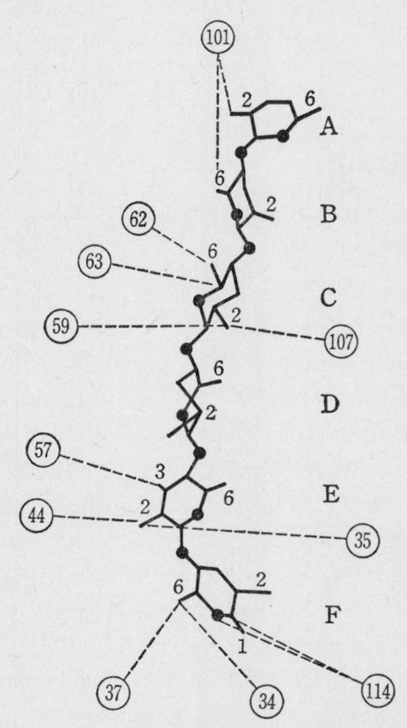

Fig. 42. Binding of substrate to the lysozyme molecule.[7)]

101: Hydrogen bond between >CO of β-carboxyl group of Asp-101 and >NH of 2-acetamide group of sugar residue at subsite A, and hydrogen bond between –OH of β-carboxyl group of Asp-101 and 6-0 of sugar residue at subsite A.

62: Hydrogen bond between >NH of indole ring of Trp-62 and 6-0 of sugar rèsidue at subsite C.

63: Hydrogen bond between >NH of indole ring of Trp-62 and 3-0 of sugar residue at subsite C.

107: Hydrogen bond between >CO of peptide bond of Ala-107 and >NH of 2-acetamide group of sugar residue at subsite C.

59: Hydrogen bond between >NH of peptide bond of Ile-59 and >CO of 2-acetamide group of sugar residue at subsite C.

57: Hydrogen bond between $-NH_2$ of amide group of Gln-57 and 3-0 of sugar residue at subsite E.

35: Hydrogen bond between >CO of peptide bond of Glu-35 and >NH of 2-acetamide group of sugar residue at subsite E.

44: Hydrogen bond between >NH peptide bond of Asp-44 and >CO of 2-acetamide group of sugar residue at subsite E.

34: Hydrogen bond between >CO of peptide bond of Phe-34 and 6-0 of sugar residue at subsite F.

37: Hydrogen bond between $-NH_2$ of amide group of Asn-37 and 6-0 of sugar residue at subsite F.

114: Hydrogen bond between $-NH_2$ of guanidyl group of Arg-114 and 5-0 and 1-0 of sugar residue at subsite F.

Groups located near Glu-35: Ala-110; C_β and C_γ of Glu-57, C_{δ_1} of Trp-108.

Groups located near Asp-52: hydrogen bond network of Asn-46, Ser-50 and Asn-59.

Subsite D: When the sugar residue takes the chair conformation, >CO of the peptide bond of Asp-52 and C-6 of the residue are in too close proximity, and Trp-108 and the 2-acetamide group of the sugar residue at subsite C come into contact. When a sugar residue with half-chair conformation is fitted to subsite D, these steric hindrances are lost, and a hydrogen bond between >CO of the peptide bond of Asp-102 and 6-OH of the residue can be formed.

As shown in the figure, the hydrogen bonds between the lysozyme molecule and the substrate play a fundamental role in the binding behavior. The hydrogen bonds formed between adjacent sugar residues (5-0 and 3-0) are also considered significant. It is probable that a degree of hydrophobic interaction between the lysozyme molecule and the substrate also contributes in the binding process.

As described above, the character of the binding subsites for sugar residues has been clarified and the binding sites for oligosaccharides of *N*-acetylglucosamine determined. Further, it is obviously impossible for an *N*-acetylmuramic acid residue to occupy subsite C, since it has a bulky lactyl group at the 3-0 position. On the other hand, *N*-acetylglucosamine, *N*,*N*′-diacetylchitobiose, glucose, xylose, etc. could all be acceptors in the transglycosylation catalyzed by lysozyme. It is therefore clear that such compounds may bind at subsites E and F. The possible positioning of the substrate binding sites thus determined is given in Table 12.

Table 12

Binding Sites of Substrate

X†[1]	Y†[2]	A	B	C	D	E	F†[2]	
				NAG		NAG		
			NAG—	NAG		NAG—	NAG	
		NAG—	NAG—	NAG		G		
	NAG—	NAG—	NAG—	NAG		G —	G	
NAG—	NAG—	NAG—	NAG—	NAG		X		
			NAM—	NAG				
		NAG—	NAM					
				NAG—	NAM			
		NAG—	NAM—	NAG				
		NAG—	NAM—	NAG—	NAM			
......	NAG—	NAG—	NAG—	NAG—	NAG—	NAG—	NAG	......†[3]
......	NAM—	NAG—	NAM—	NAG—	NAM—	NAG—	NAM	.....†[4]

†[1] Imaginary subsite.

†[2] From transglycosylation.

†[3] Chitin derivative.

†[4] Cell-wall polysaccharide.

Table 13

Binding Sites of Oligosaccharide Substrates[53)]

	A	B	C	D	E
NAG—	NAG—	NAG—	NAG †[1]		
	NAG—	NAG—	NAG †[2]		
		NAG—	NAG †[2]		
			NAG †[2]		
	NAG—	NAG—	NAG-Me †[3]		
		NAG—	NAG-Me †[3]		
			NAG-Me †[3]		
			NAG ——	G-Me †[4]	

†[1] α- and β-Anomers bound in the same way.
†[2] α- and β-Anomers bound differently.
†[3] Methyl β-*N*-acetylglucosaminide.
†[4] Methyl β-D-glucoside.

Lysozyme can slowly hydrolyze the tri- and tetra-saccharides of *N*-acetylglucosamine and the cell-wall tetrasaccharide $(NAG\text{–}NAM)_2$, as will be described later. The bond of the substrate to be hydrolyzed should be situated across the catalytic region between subsites D and E. Consequently, these oligosaccharides should occupy subsites D and E following hydrolysis. The patterns listed in Table 12 thus indicate only the possible and most probable binding sites. The oligosaccharides could also occupy, at low probability, sites other than the listed ones.

Dahlquist and Raftery[53)] and Parsons and Raftery[191)] established the binding sites of oligosaccharides by the proton magnetic resonance method, as listed in Table 13.

8.2.2. Methods for Detection of Lysozyme-Substrate Complex in Solution

It is very difficult to confirm the position of the binding site of the substrate* in solution. The following principles have been employed for the measurement of substrate binding in solution.

(i) In the binding site or the cleft, there exist three tryptophan residues, Trp-62, Trp-63 and Trp-108. Changes in the environment around these tryptophan residues caused by binding of the substrate allow the possibility of spectrophotometric detection of the binding of the substrate to the lysozyme molecule, since changes in the environment of the chromophore cause changes in its optical behavior. From such a viewpoint, the following methods have been employed for detecting changes in the optical behavior of the tryptophan residues resulting from substrate binding: *difference spectrophotometry*, which is a method for measurement of changes in the absorption spectrum; *fluorescence spectrophotometry*; *optical rotatory dispersion*, by which changes in the Cotton effect of the tryptophan residues can be measured; and *the circular dichroism method.*

(ii) Changes in the structure of the substrate due to binding with lysozyme can also be used for determining the amount of substrate bound to the enzyme molecule. The longest wave length of the absorption band of the 2-acetamide group of the substrate lies at about 220 nm. A change in this band cannot be measured by difference spectro-

* Saccharides which can specifically bind to the lysozyme molecule are here termed the "substrate", regardless of their degree of polymerization.

photometry or fluorescence spectrophotometry. The only suitable method at present is *proton magnetic resonance*, by which changes in the chemical shift of the proton resonance of the 2-acetamide group can be measured.

(iii) *The dialysis equilibrium method.* This method requires a comparatively long time period, and is therefore not applicable to easily hydrolyzable substrates. Furthermore, polysaccharide substrates are unsuited to the method, due to the nonpermeability of the substrate through the membrane.

(iv) *Adsorption on insoluble substrates.* Chitin, chitin derivatives, bacterial cell-walls and bacterial cell-bodies have all been used as agents for the adsorption of lysozyme. It should be noted however that there is frequently no parallelism between the adsorption of lysozyme on an insoluble substrate and the formation of a lysozyme complex with a soluble substrate, as detected by spectrophotometry.

(v) *Rate of hydrolysis, inhibition and transglycosylation.* In the case of lysozyme-catayzed reactions, the binding of substrate to the enzyme molecule does not necessarily entail hydrolysis of the substrate. The formation of a "nonproductive" complex, or complex which can be hydrolyzed only with difficulty, is well known. The formation of enzyme-substrate complex can only be detected kinetically by measuring the rate of hydrolysis, inhibition and transglycosylation, that is when one has sufficient knowledge about the formation of nonproductive complex.

A. Difference Spectrophotometry[86,155,163]

In the reference compartment of a spectrophotometer, two cells are placed; one containing 0.1% lysozyme solution ($OD_{280} \doteqdot 2.5$), and the other the substrate solution at a suitable concentration (for instance, 0.1–0.2% for glycol chitin). One of the two cells in the sample compartment contains a mixture of the 0.1% lysozyme and substrate solutions and the other contains buffer solution. The difference spectrum measured with this system gave positive peaks at 293, 286 and 275 nm, and was in general quite similar to that arising from the red-shift of the absorption spectrum of the amino acid tryptophan. The intensity of the difference spectrum was represented by the optical density at 293 nm, ΔOD_{293}, or the molar extinction coefficient at 293 nm in the difference spectrum, $\Delta\varepsilon_{293}$ (calculated on the basis of the molecular weight of lysozyme). When all the lysozyme had formed enzyme-substrate complex in the solution, the value of ΔOD_{293} was 0.11, and that of $\Delta\varepsilon_{293}$, 1650.

Under the normal conditions described above, difference spectra were observed only with *N,N'*-diacetylchitobiose, oligosaccharides of *N*-acetylglucosamine and chitin derivatives.* Glucosamine hydrochloride, hexoses and their oligosaccharides and polysaccharides, chitosan, etc. did not exhibit difference spectra. That is to say, difference spectra were observed only with oligosaccharides and polysaccharides of *N*-acetylglucosamine. A mixture containing glycol chitin and other proteins (or the amino acid tryptophan) did not exhibit a difference spectrum.

Fig. 43 shows the difference spectrum for lysozyme-glycol chitin complex. The results indicate that the difference spectrum was observed only when the substrate bound specifically to the binding site of lysozyme. The difference spectrum observed between the lysozyme-substrate complex and lysozyme alone entirely resembled that arising from the red-shift of the absorption spectrum of the amino acid tryptophan. This suggests that the difference spectrum of the lysozyme-substrate complex may be brought about by a red-shift of the absorption spectrum of tryptophan residues locating in or near the cleft or the binding site of the lysozyme molecule.

The next problem to be solved is the cause of the red-shift in the ab-

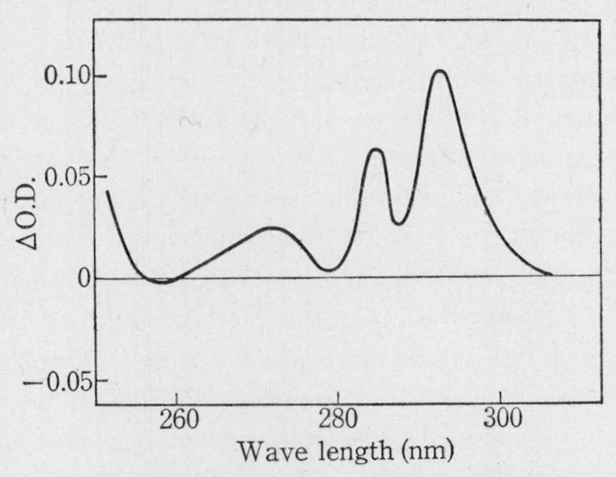

Fig. 43. Difference spectrum for lysozyme-glycol chitin complex.

* *N*-Acetylglucosamine at high concentrations, such as 0.5 M, also exhibited a difference spectrum.

sorption spectrum of the tryptophan residues in the lysozyme-substrate complex. Extensive investigations have been carried out on factors promoting such a shift in the absorption spectrum. Generally, the red-shift of the absorption spectrum of a chromophore is caused by (1) contact of the chromophore with a medium of high density, (2) burial or contact of the chromophore into or with a hydrophobic region, (3) changes in the dissociation state of nearby polar groups and (4) binding of the chromophore to other groups. The red-shift of the absorption spectrum of tryptophan residues observable in the lysozyme-substrate complex formation is therefore considered to arise from one or other of the following factors or an appropriate combination of them.
(i) Contact of a certain tryptophan residue with the substrate.
(ii) Conformational change of the lysozyme molecule accompanied by binding of the substrate, resulting in the burial of the tryptophan residue into the interior of the molecule with higher hydrophobicity.
(iii) Covering of a certain tryptophan residue by the substrate, so keeping the residue out of contact with the medium.

As described above, the lysozyme molecule contains six tryptophan residues, of which two locate in the interior of the molecule and are inaccessible to the medium. In the lysozyme-substrate complex, however, three of the tryptophan residues are inaccessible to the medium. This suggests that one originally accessible tryptophan residue is buried into the interior of the lysozyme molecule in the course of complex formation. Such burial of one of the tryptophan residues is not the result of mere covering by bound substrate. As mentioned previously, the tryptophan residue contributing most to the difference spectrum of lysozyme-substrate complex is Trp-62.

Blake *et al.*[7] confirmed that >NH of the indole nucleus of Trp-62 forms a hydrogen bond with 6-0 of the *N*-acetylglucosamine residue of the substrate at subsite C, and that the indole nucleus of Trp-62 then moves about 0.75 Å towards the binding site or the cleft. Thus, it is obvious that the red-shift of the absorption spectrum of one of the tryptophan residues in the process of complex formation, which is precisely observable by difference spectrophotometry, is caused by this hydrogen bonding between the indole >NH of Trp-62 and 6-0 of the *N*-acetylglucosamine residue.

When the intensity of the difference spectrum was plotted as a function of the concentration of the substrate, an asymptote was obtained.

The equilibrium constant in the formation of the lysozyme-substrate complex can be calculated from this asymptote.

Fig. 44[147b)] illustrates the relation between the molar extinction coefficient at 293 nm in the difference spectrum and the concentration of substrate, with glycol chitin as substrate. Fig. 45[47)] illustrates the relation between the amount of lysozyme-substrate complex measured

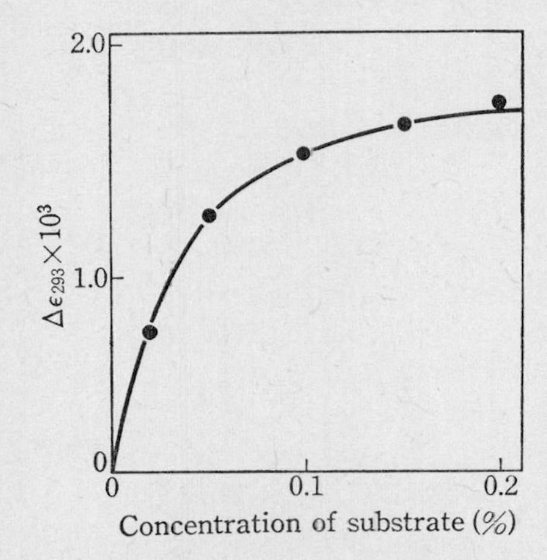

Fig. 44. Relation between the molar extinction coefficient at 293 nm in the difference spectrum and substrate concentration.[147b)] Substrate, glycol chitin; lysozyme, 0.1%; 0.1 M acetate at pH 4.5, 25 °C.

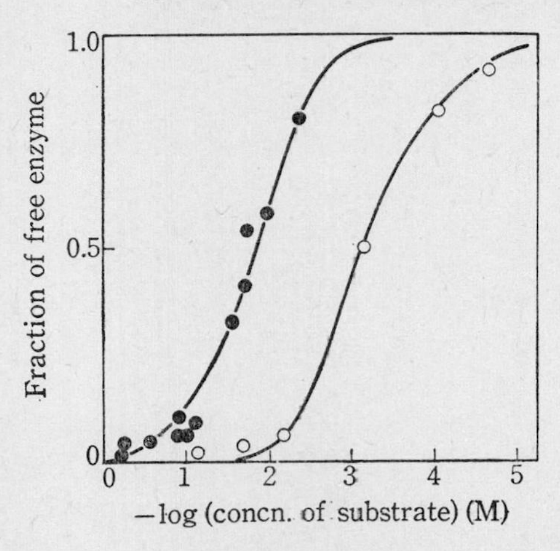

Fig. 45. Relation between amount of lysozyme-substrate complex measured from the difference spectrum and substrate concentration.[47)] ●, NAG; ○, $(NAG)_2$.

from the difference spectrum and the concentration of substrate, with *N*-acetylglucosamine (NAG) or $(NAG)_2$ as substrate.

B. Fluorescence Spectrophotometry

The fluorescence spectrophotometry technique for measurement of the amount of lysozyme-substrate complex in solution was established by Sharon and his co-workers,[192,193] and by Lehrer and Fasman.[48]

Although the measurement of the absolute intensity of the fluorescence spectrum is a difficult task and the relative intensity of the spectrum is very sensitive to the conditions of measurement, a fine conformational change, such as one not causing a change in the absorption spectrum, can be measured precisely. This is due to the fact that the fluorescence spectrum can reflect fine environmental changes around the fluorescent residues (Trp and Tyr) in the protein molecule.

Shinitzky *et al.*[192] measured the fluorescence spectra of lysozyme complexes with various oligosaccharides. The results are shown in Table 14. As can be seen from the table, the shift in wave length at the peak of the fluorescence spectrum, $\Delta\lambda_{max}$, showed a certain regularity, whereas the change in quantum yield of the peak, ΔQ, changed at random from substrate to substrate. These results appear to indicate that each series of substrate provides a different mode of binding. However, it was found that for each substrate there is a certain relationship between the fluorescence intensity and concentration. In the case of

TABLE 14
Fluorescence Spectrum of Lysozyme-Substrate Complex[192]

Substrate	$\Delta\lambda_{max}$ (nm)†1	ΔQ (%)†2
$(NAG)_4$	−10	40
$(NAG)_2$	−10	25
$(NAG–NAM)_2$	−10	−12
NAG–NAM–peptide	−10	7
Gal–NAG	0	18
NAG	0	−7
NAG–NAM	0	−12
NAG–NAM–methyl ester	0	−12
Cellobiose	0	0

†1 Shift of wave length at peak in fluorescence spectrum.
†2 Change in maximum intensity in the spectrum.

oligosaccharide containing *N*-acetylglucosamine (NAG) and *N*-acetylmuramic acid (NAM), the measurements were performed at 370 nm, where the largest change in intensity was found. With oligosaccharide of *N*-acetylglucosamine, the intensity at 325 nm was measured. Fig. 46[193] gives an example of the fluorescence spectrum of lysozyme with and without NAG–NAM substrates.

When log $\{(F_0 - F)/(F - F_\infty)\}$ was plotted against log $(S)_0$, the gradient of the resultant straight was found to be unity, indicating the formation of a 1:1 type complex. Here, F_0, F, and F_∞ represent the fluorescence intensity of lysozyme alone, of lysozyme plus substrate, and of lysozyme saturated by substrate, respectively. $(S)_0$ represents the total concentration of substrate.

It is obvious that a substrate concentration which satisfies the equation, log $\{(F_0 - F)/(F - F_\infty)\} = 0$, is equal to the equilibrium constant K, which is defined by Eq. (13).

$$K = (ES)/(E)\,(S), \tag{13}$$

where (ES), (E), and (S) represent the molar concentrations of the lyso-

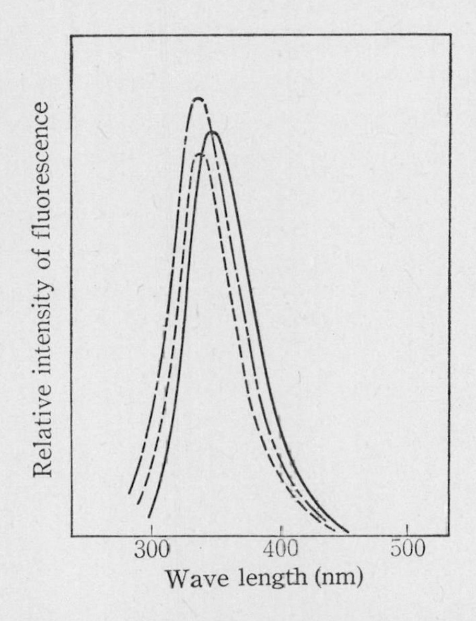

Fig. 46. Fluorescence spectrum of lysozyme with and without NAG-NAM substrates.[193] —— Lysozyme solution at 0.081 mg/ml; ---- lysozyme solution plus 7.4×10^{-3} M $(NAG\text{-}NAM)_2$; —·— lysozyme solution plus 2×10^{-3} M $(NAG)_2$.

zyme-substrate complex, the free enzyme, and the free substrate, respectively.

Lehrer and Fasman measured the intensity of the fluorescence spectrum of a mixture consisting of 0.005–0.01% lysozyme solution and 0.06–3.0% substrate solution, and calculated the equilibrium constant. When variable β is represented by the equation

$$\beta=(F-F_0)\,/\,(F_\infty-F_0)\,, \tag{14}$$

the equilibrium constant is given by

$$K=\beta/(1-\beta)\cdot 1/(\mathrm{S})\,, \tag{15}$$

where

$$(\mathrm{S})=(\mathrm{S})_0-\beta(\mathrm{E}) \tag{16}$$

The intercept of the straight line from plotting $1/(F-F_0)$ against $1/(\mathrm{S})_0$ gives the value of $(F_\infty-F_0)$. Under conditions satisfying $\beta=1/2$, the value of K becomes $1/\{(\mathrm{S})_0-\frac{1}{2}(\mathrm{E})_0\}$, where $(\mathrm{E})_0$ is the total concentration of the enzyme.

Thus, when a certain substrate is chosen and the relation between the concentration of the substrate and the fluorescence intensity at a suitable wave length is obtained, the equilibrium constant or the amount of bound substrate can be calculated quantitatively from the above relationships.

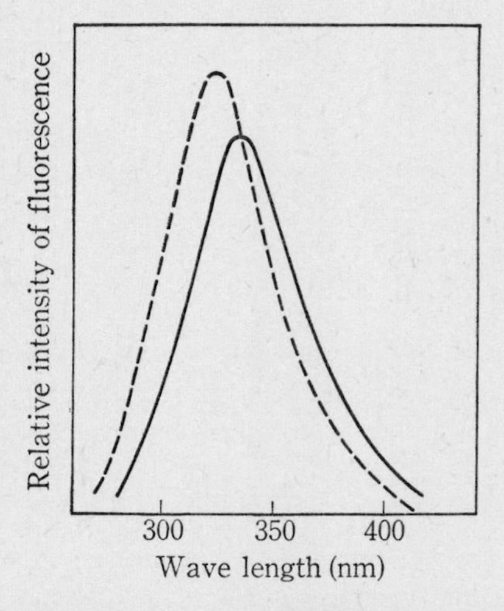

Fig. 47. Fluorescence spectrum of lysozyme with and without glycol chitin substrate.[194] —— Lysozyme solution at 0.33 mg/ml; ---- lysozyme solution plus glycol chitin at 0.25 mg/ml; pH 5.6

Hayashi *et al.*[194] observed the fluorescence spectrum of lysozyme in the presence of glycol chitin under various conditions. An example of the resultant spectra is shown in Fig. 47.

C. Optical Rotatory Dispersion

Adkins and Yang[195] measured the difference in optical rotatory dispersion between lysozyme and lysozyme-substrate complex. They found that the magnitude of this difference could be represented by a function of the concentration of the substrate.

X-ray crystallographic studies of lysozyme-substrate complex have demonstrated that the main chain structure of the lysozyme molecule remains unchanged during complex formation. Also, in solution, changes in the main chain structure owing to complex formation have not been observed. Therefore, the difference in optical rotatory dispersion arising from complex formation appears to stem from some contribution due to a change in the side chain structure.

D. Circular Dichroism[72]

The circular dichroism of lysozyme solution exhibits four positive bands in the range from 270 to 300 nm. These are changeable on the formation of lysozyme-substrate complex. The extent of the change in intensity of the positive peak at 294 nm is the most precise index of the amount of complex formed in solution.

When the difference in peak intensities between the lysozyme solution and the mixture containing lysozyme and substrate is represented by the difference in ellipticity, $\Delta[\theta]$, the reciprocal of the difference can be represented by Eq. (17).

$$\frac{1}{\Delta[\theta]} = \frac{1}{\Delta[\theta]_\infty}\left\{\frac{1}{K(\mathrm{S})_0}+1\right\}, \tag{17}$$

where $\Delta[\theta]_\infty$ represents the difference in ellipticity between the lysozyme solution and a mixture containing lysozyme plus a large excess of substrate (in which all the lysozyme is in the form of lysozyme-substrate complex). Plotting $\frac{1}{\Delta[\theta]}$ *vs.* $\frac{1}{(\mathrm{S})_0}$ may give a straight line, and the intercept and gradient of the line then give the reciprocal values of $\Delta[\theta]_\infty$ and K, respectively.

When lysozyme forms the 1:1 type complex with the substrate, Eq. (18) may be obtained.

$$\log\left\{\frac{[\theta]-[\theta]_0}{[\theta]_\infty-[\theta]}\right\}=\log K+\log(\mathrm{S})_0 \tag{18}$$

The plotting of log $([\theta]-[\theta]_0/[\theta]_\infty-[\theta])$ *vs.* log $(\mathrm{S})_0$ gives a straight line with a slope of unity. $[\theta]_0$ and $[\theta]_\infty$ represent the ellipticities of lysozyme and lysozyme saturated by substrate, respectively, and $[\theta]$ is the measured ellipticity.

E. Proton Magnetic Resonance

The detection of lysozyme-substrate complex in solution by proton magnetic resonance was first reported by Thomas.[196] Application of the method was further developed by Raftery *et al.*[49,53] They measured the chemical shift in proton resonance of the 2-acetamide group of the substrate at 100 MHz, 30°C and pH 5.5 in a mixture consisting of 3×10^{-3} M acetate-free lysozyme solution and 10^{-1}–10^{-2} M substrate solution.

The measured chemical shift, δ, of the 2-acetamide group of the substrate may be written

$$\delta=P\Delta\,, \tag{19}$$

where Δ represents the chemical shift of the 2-acetamide group of the substrate bound completely to the lysozyme molecule, and P is the fraction of bound substrate to total substrate. Since δ is also written as

$$\delta=(\mathrm{ES})/(\mathrm{S})_0\cdot\Delta\,,$$

or

$$(\mathrm{ES})=\delta(\mathrm{S})_0/\Delta \tag{20}$$

the equilibrium constant K may be calculated from Eq. (21).

$$K=(\mathrm{E})_0\left(\frac{\Delta-\delta}{\delta}\right)-\frac{\delta}{\Delta}(\mathrm{S})_0\left(\frac{\Delta-\delta}{\delta}\right) \tag{21}$$

When the conditions satisfying $\delta<\Delta$ and $K\sim(\mathrm{S})_0$ are adopted for the measurement, Eq. (21) simplifies to Eq. (22).

$$(\mathrm{S})_0=\frac{(\mathrm{E})_0\Delta}{\delta}-K-(\mathrm{E})_0 \tag{22}$$

The intercept and gradient of the straight line which results from plotting $(\mathrm{S})_0$ *vs.* $1/\delta$ give the values of $-\{K+(\mathrm{E})_0\}$ and $\Delta(\mathrm{E})_0$, respectively.

F. Equilibrium Dialysis[193]

A mixture consisting of *ca.* 3% lysozyme solution and 10^{-2}–10^{-3} M substrate solution was dialyzed against substrate solution of the same concentration as in the mixture. When the semipermeable tube was rotated mechanically, equilibrium was attained in a few days.

The equilibrium constant K for the formation of lysozyme-substrate complex was then calculated from Eq. (23).

$$K = b/f \cdot (E) \tag{23}$$

where b is the difference in substrate concentration between the inner and outer solutions of the dialyzing tube, and f is the substrate concentration in the outer tube.

G. Adsorption on Insoluble Substrates[165,197]

When lysozyme dissolved in buffer solutions with an ionic strength of about 0.1 was applied to a column of chitin or insoluble chitin derivatives, a strong adsorption was observed in the range from pH 6 to 10 (as illustrated in Fig. 48).[197] It is interesting that adsorption did not take place to any appreciable extent in the range from pH 2 to 6, i.e. where lysozyme shows β-1,4-*N*-acetylglucosaminidase activity. However, the increase in ionic strength of the lysozyme solution resulting from addition of neutral salts caused a profound increase in the adsorption of lysozyme in the range from pH 2 to 6. The adsorption of lysozyme onto insoluble substrate was specific, and was not caused by simple electrostatic interaction or ion exchange. Insoluble substrate adsorbed about 25% its weight of lysozyme at neutrality. Generally, inactivated lysozyme was not adsorbed onto the insoluble substrate column. For example, NBS-oxidized lysozyme was not adsorbed at all in the whole pH region, even though buffer solutions with high ionic strength were used. The only exception was observed with iodine-oxidized lysozyme (where Trp-108 is selectively oxidized), the adsorption behavior of the iodine-oxidized lysozyme being the same as that of the native lysozyme (see Fig. 48).

After short intervals, the adsorbed lysozyme was easily eluted by 0.2 N acetic acid. However, when lysozyme had been adsorbed and kept for a longer time in the column bed (for instance, overnight), the adsorption became partly irreversible.

Lysozyme forms productive complex with polysaccharide substrates in solution, at pH 3–7; this can be detected by difference spectrophotometry. On the other hand, the formation of nonproductive complex in alkaline media is evidenced by a complete restraint in the isoelectric crystallization of lysozyme in the presence of one tenth its weight of

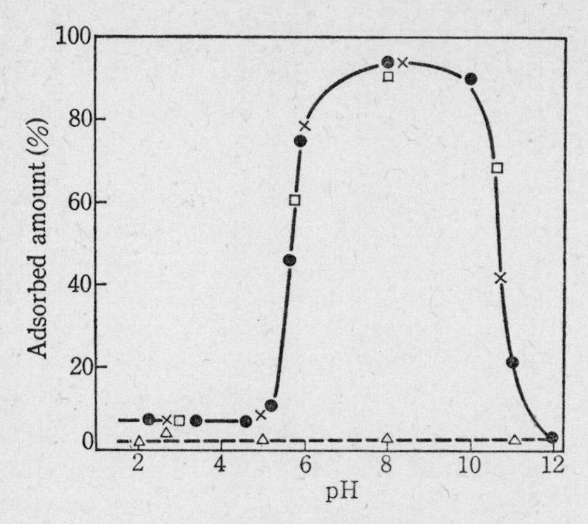

Fig. 48. pH dependence of adsorbed amount of lysozyme on a column of insoluble carboxymethyl chitin.[197] ●, untreated lysozyme, 2 mg; ×, acetylated lysozyme, 2 mg; △, NBS-oxidized lysozyme, 3 mg; □, iodine-oxidized lysozyme, 2.5 mg. Column size, 1.5 × 4.0 cm.

substrate. It is interesting that the adsorption of lysozyme on an insoluble substrate column (for example, 6-0 derivatives) can be used as an index of the amount of nonproductive complex formed in neutral solutions of low ionic strength. Furthermore, it should be emphasized that the adsorption of lysozyme on insoluble substrate columns depends strongly upon the ionic strength of the solution.

Loquet *et al.*[198] have measured the adsorption of lysozyme of various origins onto the cell-bodies of *M. lysodeikticus*.

H. Competitive Inhibition[199,200]

The cationic detergent dimethylbenzylmyristylammonium chloride (DBMA) binds to the cleft of the lysozyme molecule, inhibiting its activity competitively. The binding of DBMA causes a remarkable spectral change in the tryptophan residues locating in the cleft, as detected by difference spectrophotometry. Hayashi *et al.*[199,200] found that the molar extinction coefficient at 297 nm in the red-shift difference spectrum, $\Delta\varepsilon_{297}$, arising from the binding of the detergent, increased exponentially to a certain limiting value when plotted as a function of the concentration of DBMA (Fig. 49). When the lysozyme concentration was fixed at 0.1%, a DBMA concentration above 0.2% gave a

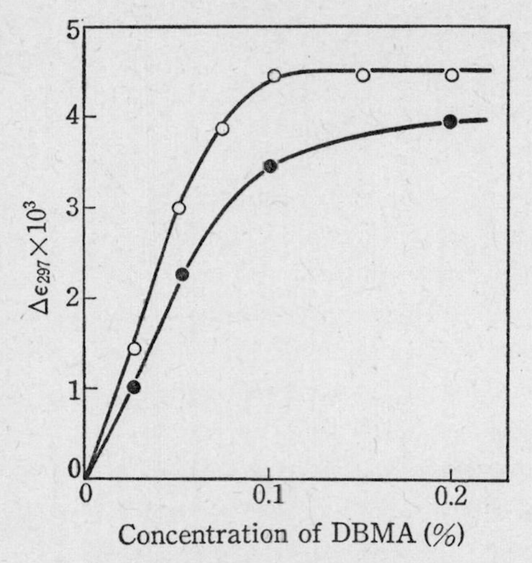

Fig. 49. Interaction of lysozyme with DBMA.[199, 200] ●, 0.1% lysozyme; ○, 0.1% acetylated lysozyme; pH 6.5.

constant value for $\Delta\varepsilon_{297}$ (*ca.* 4000). When substrate solution was added to a mixture consisting of lysozyme and DBMA solutions, the value of $\Delta\varepsilon_{297}$ decreased gradually with time, and after 100 min at 30°C reached a constant value. Conversely, when DBMA solution was added to a mixture consisting of lysozyme and substrate solutions, the value of $\Delta\varepsilon_{297}$ increased gradually, attaining the same value after the same time as above. These results indicate that DBMA or substrate previously bound to the cleft of the lysozyme molecule is repelled through the addition of substrate or DBMA, respectively, until a state of equilibrium is reached. In a mixture consisting of equal concentrations (0.1%) of lysozyme, substrate and DBMA, one third of the lysozyme molecules on average were found to have bound DBMA and the remainder to have bound substrate, in the equilibrium state. One characteristic feature was that the exchange reaction at the binding site between the substrate and the competitive inhibitor DBMA extended over a comparatively long period, whereas the binding of the substrate to the enzyme took place almost instantaneously, and the binding of the inhibitor needed only about 10 min for completion. The rate of decrease in the value of $\Delta\varepsilon_{297}$, or the time consumed before attainment of equilibrium in a mixture of lysozyme and DBMA solutions to which substrate had been added, may thus give an indication of the relative magnitude of the binding force of the chosen substrate. If the substrate has a greater capacity for binding to the enzyme, then the time taken to establish an equilibrium state should become much shorter.

I. Transglycosylation[201]

Information on the binding of substrate at subsites A, B and C of the lysozyme molecule has been obtained from complexes with oligosaccharide substrates. However, as mentioned above, it is almost impossible to obtain comparable information on subsites E and F by mere binding experiments with oligosaccharides.

Transglycosylation occurs between subsites D and E, as well as hydrolysis. An acceptor molecule must therefore occupy each of these subsites. Pollock *et al.* measured the transglycosylation reaction by using the cell-wall tetrasaccharide $(NAG–NAM)_2$ as substrate and *N*-acetylglucosamine, *N,N'*-diacetylchitobiose, glucose, maltose or xylose as acceptor. Since the capability of each compound to function as an acceptor can be assumed to be proportional to its capacity for binding at subsites E and F, characterization of these subsites is possible on the basis of transglycosylation experiments.

8.2.3. Binding Forces of Lysozyme-Substrate Complex in Solution

Before discussing the binding forces between the lysozyme and substrate molecules, it is worthwhile to consider the rate at which lysozyme-complex formation occurs. Generally, the rate of formation of enzyme-substrate complexes is very high. Sykes[202] measured the rate of formation of lysozyme-substrate complex by the proton magnetic resonance method, and obtained the data listed in Table 15. Although the methyl *N*-acetylglucosaminides are not the true substrate, the order of the rate of complex formation with true substrate can be judged roughly from the table. When the equilibrium constant K in the complex formation, $E + S \rightleftharpoons ES$, is measured by one of the various methods described above, the standard free energy change, ΔF^0, for substrate binding may be calculated from Eq. (24).

$$\Delta F^0 = -RT \ln K, \tag{24}$$

where R and T represent the gas constant and absolute temperature. In the case of polysaccharide substrates, the precise equilibrium constant

TABLE 15

Rates of substrate binding[202]

$$E + S \underset{k_{-1}}{\overset{k_{+1}}{\rightleftharpoons}} ES$$

Substrate	k_{+1} (M^{-1}. sec^{-1})	k_{-1}(sec^{-1})
Methyl *N*-acetyl-α-D-glucosaminide	1.4×10^5	2.1×10^3
Methyl *N*-acetyl-β-D-glucosaminide	1.6×10^5	1.7×10^3

was not obtained, due to the difficulty in measuring the exact molecular weight of the substrate, and the number of lysozyme molecules bound per molecule of substrate was not calculated.

In Tables 16–21, values of K and ΔF^0 independently obtained by various methods are listed. Adkins and Yang[195] reported a K value of 50 for *N*-acetylglucosamine, obtained by the difference optical rotatory dispersion method. The data obtained by the different methods are in good agreement with one another. The standard free energy change for the binding of sugar residues at each subsite is summarized in Table 22. In the calculations it was assumed that the substrates bind only at specific binding sites (as listed in Table 12 and 13). That is to say, tetrasaccharide of *N*-acetylglucosamine binds at subsites Y,A,B, and C, while *N,N'*-diacetylchitobiose binds at subsites B and C. As discussed already, the tetrasaccharide is hydrolyzed by lysozyme, even though the rate of hydrolysis is very low, and it also exhibits a capacity

TABLE 16

Free energy change in substrate binding[51]

Substrate	$-\Delta F^0$ (kcal/mole)
NAG	4.5
NAM	4.2
NAG–NAM	3.5
$(NAG)_2$	7.2
$(NAG)_3$	9.5
Cellobiose	2.2
Acetamide derivatives	2–3

to act as a donor in transglycosylation. Accordingly, some part of the tetrasaccharide must bind at certain other binding sites, including subsites D and E. Furthermore, the fact that N,N'-diacetylchitobiose is able to bind at subsites E and F has been shown by transglycosylation experiments.

It seems unlikely that the values of K and ΔF^0 calculated from the above assumptions deviate considerably from the real values. However,

TABLE 17

Binding of Substrate (Difference Spectrophotometry)[52]

Substrate	K
NAG	20
$(NAG)_2$	5.7×10^3
$(NAG)_3$	
$\mu=0.1$	1.5×10^5
$\mu=0.2$	9.3×10^4
$\mu=0.3$	1.3×10^5
$(NAG)_4$	1.1×10^5
$(NAG)_5$	1.1×10^5
$(NAG)_6$	1.6×10^5

TABLE 18

Binding of Substrate[193]

Substrate	K	$-\Delta F^0$ (kcal/mole)
NAG	15–20	1.6
$(NAG)_2$	5.5×10^3	5.1
$(NAG)_3$	1.1×10^5, 2×10^5†1	6.9
$(NAG)_4$	—	6.8
$(NAG)_5$	—	6.9
NAM–NAG	1.1×10^4	5.5
NAG–NAM	20	1.7
NAG–NAM–NAG	2.8×10^5	7.4
NAG–NAM–NAG–NAM	2.1×10^3	4.5
NAG–NAM–NAG–NAM dimethyl ester	1.9×10^3	—
NAG–NAM–NAG–NAM–pentapeptide	7×10^3	—
NAG–NAM	—	1.7†

†1 Ref. 48. †2 Value of subsites C and D.

Table 19

Difference in Free Energy Changes of Two Substrates[193]

Substrate 1	Substrate 2	$\Delta(-\Delta F^0)$ (kcal/mole)†[1]	Site†[2]
NAG–NAG–NAG–NAG	NAG–NAG–NAG	0	Y
NAG–NAG–NAG	NAG–NAG	1.8	A
NAG–NAM–NAG	NAM–NAG	1.9	A
NAG–NAG	NAG	3.5	B
NAM–NAG	NAG	3.9	B
NAG–NAM–**NAG**	NAG–NAM	5.7	C
NAG–NAM–NAG–**NAM**	NAG–NAM–NAG	−2.9	D

†[1] Contribution of residue indicated in bold type.
†[2] Subsite occupied by residue indicated in bold type, which is decided by referring to the results in Table 12 and 13.

Table 20

Binding of Substrate (NMR Method)[49]/[53]

Substrate	K
α-Methyl NAG	19
β-Methyl NAG	30
α-NAG	67
β-NAG	30

Table 21

Binding of Lysozyme to Cell-bodies of *M. lysodeikticus* [198]

Lysozyme	Relative equilibrium constant
Human leucocyte	90
Human serum	100
Human milk	110
Hen egg-white	115
Duck egg-white I	150
Duck egg-white II	200
Goose egg-white	400

particular caution must be exercised in assigning values to subsite D, since precise experimental measurement of the values at this subsite is difficult.

8.2.4. Splitting Mechanism of Hydrolyzed Products

The β-1,4-*N*-acetylglucosaminide linkage of the substrate to be hydrolyzed is that lying between subsites D and E. Considering the case where hexasaccharide of *N*-acetylglucosamine (which most probably occupies all the subsites, A to F) is the substrate, the standard free energy change of the binding of the hexasaccharide is calcluated to be −10.2 kcal/mole from Table 22. This substrate may be hydrolyzed to tetrasaccharide (occupying subsites A to D) and *N,N'*-diacetylchitobiose (occupying subsites E and F). The standard free energy changes of the products, that is tetrasaccharide and disaccharide, are calculated to be −7.5 and −3.8 kcal/mole, respectively. On the other hand, the tetrasaccharide and disaccharide are known to bind most stably at subsites Y,A,B and C, and B and C, respectively. Such binding gives standard free energy changes of −10.2 kcal/mole for the tetrasaccharide and −8.6 kcal/mole for the disaccharide. Consequently, it is obvious that the split products tend to leave the original sites and to move to the most stable binding sites. In other words, the hydrolyzed products split away from the originally occupied binding sites.

It should be noted that the substrate, the hexasaccharide and the

TABLE 22

Free Energy Change at Subsites for the Binding of Sugar Residues

Subsite	$-\Delta F^0$ (kcal/mole)
X	0
Y	0
A	1.8–2.3
B	2.7–4.2
C	4.5–5.7
D	−2.9 – −6.0
E	1.2–4.0
F	0.5–2.0
Sum	7.8–12.6
Average	10.2

hydrolyzed products compete with one another for the binding site in the cleft, according to their value of ΔF^0. This may result in product inhibition.

In reality, the bond cleavage taking place between subsites D and E accompanies both transglycosylation and hydrolysis. As described above, the hexasaccharide, tetrasaccharide and disaccharide seem to occupy not only the specific sites listed in Table 12 and 13, but also various other sites, at a state of equilibrium. Therefore, the real splitting mechanism must follow a comparatively complicated pattern.

The standard free energy change for the binding of *N*-acetylglucosamine residue at subsite D has a positive sign. This is consistent with the X-ray crystallographic finding that the *N*-acetylglucosamine residue at subsite D is distorted, exhibiting the half-chair conformation. The binding of sugar residue at subsite D cannot take place unless the negative free energy change gained at other subsites just cancels or exceeds the positive value at subsite D.

Generally, even though the splitting mechanism of products for various enzymes has been widely discussed, clear evidence in support of the proposed theories has not yet been obtained. In the case of lysozyme, several nonspecific forces such as electrostatic repulsion appear to contribute in the splitting of products to a variable extent. Nevertheless, it must be emphasized that the splitting of products from lysozyme-product complex is probably promoted mainly by the distortion of the sugar residue at subsite D and by the tendency towards attaining thermodynamic equilibrium.

8.2.5. Productive Complex, Nonproductive Complex, and Inhibition by Substrate Analogs

If substrate binds only at the sites indicated in Table 12 and 13, oligosaccharide substrates which do not occupy subsites D and E could not be hydrolyzed. However, oligosaccharides of polymerization degree greater than two can be hydrolyzed and undergo transglycosylation, clearly indicating that oligosaccharides are also able to bind at several sites other than those listed in the tables.

The complex in which the substrate occupies sites other than subsites D and E is called "nonproductive" complex. The hydrolyzable complex, in which the glycosidic bond susceptible to cleavage lies across the catalytic region between subsites D and E, is called "productive" complex. The extent of formation of productive complex increases with increase in the degree of polymerization of the substrate.

The pH dependence of substrate binding exhibits a broader profile than that related to either the hydrolysis or transglycosylation of the substrate. Under certain conditions, therefore, complex in which the substrate occupies subsites D and E may itself not be hydrolyzed. It should therefore be designated as nonproductive complex. Such nonproductive complex has frequently been observed in the case of polysaccharide substrates. For instance, complexes with glycol chitin, formed at high temperatures, a pH of 8 to 10 or in concentrated salt solutions, probably belong to this type of complex. The catalytic groups located near subsites D and E appear to have a stereostructure which is unfavorable for promotion of the cleavage of glycoside linkages.

8.2.6. Stability of Lysozyme-Substrate Complex in Solution

A. pH Dependence of the Binding of Substrate

The pH dependence of the standard free energy change due to the binding of trisaccharide of *N*-acetylglucosamine to the lysozyme molecule is illustrated in Fig. 50.[47] As can be seen, the curve shows a gently

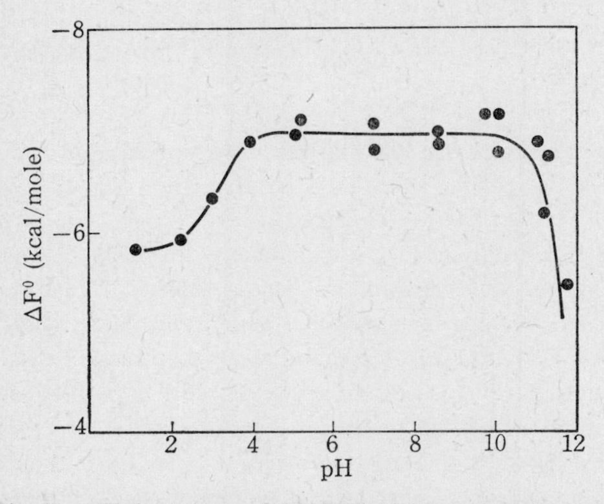

Fig. 50. pH dependence of the standard free energy change of binding of $(NAG)_3$ measured by difference spectrophotometry.[47]

decreasing slope or plateau in the range from pH 4 to pH 10. The maximum value lies close to pH 5, indicating that two groups in the lysozyme molecule with p*K*'s of 6.5 and 4.2 are primarily responsible for the binding of the substrate. The p*K* value of the former suggests that this group may be the β-carboxyl of Glu-35, while the latter may be the γ-carboxylate of either Asp-101 or Asp-103. A similar result has also been reported by Dahlquist *et al.*[52]

Chipman *et al.*[193] measured the pH dependence of the binding of cell-wall tetrasaccharide by fluorescence spectrophotometry and obtained the results given in Fig. 51. As can be seen, the curve exhibited an even

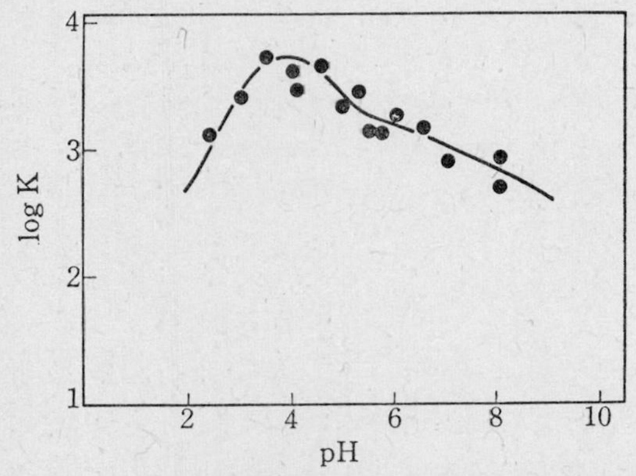

Fig. 51. pH dependence of the equilibrium constant of lysozyme–(NAG–NAM)$_2$ complex measured by fluorescence spectrophotometry.[193]

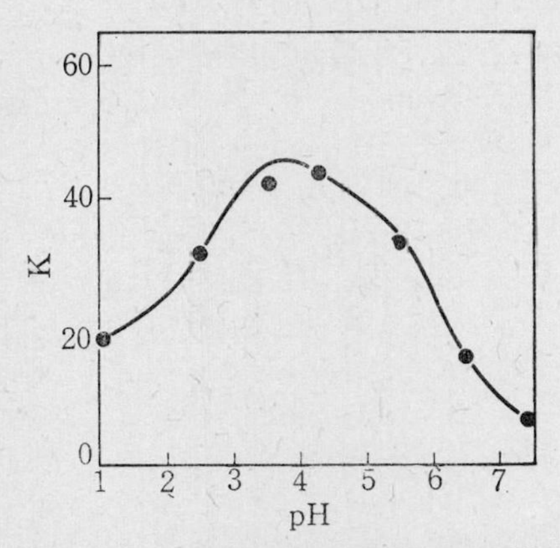

Fig. 52. pH dependence of the equilibrium constant of lysozyme–NAG complex measured by CD.[72]

downward slope on the alkaline side of pH 4. The pH dependence of the binding of *N*-acetylglucosamine, as measured by circular dichroism, is illustrated in Fig. 52.[72] The strong pH dependence of binding with polysaccharide substrate (glycol chitin) was investigated by Imoto *et al.*,[165,197] yielding the results in Fig. 53. The data for partially deacetylated chitins (chitosan) indicate that this type of substrate undergoes complex formation much more readily than glycol chitin at alkaline pH's.

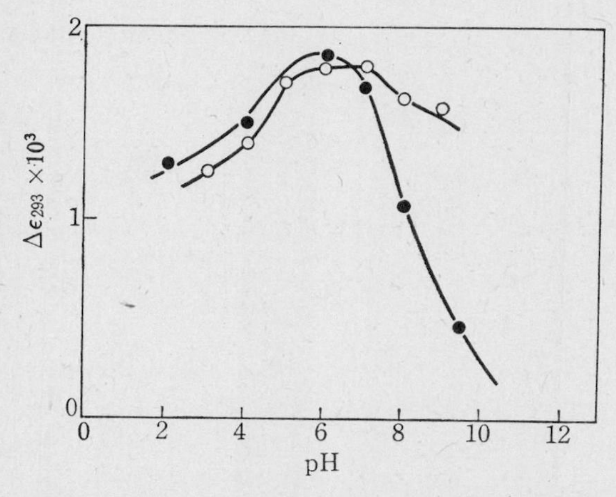

Fig. 53. pH dependence of complex formation between lysozyme and chitin derivatives measured by difference spectrophotometry.[165, 197] ●, Glycol chitin; ○, partially deacetylated glycol chitin.

B. Temperature Dependence of the Binding of Substrate

It has been confirmed by various techniques that the native structure of lysozyme is stable up to 75°C.[125,203,204] Maximum activity, however, is observed at 50°C.

The decrease in activity of various enzymes in the region above the optimum temperature has been explained on the basis of thermal denaturation of the enzyme molecule, which is presumed to cause a loss in the ability to bind to substrate. In the case of lysozyme, however, the decrease in activity in the temperature region 50–75°C must be attributed to some other factor, since the enzyme is not denatured at such temperatures.

Dahlquist and Raftery[53] reported that the binding of methyl-β-*N*-acetylglucosaminide decreased linearly with rise in temperature in the

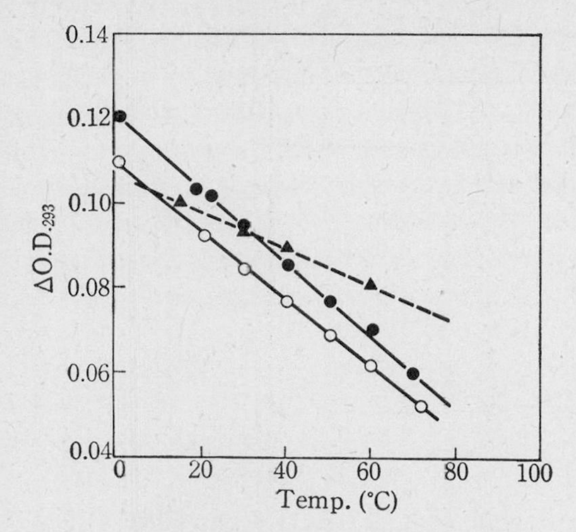

Fig. 54. Temperature dependence of complex formation between lysozyme and chitin derivatives measured by difference spectrophotometry.[164] ●, Glycol chitin; ○, glycol chitin hydrolyzed previously by lysozyme; ▲, solvent effect of 30% glucose solution.

range from 30 to 55°C. Hayashi *et al.*[164] measured the binding of glycol chitin by difference spectrophotometry, obtaining the results in Fig. 54. The intensity of the difference spectrum decreased linearly with rise in temperature of the solution, falling by about 50% over the range from 0 to 70°C. Comparable results were obtained with the oligosaccharide, i.e. the hydrolyzed product of glycol chitin with lysozyme. On the other hand, the change in intensity of the difference spectrum of lysozyme due to the solvent effect of 30% glucose solution was also found to be a linear decrease with rise in temperature of the solution (the broken line in Fig. 54). This fact suggests that the decrease in intensity of the difference spectrum of the lysozyme-substrate complex at higher temperatures does not necessarily imply any decrease in the amount of complex formed, but may well be due to the lowering of the density of the medium. Thus, it seems very likely that complex can be fully formed even when the temperature is not below 75°C, and that the observed decrease in activity results from a change in the fine stereostructure of catalytic groups.

C. Effect of Neutral Salts

Neutral salts at high concentrations such as 1.0 M cause a strong inhibition of activity, but, on the contrary, enhance the formation of

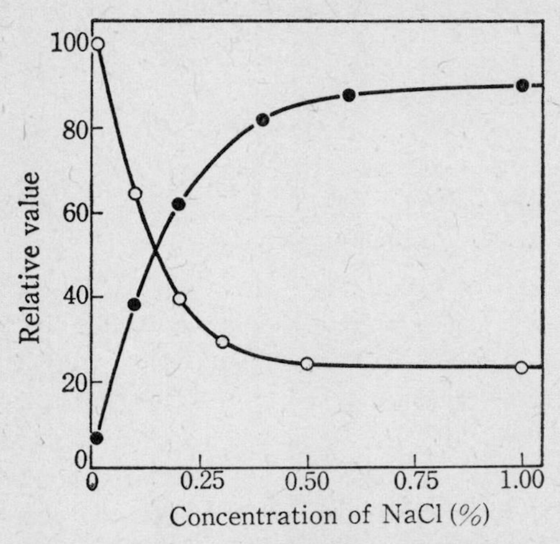

Fig. 55. Effect of sodium chloride on lysozyme activity (○) and the formation of lysozyme–glycol chitin complex (●).[165]

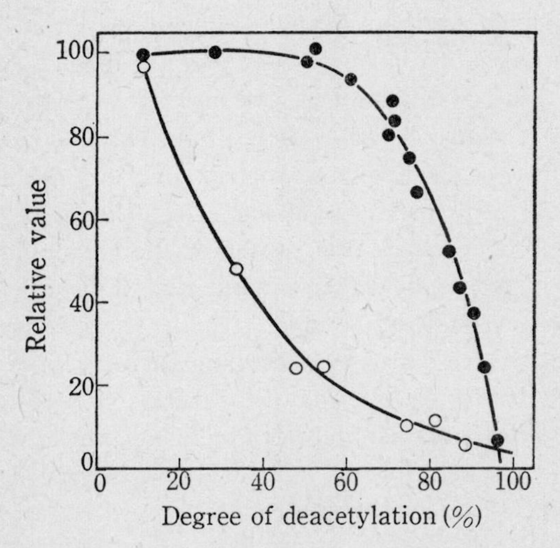

Fig. 56. Effect of deacetylation of chitin on its ability to form complex (●) and its digestibility by lysozyme (○).[147b]

lysozyme-substrate complex in solution, as shown in Fig. 55.[165] The mechanisms by which neutral salts display these effects as yet remain unknown.

D. Effects with Various Substrates

Generally speaking, most 6-0 derivatives of chitin (*ca.* 50% substitution) show almost the same capability for forming complex, i.e. it is irrespective of the nature of the substituent. Deacetylation of the 2-acetamide group of chitin causes a considerable increase in its resistance to digestion by lysozyme. The capability to form complex, however, remains unchanged up to 50% deacetylation. Full deacetylation results in a complete resistance to digestion by lysozyme and also a complete loss in the ability to form complex, as shown by Fig. 56.[147b] From these results, it is assumed that only the 2-acetamide group at subsite C is essential for the binding of polysaccharide substrate, and that the protonated amino group of the sugar residue at subsite D does not affect the binding behavior of the substrate.

E. Adsorption on Insoluble Substrates

The pH dependence of lysozyme adsorption on insoluble substrates is very characteristic (Fig. 48). The amount of sodium chloride is also critical, the degree of adsorption being enhanced in the presence of salt at high concentrations, as shown in Fig. 57.[165]

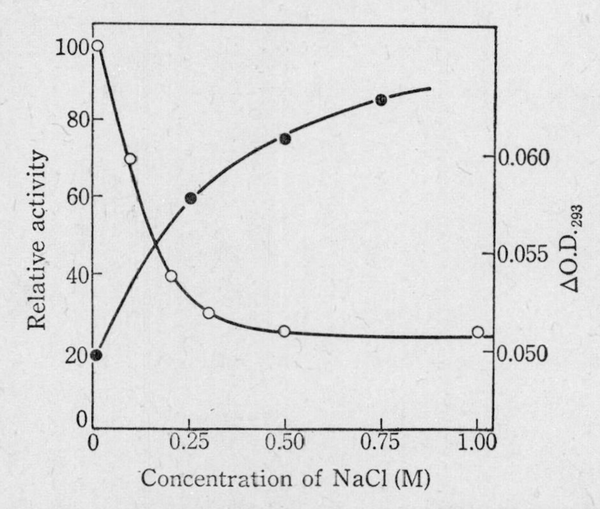

Fig. 57. Effect of sodium chloride on lysozyme adsorption onto a column of carboxymethyl chitin.[165] ●, Adsorption at pH 4.5; ○, activity.

8.3. Mechanism of Lysozyme-catalyzed Hydrolysis of the β-1,4-*N*-Acetylglucosaminide Linkage

8.3.1. Catalytic Hydrolysis

Before describing the mechanism of lysozyme-catalyzed hydrolysis of the β-1,4-*N*-acetylglucosaminide linkage, it is valuable to give a brief description of the general mechanism of hydrolytic reactions taking the ester bond as an example.

A. Tetrahedral Intermediate

Tetrahedral intermediate, in which the sp^2 carbonyl carbon of an ester is converted to sp^3 carbon, has been proposed as the most common and critical intermediate of hydrolytic reactions involving the ester bond. The role of the catalyst for the reaction, irrespective of its mode of action, is to accelerate the formation or stabilization of tetrahedral intermediate. The real structures of tetrahedral intermediates in hydrolytic reactions catalyzed by acids and bases are assumed to be as follows.

Base:

$$\mathrm{R{-}\overset{\displaystyle O}{\overset{\|}{C}}{-}O{-}R' + B{:} \longrightarrow \left[R{-}\overset{\displaystyle O^-}{\overset{|}{\underset{\underset{\displaystyle B}{|}}{C}}}{-}O{-}R' \right]}$$

$$\mathrm{\xrightarrow{+OH^-} R{-}\overset{\displaystyle O}{\overset{\|}{C}}{-}OH + B{:} + R'{-}O^- \xrightarrow{+H^+} R'{-}OH}$$

Acid:

$$\mathrm{R{-}\overset{\displaystyle O}{\overset{\|}{C}}{-}O{-}R' + AH + H_2O \longrightarrow \left[R{-}\overset{\displaystyle OH}{\overset{|}{C}}{-}O{-}R' \quad (C{-}O(H){\cdots}H{-}O(H){\cdots}H^+) \right] A{:}}$$

$$\mathrm{\longrightarrow R{-}\overset{\displaystyle O}{\overset{\|}{C}}{-}OH + AH + R'{-}OH}$$

In these structures, the leaving groups are alkyloxy (—O—R), base (—B) and hydroxyl (—OH) groups. The removal of an alkyloxy group leaves a carbonium ion at the carbonyl carbon. Substitution of a base group or nucleophile attached at the carbonyl carbon by a hydroxyl group in base catalysis, and the transformation of the carbonyl carbonium ion to carbonyl carbon in acid catalysis, will then terminate the hydrolytic reaction. Removal of the leaving groups, —B and —OH, reverts the intermediate to the original ester.

B. Position of Bond Fission (Fission Type)

Bond fission in ester hydrolysis takes place at two different positions; namely, at the bond between the carbonyl carbon and ester-oxygen, and at the bond between the ester-oxygen and alkyl carbon of the alcohol moiety. The former (a) is called "acyl fission", and the latter (b) is called "alkyl fission".

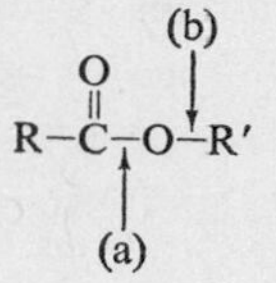

$$\underset{\uparrow (a)}{\mathrm{R{-}\overset{\overset{\displaystyle O}{\|}}{C}{-}}}\mathrm{O}\overset{(b)\downarrow}{-}\mathrm{R'}$$

The fission type depends upon the position of the carbon atom at which the carbonium ion exists preferentially in the intermediary step. When carbonyl carbon forms the carbonium ion, fission is of the acyl fission type; when the carbon atom of the alcohol moiety forms the carbonium ion, alkyl fission occurs. Generally, carbonyl carbon has a much greater tendency to form the carbonium ion than the carbon atom of the alcohol moiety, resulting in acyl fission. Therefore, the ester-oxygen is usually present as the hydroxyl group of the alcohol molecule after hydrolysis. A typical example of alkyl fission is seen in the hydrolysis of tertiary butyl alcohol ester; here, the tertiary carbon of the butyl alcohol forms a particularly stable carbonium ion.

C. Mode of Action of Catalyst

a. Specific Catalysis

Specific catalysis refers to reactions catalyzed by lyate species, hydrogen ions, water, or hydroxide ions. Hydrolytic reactions catalyzed by simple acids or bases, such as hydrochloric acid or sodium hydroxide solution, are of course also included under the term specific catalysis.

b. Nucleophilic Displacement and Nucleophilic Catalysis

Nucleophilic reagents displace the alkyloxy group of an ester. When the displaced carbonyl compound, such as an amide, is very stable and not susceptible to hydrolysis within the medium, the reaction is referred to simply as "nucleophilic displacement". When the displaced carbonyl compound can react further, e.g. with hydroxide ions in solution, regenerating the nucleophilic group, the reaction is termed "nucleophilic catalysis".

c. General Catalysis

In the case of general catalysis, the catalyst does not bind covalently to the substrate (reactant); i.e. the catalyst does not participate directly in the tetrahedral intermediate. The catalysts in this group may assist attack by water and nucleophiles, or may enhance the activity of a separate catalyst by donating or withdrawing a hydrogen ion (i.e. acting as acid or base, respectively). The following equations explain in more detail the various modes of action of catalysts of the above three categories.

(a) Specific catalysis

H_3O^+, H_2O, OH^-

(b) Nucleophilic catalysis

(1) Nucleophilic displacement

$$R{-}\overset{\overset{\displaystyle O}{\|}}{C}{-}O{-}R' + B: \longrightarrow R{-}\overset{\overset{\displaystyle O}{\|}}{C}{-}B + R'{-}OH$$

(2) Nucleophilic catalysis

$$R{-}\overset{\overset{\displaystyle O}{\|}}{C}{-}O{-}R' + B: \longrightarrow R{-}\overset{\overset{\displaystyle O}{\|}}{C}{-}B + R'{-}OH$$

$$R{-}\overset{\overset{\displaystyle O}{\|}}{C}{-}B + OH^- \longrightarrow R{-}\overset{\overset{\displaystyle O}{\|}}{C}{-}OH + B:$$

(c) General catalysis

(1) Hydrolysis

(i) General acid

$$R{-}\overset{\overset{\displaystyle O}{\|}}{C}{-}O{-}R' + AH + H_2O \longrightarrow \left[\begin{array}{c} H{-}A \\ O \\ \| \\ R{-}C{-}O{-}R' \\ O \\ H \quad H \end{array}\right]$$

$$\longrightarrow \left[R-\overset{\displaystyle OH}{\underset{\displaystyle OH}{\overset{|}{\underset{|}{C}}}}-O-R' \right] + A: + H^+$$

$$\longrightarrow R-\overset{\displaystyle O}{\overset{\|}{C}}-OH + R'-OH + AH$$

(ii) General base

$$R-\overset{\displaystyle O}{\overset{\|}{C}}-O-R' + B: + H_2O \longrightarrow \left[R-\overset{\displaystyle O}{\overset{\|}{C}}-O-R' \atop {\vdots \atop {O \atop H \quad H \cdots B}} \right]$$

$$\longrightarrow \left[R-\overset{\displaystyle O^-}{\underset{\displaystyle OH}{\overset{|}{\underset{|}{C}}}}-O-R' \right] + BH \longrightarrow R-\overset{\displaystyle O}{\overset{\|}{C}}-OH + R'-OH + B:$$

(2) Nucleophilic displacement
(i) General acid

$$R-\overset{\displaystyle O}{\overset{\|}{C}}-O-R' + B: + AH \longrightarrow \left[R-\overset{\displaystyle O}{\overset{\|}{C}}-O-R' \atop {\vdots \atop B:} \right] \nwarrow AH$$

$$\longrightarrow \left[R-\overset{\displaystyle OH}{\underset{\displaystyle B}{\overset{|}{\underset{|}{C}}}}-O-R' \right] + AH \longrightarrow R-\overset{\displaystyle O}{\overset{\|}{C}}-B + R'-OH + AH$$

(ii) General base

$$R-\overset{\displaystyle O}{\overset{\|}{C}}-O-R' + B: + B: \longrightarrow \left[R-\overset{\displaystyle O}{\overset{\|}{C}}-O-R' \atop {\vdots \atop {B: \atop \cdots B:}} \right]$$

$$\longrightarrow \left[R-\overset{\displaystyle O^-}{\underset{\displaystyle B}{\overset{|}{\underset{|}{C}}}}-O-R' \right] + B: \longrightarrow R-\overset{\displaystyle O}{\overset{\|}{C}}-B + R'-OH + B:$$

D. Anchimeric Assistance

Intramolecular hydrolysis or enzymatic hydrolysis (in which the catalytic group and the bond susceptible to cleavage are in close proximity, or juxtaposition) proceeds with great efficiency, and frequently occurs by a mechanism different from that found for simple bimolecular hydrolysis. The intramolecular hydrolysis of suitable model compounds has been investigated intensively in many laboratories in an attempt to explain the extraordinary efficiency of the enzymatic reaction. For further information, the reader is recommended to consult the available literature on intramolecular catalysis.

E. Cooperative Action of Catalysts

When the catalyst is of poly(multi-)functional type, and each functional group has potential as a catalyst, the functional groups may catalyze a single reaction cooperatively. In the case of a polymeric catalyst such as an enzyme, the catalyst molecule contains a number of functional groups, though their reactivities may of course vary. A single molecule of a polymeric catalyst may contain both acidic and basic groups without neutralization or salt formation between them; this is one of the characteristics of enzymes.

During the catalytic hydrolysis of an ester, acidic and basic groups in the enzyme molecule may simultaneously attack the carbonyl group in a push-pull way, forming tetrahedral intermediate. This type of cooperative action is called a "concerted reaction" (a). Two or more catalytic groups may sequentially attack the same carbonyl group of the ester; this is known as a "consecutive reaction" or "double (multi-) displacement reaction" (b). Cooperative catalysis by catalytic groups leads to a considerable lowering of the activation energy. Examples of cooperative catalysis are given in the following schemes.

(a) Concerted reaction (general acid catalysis and nucleophilic catalysis)

$$R{-}\overset{\overset{\displaystyle O}{\|}}{C}{-}O{-}R' \;+\; \text{B:} \qquad \text{AH} \longrightarrow R{-}\overset{\overset{\displaystyle O\cdots H{-}A}{\|}}{C}{-}O{-}R' \atop \text{B:}$$

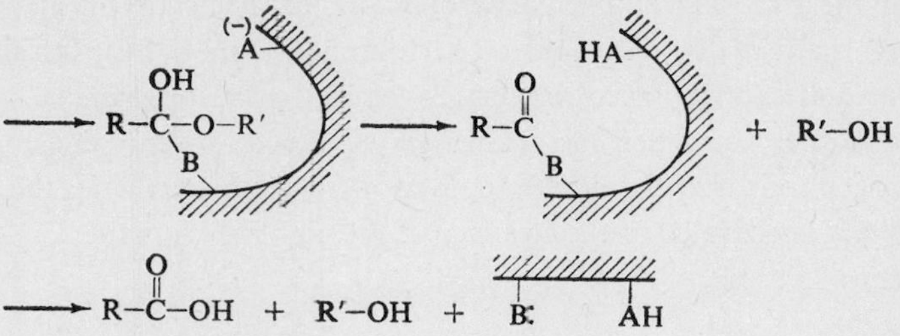

→ R–C(OH)(–B)–O–R′ [(–)A] → R–C(=O)–B [HA] + R′–OH

→ R–C(=O)–OH + R′–OH + B: AH

(b) Consecutive reaction

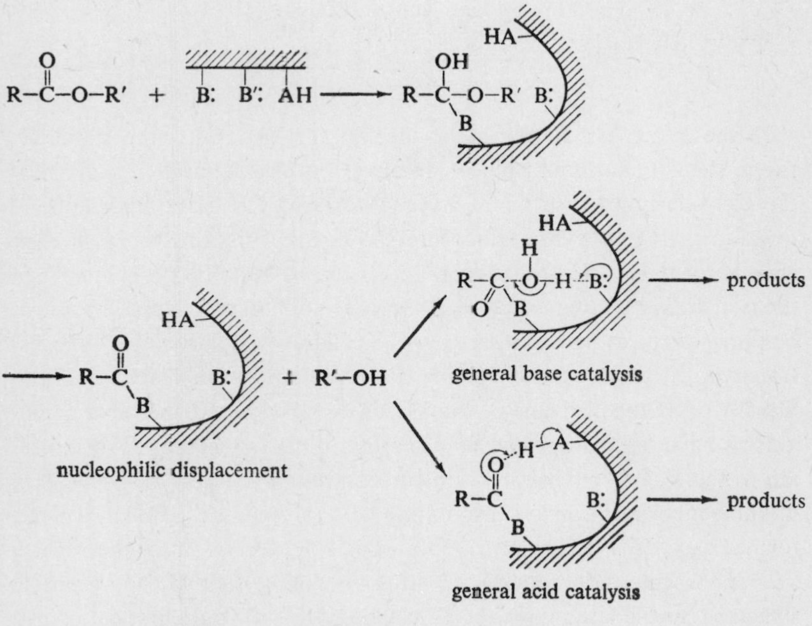

R–C(=O)–O–R′ + B: B′: AH → R–C(OH)(–B)–O–R′ B: [HA]

→ R–C(=O)–B [HA, B:] + R′–OH

nucleophilic displacement

R–C(=O)(–B)···O(H)–H···B: [HA] → products

general base catalysis

R–C(=O···H–A)–B [B:] → products

general acid catalysis

8.3.2. Hydrolysis of the Glycoside Linkage by Specific Catalysts

In general, the glycoside linkage is hydrolyzed easily at moderate concentrations of acid or hydrogen ions, while it shows a relatively strong resistance to the catalytic action of bases or hydroxide ions. However, as is well known, aryl glycosides and esters of sugars and acids are hydrolyzed by bases and hydroxide ions. The resistance of the glycoside linkage to hydrolysis by bases may be explained by the fact that in the glycoside linkage the carbon atom connected to the glycoside-oxygen is already in a form which corresponds to that of the tetrahedral intermediate appearing in the process of hydrolysis of esters.

Let us next consider the hydrolysis of the β-1,4-glucoside linkage. Hydrogen ions may accelerate the hydrolysis by neutralizing the negative charge generated in the cleavage of the glucoside linkage, C-1—O—C-4 (Eq. 25).

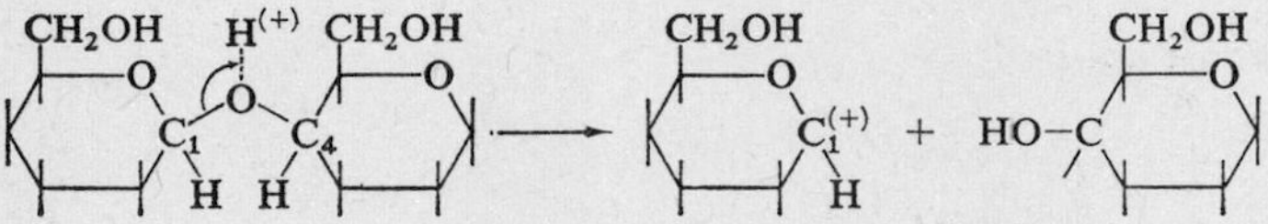

$$\text{(glucose ring, } CH_2OH\text{)}-C_1(H)-\overset{H^{(+)}}{O}-C_4(H)-\text{(glucose ring, } CH_2OH\text{)} \longrightarrow \text{(glucose ring, } CH_2OH\text{)}-C_1^{(+)}(H) + HO-C\text{(glucose ring, } CH_2OH\text{)} \quad (25)$$

In the hydrolysis of the glucoside linkage, generally, the bond between the C-1 atom and glucoside-oxygen is cleaved and σ-electrons of the C-1—O bond shift to the oxygen, forming C-1 carbonium ion and oxyanion. *t*-Butyl pyranoglucoside shows alkyl fission,[205] due to the ease of formation of *t*-butyl carbonium ion, as described previously. In the case of sugar esters such as phosphate and sulfate esters, the sulfur or phosphorus atom carries a positive charge by polarization. The fission type in these esters depends markedly upon the reaction conditions.[206] For the cleavage of the glucoside linkage, two possible mechanisms have been proposed. One is a bimolecular reaction corresponding to the S_N2 mechanism of substitution reactions, and the other is a unimolecular reaction corresponding to the S_N1 mechanism. In the former case, hydroxide ions would attack the C-1 atom at the reverse side of the glucoside linkage, resulting in an inversion of the anomeric structure. In the latter case, the C-1 carbonium ion would first be formed by attack of hydrogen ions on the glucoside-oxygen. Hydroxide ions would then attack the C-1 carbonium ion, resulting in the formation of equal amounts of α- and β-anomers. (The C-1 carbonium ion has a planar structure, which permits attack by hydroxide ions from either the front or the back of the plane.)

In the reaction mechanism involving C-1 carbonium ion intermediate, stabilization of the carbonium ion is considered to be the rate-determining step. Stabilization of the carbonium ion intermediate in solution may be attained in various ways. First, the carbonium ion may be stabilized by interaction with the solvent molecule, by salt formation with any other anion in the medium, by ion-pair interaction with an anion, or by covalent binding with a nucleophile group in the medium. Second, when the sugar has a ring structure (a furanose or pyranose ring) the electron could move from the ring oxygen to the C-1 carbonium ion; the resultant cation would then become stabilized according to the formula (VIII).

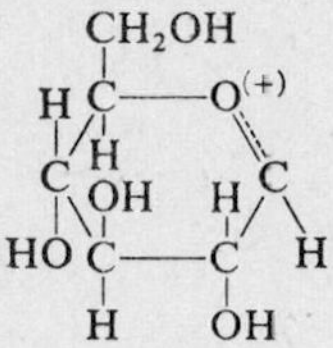

(VIII)

This type of stabilization requires a change in the ring structure; i.e. for the formation of a pyranoxonium ion, the ring structure would be transformed from the chair conformation to the half-chair conformation. This is an energy-consuming process, and is therefore unfavorable for overall hydrolysis of the glucoside linkage. Third, the carbonium ion may be stabilized by ring formation with a substituent or C-6–OH group. The resultant structures with an acetyl group at the C-2 position, with a methyl group at the C-2 position, and with a C-6–OH group are (IX), (X), and (XI), respectively. Finally, there is a limited possibility that the carbonium ion might be stabilized as a glucal-like structure (XII), or by ring formation with 2-0 (XIII).

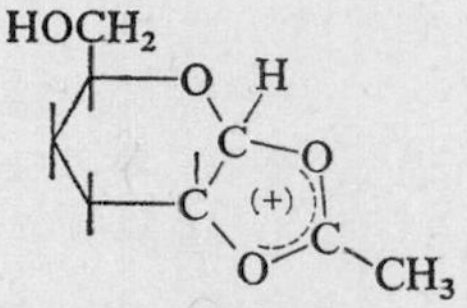

(IX) 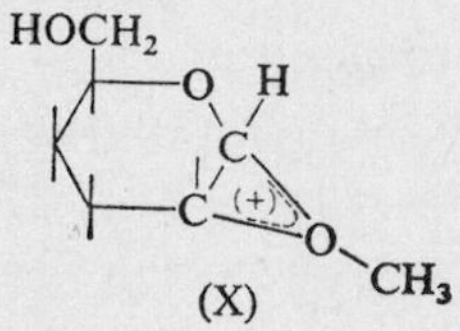 (X)

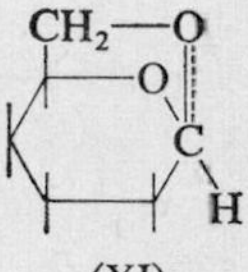

(XI) 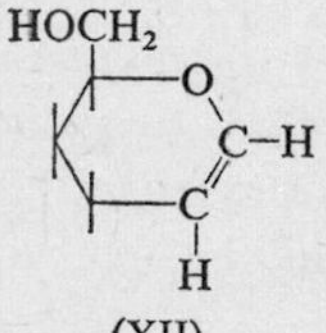 (XII) 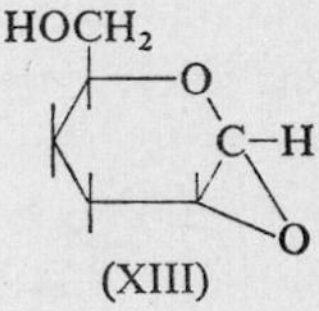 (XIII)

Countless experiments have been carried out under various conditions to distinguish which of these mechanisms actually operate in hydrolytic reactions. For hydrolysis of the *N*-acetylglucosaminide linkage, it is reasonable to assume that the C-1 carbonium ion is stabilized by the formation of a ring structure with the substituent acetamide group at the C-2 position. The carbonium ion in the intermediary step has a planar structure, unless it forms a glycosyl compound with some nearby nucleophile or forms a ring structure with the substituent. Thus, attack from hydroxide ions can occur from both the front and the back of the plane of the carbonium ion, resulting in the formation of equal amounts of α- and β-anomers.

Vernon[207] has concluded from the experimental results and a consideration of all possible mechanisms that the hydrolysis of glucoside linkages by a simple acid catalyst may proceed via carbonium ion intermediate, which may react with hydroxide ion rapidly, before the pyranose ring containing the C-1 carbonium ion goes to the half-chair conformation. Thus, the acid- or proton-catalyzed hydrolysis of glucoside linkages seems to proceed by a mixed or intermediary type of $A1$ (S_N1) and $A2$ (S_N2) mechanisms.

Since enzymes are polyfunctional catalysts in which the functional groups are able to act sequentially as unit catalysts or cooperatively as clustered catalysts, the mechanism of hydrolysis catalyzed by enzymes may be very complex, if any of the mechanisms proposed by Vernon are involved in enzymatic hydrolysis.

8.3.3. Enzymatic Hydrolysis of the Glycoside Linkage

Many enzymes catalyzing the hydrolysis of glycoside or sugar ester linkages are known and their modes of action have been intensively investigated. The detailed mechanism by which hydrolysis of glycoside linkages proceeds has not been elucidated, however, although a few proposals for glycosidase mechanisms appear in references. In this section, the proposed mechanism for the action of amylases will be introduced in order to outline the progress in this field.

Amylases are enzymes hydrolyzing the α-1,4-glucoside linkage of polysaccharide starch and are classified into α- and β-amylases, based on the anomeric structure at the reducing end of the products. α-Amylase forms a product with α-anomeric structure, retaining the same sign of optical activity as the cleaved bond of the substrate. β-Amylase forms β-anomer, resulting in the inversion of the anomeric structure. Both amylases have been shown to cleave the bond between C-1 and glucoside-oxygen by the incorporation of ^{18}O at the reducing end from $H_2{}^{18}O$ solvent.

Koshland Jr.[208] proposed a mechanism for β-amylase as follows. First, β-amylase catalyzes the attack of hydrogen ion on the glucoside-oxygen, accelerating the formation of oxonium ion or conjugate acid. At the same time, hydroxide ion attacks the C-1 atom from the back, with respect to the position of the glucoside-oxygen. The binding of hydroxide ion to the C-1 atom coincides with the splitting of the bond between C-1 and the glucoside-oxygen, bringing about the inversion of optical activity or anomeric structure. This mechanism can be categorized as S_N2, being a bimolecular mechanism denoted as *A*2.

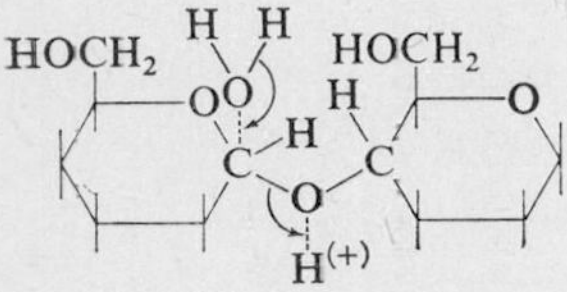

(XIV)

In this case, the C-1 carbonium ion intermediate, and hence the half-chair conformation, need not be considered. If the hydrolytic reaction catalyzed by β-amylase proceeds by Koshland's mechanism, the β-amylase molecule should provide catalytic groups able to accelerate the attack of hydrogen ion and hydroxide ion by general catalysis. An acidic group of β-amylase might act as a general acid catalyst for the attack of hydrogen ions on glucoside-oxygen, and a basic group as a general base catalyst for the attack of hydroxide ion on the C-1 atom.

Simple *A*2 mechanisms cannot be involved in the hydrolysis of α-1,4-glucoside linkages catalyzed by α-amylase, because retention of anomeric structure is observed in this reaction. If the hydrolytic reaction proceeds by an *A*1 mechanism, the products should be a mixture of equal amounts of α- and β-anomers at the reducing ends. A simple *A*1 mechanism may therefore be ruled out in considering possible mechanisms for the action of α-amylase. These conclusions thus suggest that the real mechanism for hydrolysis catalyzed by α-amylase is rather complex.

Koshland Jr.[208,209] proposed a consecutive combination of *A*2 mechanisms or the double displacement mechanism as the overall mechanism for hydrolysis catalyzed by α-amylase. The acidic group ($—NH_3^+$) in the α-amylase molecule donates a hydrogen ion as a general acid, accelerating the formation of oxonium ion at the glucoside-oxygen. Then a nucleophile (carboxylate anion) of the enzyme attacks the C-1 atom, resulting in a glucosyl amylase, as shown in Eq. (26).

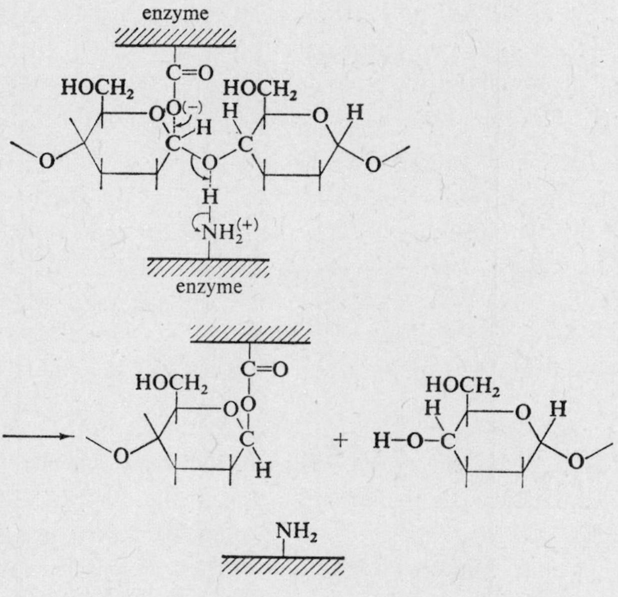

(26)

The enzyme moiety in the intermediate glucosyl amylase is then replaced by hydroxide ion from the solvent (water). The attack of the water molecule is enhanced by the general base catalysis of the neutral amino group. Thus, the anomeric structure at the reducing end is held unchanged after the hydrolysis of the glucoside linkage (Eq. (27)).

Contrary to Koshland's proposal, Mayer and Lahner[210] and Halpern and Leibowitz[211] reported *A*1 mechanisms for both α- and β-amylases. This mechanism contains a sequence of reactions. First, the oxonium ion is formed on the glucoside-oxygen, then the glucoside linkage is cleaved and the carbonium ion at the C-1 atom formed. This C-1 car-

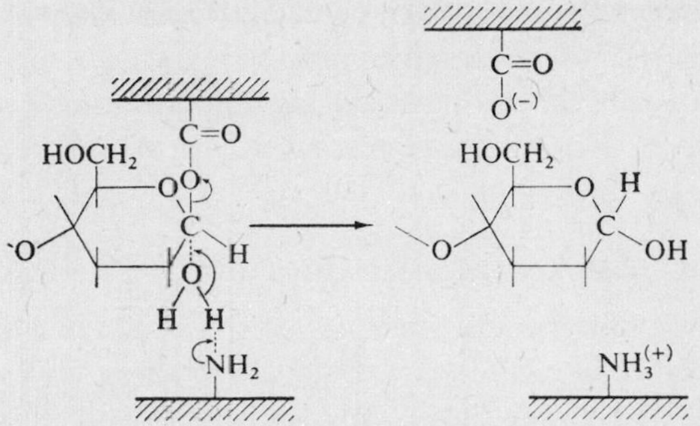

(27)

bonium ion is stabilized in a pyranose ring by the formation of the half-chair pyranoxonium ion. The positive charge seems to be further stabilized by an interaction (ion-pair formation) with a negative charge of the enzyme molecule. The difference in action seen between α- and β-amylases is essentially due to the direction of rotation of the pyranose ring to form the half-chair conformation. In the case of α-amylase, the rotation of the pyranose ring to form the pyranoxonium ring might be permissible only in a definite direction, and as a result the groups attached to the C-2, C-3, C-4 and C-5 atoms take the axial positions (XV). In the case of β-amylase, the direction of the rotation is the reverse, resulting in equatorial positions for the groups on C-2, C-3, C-4 and C-5 (XVI).

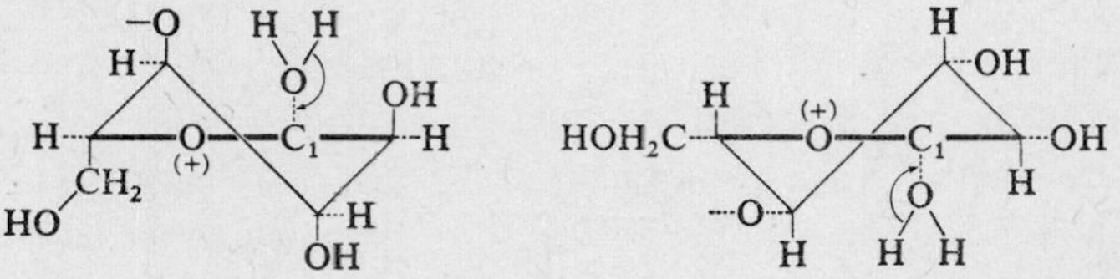

(XV) (XVI)

These restrictions on the direction of rotation of the pyranose ring to form the half-chair conformation may arise from the surface structure or topography of each enzyme. The resulting ring intermediate with half-chair conformation may then be fixed in the active site of the enzyme. Subsequently, the attack of a water molecule at the C-1 atom may be permitted from a definite direction according to the topography of the enzyme. Thus, α-amylase produced only the α-anomeric structure at the reducing end, and β-amylase only β-anomer.

In general, the mechanism of hydrolysis of glycoside linkages catalyzed by enzymes has not been fully understood. However, it seems clear that most of the mechanisms of enzyme action involve the formation of a carbonium ion intermediate. The problems remaining are those concerning the stabilization mechanism of the carbonium ion intermediate.

8.3.4. Catalytic Mechanism of Lysozyme Action

A. Fission Type and the Position of the Linkage Susceptible to Cleavage

Bond cleavage during the hydrolysis of β-1,4-*N*-acetylglucosaminide linkages catalyzed by lysozyme is known to take place between C-1 and glucosaminide-oxygen.[47] The next problem is to determine which bond among the six *N*-acetylglucosamine residues of the substrate bound in the cleft of the lysozyme molecule is cleaved by catalytic hydrolysis. In order to decide the position of the *N*-acetylglucosaminide linkage susceptible to cleavage, it will be preferable to summarize the experimental results obtained so far.

(i) A lysozyme-substrate complex in which the sequence of subsites D, E and F is occupied by *N*-acetylglucosamine residues has not been isolated.

(ii) The glucoside linkage between *N*-acetylglucosamine and *N*-acetylmuramic acid residues of cell-wall tetrasaccharide NAG–NAM–NAG–NAM was hydrolyzed by lysozyme. Since *N*-acetylmuramic acid has a lactic acid residue at the C-3 position, it cannot occupy subsite C because of the steric effect of the bulky substituent. Therefore, the subsite occupied by *N*-acetylmuramic acid residue must be one of the subsites A, B, and D, and the linkage to be cleaved may be any one of those located between A and B, between B and C, and between D and E.

(iii) For a suitable orientation of *N*-acetylglucosamine residues at subsite D, the residue must assume the half-chair conformation in which C-1, C-2, C-5 and 5-O locate in the same plane. When the C-1 atom in the pyranose ring forms a carbonium ion in the hydrolysis of *N*-acetylglucosaminide linkages, the pyranose ring in the chair conformation transforms to a half-chair conformation. The distortion of the pyranose ring to a half-chair conformation occurring in the binding process might accelerate carbonium ion formation at the C-1 atom in the process of bond cleavage.

(iv) The free energy change arising from the binding of *N*-acetylglucosamine residue at subsite D has a positive sign. This process thus necessitates the consumption of free energy, which is consistent with the observation that the *N*-acetylglucosamine residue at subsite D takes a half-chair conformation which consumes more than 10 kcal/mole free energy in the transformation from the chair conformation.

(v) One of the catalytic groups in the lysozyme molecule is considered to be a protonated carboxyl group. In fact, Glu-35 and Asp-52 are located near the *N*-acetylglucosaminide linkage between the subsites D and E. The former group exhibits a p*K* value of about 6.0.

These findings suggest that the *N*-acetylglucosaminide linkage susceptible to hydrolysis is located between subsites D and E.

B. Hydrolytic Activity of Lysozyme

The catalytic groups in lysozyme are the protonated γ-carboxyl group of the side chain of Glu-35 and the β-carboxylate anion of the side chain of Asp-52. Before discussing the detailed mechanism of hydrolysis of the *N*-acetylglucosaminide linkage, it may be valuable to summarize the findings so far obtained.

(i) The distortion of the pyranose ring of *N*-acetylglucosamine residue at subsite D is favorable for the formation of the C-1 carbonium ion intermediate.

(ii) The C-1 carbonium ion may be surrounded with a hydrophobic region. This may be associated with the mechanism of stabilization of the C-1 carbonium ion.

(iii) The anomeric structure at the C-1 atom remained unchanged throughout the hydrolysis or transglycosylation. The percentage of β-anomer in methyl *N*-acetylglucosaminide, which was formed by transglycosylation in the presence of methanol, was 99.5% for human lysozyme and 99.7% for hen egg-white lysozyme. In the case of papaya lysozyme, transglycosylation was not observable and a product with α-anomeric structure was formed to some extent.[212]

(iv) The great efficiency of transglycosylation catalyzed by lysozyme suggests that the C-1 carbonium ion must be stabilized. However, there is no evidence that glucosaminyl lysozyme is an intermediate, though its formation from the C-1 carbonium ion and the carboxylate anion of Asp-52 might seem possible.

(v) Glu-35 locates in a hydrophobic region of the molecule, exhibiting abnormal dissociation with a p*K* value near 6.0. Asp-52 is contained in a network of hydrogen bonds, existing as a carboxylate anion in the pH region where lysozyme shows *N*-acetylglucosaminidase activity.

(vi) There is no difference in the hydrolytic rates of $(NAG)_n$ type substrate and $(NAG\text{–}NAM)_{n12}$ substrate.

(vii) The hydrolysis of *N*-acetylglucosaminide linkage catalyzed by lysozyme is observable in the region from pH 2 to 8, with the maximum rate at pH 4.5–5.5. The pH dependence of transglycosylation is completely parallel to that of the rate of hydrolysis.

(viii) Organic solvents inhibit the activity. Neutral salts exhibit a similar effect but do not inhibit the formation of lysozyme-substrate complex. Glucose solution at a high concentration such as 30% enhances the activity.

C. Mechanism of Hydrolysis

From the above findings it can be inductively concluded that the β-1,4-acetylglucosaminide linkage is hydrolyzed by a cooperative catalysis of Glu-35 and Asp-52. Glu-35 appears to act as general acid catalyst, though the real mode of action of Asp-52 is not yet well understood. More details of the mechanism of lysozyme-catalyzed hydrolysis of the *N*-acetylglucosaminide linkage are being sought in many laboratories.

Piszkiewicz and Bruice[213–15] studied the hydrolysis of phenyl and methyl *N*-acetylglucosaminides in connection with the mechanism of hydrolysis catalyzed by lysozyme, emphasizing especially the anchimeric role of the 2-acetamide group of *N*-acetylglucosamine residues for the rate of hydrolysis.

a. 2-Acetamide Group

First, the rates of hydrolysis of *p*- and *o*-nitrophenyl *N*-acetylglucosaminides were measured in the region from pH 0.75 to 11.72 at 78.2°C, following the liberation of nitrophenol by spectrophotometry. α-Anomers of these *N*-acetylglucosaminides can be hydrolyzed by specific acid and specific base catalysts. On the other hand, the rate of hydrolysis of the β-anomers exhibited a plateau in the neutral region. In particular, nitrophenyl β-*N*-acetylglucosaminides (NP–β-NAG) showed a constant rate of hydrolysis throughout nearly all pH values even in the absence of buffer solutions to keep the pH constant. This means that NP–β-NAG was hydrolyzed spontaneously by solvent water molecules. Although *p*- and *o*-nitrophenyl β-glucopyranosides also showed a plateau, their rates of hydrolysis were considerably lower than that of NP–β-

NAG, and the pH region for the plateau was considerably narrower. This experimental result indicates that when the 2-acetamide group and aglycone take the *trans*-1,2-disposition, the 2-acetamide group can accelerate the cleavage of the *N*-acetylglucosaminide linkage. The anchimeric effect of the 2-acetamide group on the spontaneous hydrolysis seems to take place by two kinetically indistinguishable pathways. In one case, the neutral 2-acetamide group acts as intramolecular nucleophile, forming a protonated oxazolium ring as an intermediate. In the other case, the ionized 2-acetamide group acts as intramolecular nucleophile for the glucosaminide linkage with the oxonium ion, forming the oxazoline intermediate. Since the pK value of 2-methyl–Δ^2-oxazoline is 5.5, the effect of the 2-acetamide group on the spontaneous hydrolysis of NP–β-NAG may be manifested via both mechanisms. For the hydrolysis of nitrophenyl *N*-acetylglucosaminides by specific hydrogen ion catalysis, the effect of the 2-acetamide group was not observed.

(28)

β-Anomer of nitrophenyl D-glucopyranoside was hydrolyzed much less rapidly by specific base catalysis than the α-anomer. In this type of hydrolysis, it has been postulated that the oxyanion at the C-2 position does not participate in the cleavage of glucoside linkages. The effect of the 2-acetamide group was also not observable in the specific base-catalyzed hydrolysis of *N*-acetylglucosaminide linkages.

b. Anchimeric Assistance of the Carboxyl Group of Aglycone

The pH dependence of the hydrolysis of *o*-carboxyphenyl β-*N*-acetylglucosaminide (*o*-CP–β-NAG) suggested that a group with a p*K* value of 4.2 may contribute to the high rate of hydrolysis. At pH values below this p*K*, the rate of hydrolysis is greater than the value calculated by assuming that the hydrolysis proceeds only by specific acid catalysis, by as much as 8×10^4. Thus, the carboxyl group at the *ortho* position of aglycone phenol anchimerically assists the hydrolysis of *N*-acetylglucosaminide linkages. The possible mechanisms for anchimeric assistance of the *o*-carboxyl group are illustrated below. One is general catalysis of the *o*-carboxyl group (XVII), and the other is intramolecular nucleophilic displacement by the carboxyl group (XVIII).

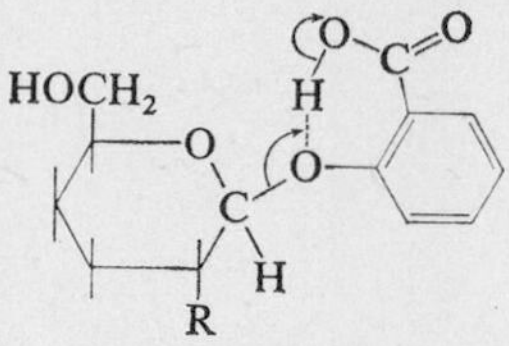

(XVII) 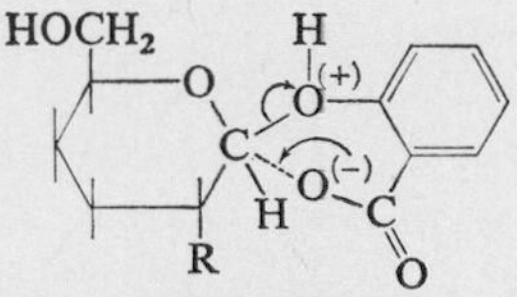 (XVIII)

The hydrolysis of *o*-CP–β-NAG in acid media proceeded more rapidly than that of *o*-carboxylphenyl β-D-glucopyranoside (*o*-CP–β-G). This implies that the 2-acetamide group can accelerate the hydrolysis of *o*-CP–β-NAG. The *o*-carboxyphenyl and 2-acetamide groups may in fact catalyze cooperatively the cleavage of *N*-acetylglucosaminide linkages as a general acid and a nucleophile, as shown in Eq. (29).

(29)

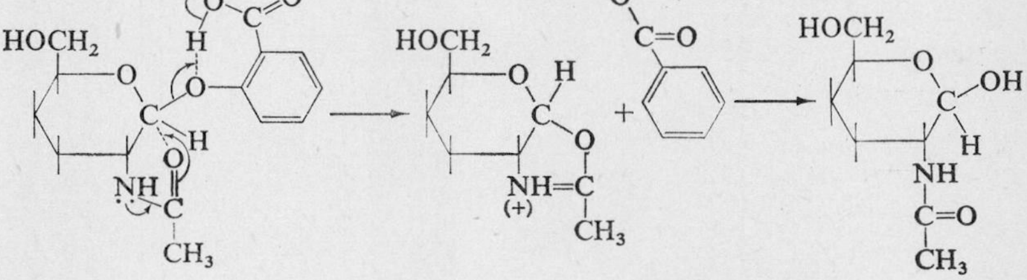

D. Effect of the 2-Acetamide Group in Specific Acid Catalysis

No effect of the 2-acetamide group on the hydrolysis of NP–β-NAG catalyzed by specific acid or proton was observed. On the other hand, for the hydrolysis of methyl β-*N*-acetylglucosaminide (Me–β-NAG) catalyzed by specific acid, it was confirmed that the 2-acetamide group can accelerate the rate of hydrolysis of the *N*-acetylglucosaminide linkage. The mechanism of participation of the 2-acetamide group was considered to be as follows. First, hydrogen ion acts either on the 2-acetamide group or on glucosaminide-oxygen, resulting in protonation of the 2-acetamide group or oxonium ion of glucosaminide-oxygen. In the former case, the 2-acetamide group acts as intramolecular general acid and the C-1 carbonium ion formed is stabilized in the structure of the pyranoxonium ion; in the latter case, the C-1 carbonium ion is stabilized as a protonated oxazoline ring, as shown in Eq. (30), where the 2-acetamide group acts as intramolecular nucleophile.

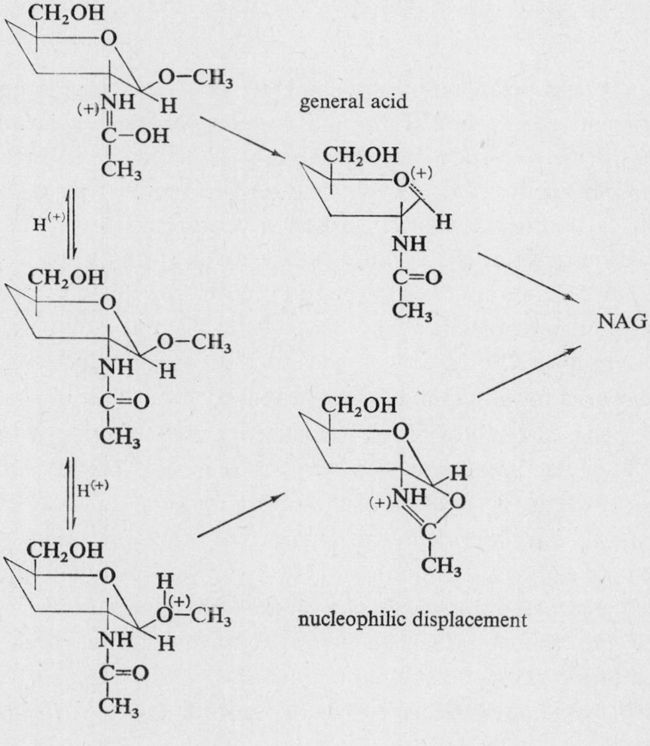

general acid

NAG (30)

nucleophilic displacement

The two mechanisms were indistinguishable kinetically. However, it was concluded from a comparison of the rates of hydrolysis of methyl β-D-glucopyranoside and Me–β-NAG that the 2-acetamide group acted as an intramolecular nucleophile for hydrolysis of the glucosaminide linkage in solutions at low pH.

From their experimental results with model compounds, Piszkiewicz and Bruice have proposed a mechanism for lysozyme action in which a concerted reaction of the general catalyst Glu-35 and intramolecular nucleophilic assistance of the 2-acetamide group are involved (Eq. (31)).

Glu–35
C=O
O
H
O–
CH_2OH
O
NH
H
C=O
CH_3
→
CH_2OH
O
H
NH
(+)
O
C
CH_3
→ product (31)

Vernon[207] has postulated a mechanism based upon a large number of experiments. He suggested that the first step for lysozyme-catalyzed hydrolysis is the distortion of a hydrolyzable bond of the substrate. In the case of productive complex, the *N*-acetylglucosamine residues bound at subsite D are distorted to the half-chair conformation, i.e. the same conformation as that of the intermediate pyranoxonium ion. Consequently, the formation of the carbonium ion intermediate may be accelerated by the distortion of the *N*-acetylglucosamine residue. The γ-carboxyl group of Glu-35 acts as general acid on the glucosaminide-oxygen between subsites D and E, leading to formation of the C-1 carbonium ion. The carbonium ion so produced cannot from the glucosaminyl lysozyme unless the β-carboxylate anion of Asp-52 shifts remarkably towards the position of the carbonium ion, as X-ray studies clearly show. The carbonium ion intermediate therefore seems to be stabilized by electrostatic interaction with the β-carboxylate anion of Asp-52. The attack of hydroxide ion on the carbonium ion occurs only from the direction opposite to that of the leaving group, resulting in retention of the anomeric structure.

Lowe[216] has studied the hydrolysis of *o*-nitrophenyl β-*N*,*N*′-diacetylchitobioside and proposed the following mechanisms (Eq. (32)) by reference to the results of X-ray crystallographic studies.

(a) (32)

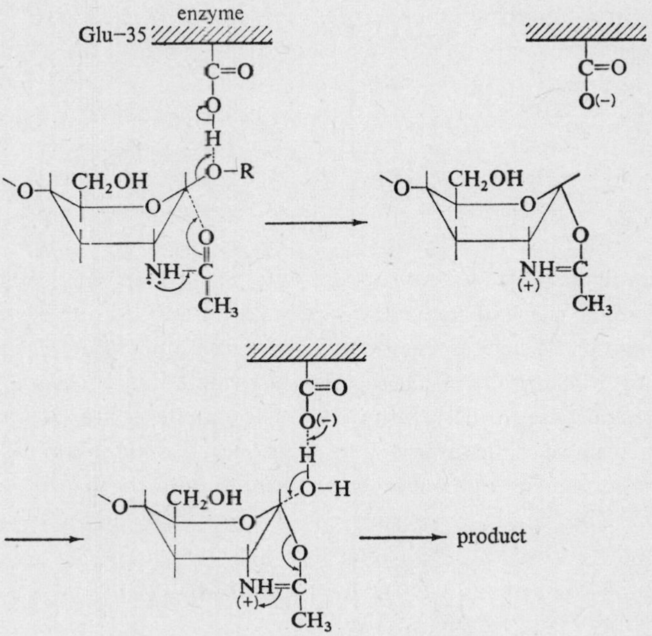

(b)

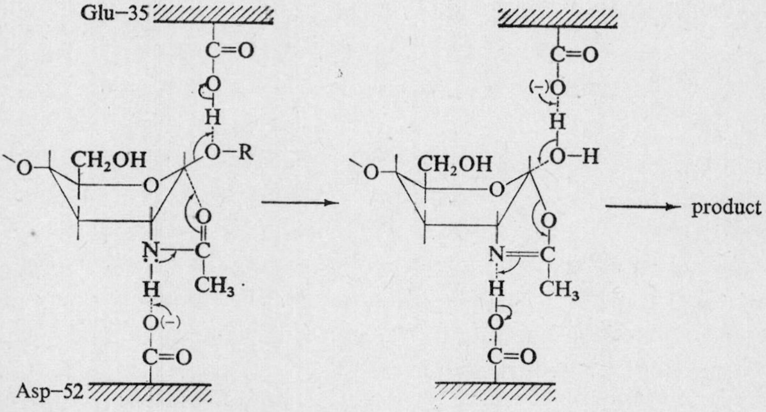

(c)

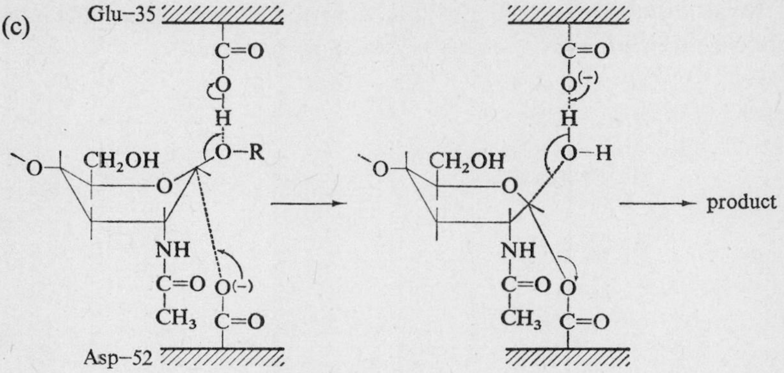

The characteristic of mechanism (a) is that the 2-acetamide group acts as an intramolecular nucleophile for the C_1 atom, and the γ-carboxyl group of Glu-35 acts as a general acid on the glucosaminide oxygen. Mechanism (b) provides that the β-carboxylate anion of Asp-52 assists as a general base in the action of the 2-acetamide group, as an intramolecular nucleophile in displacing the protonated glucosaminide linkage. In mechanism (c) the γ-carboxyl group of Glu-35 acts as a general acid on glucosaminide-oxygen, and β-carboxylate anion of Asp-52 attacks the glucosaminide linkage as a nucleophile. As pointed out by Vernon, mechanism (c), which contains the intermediate *N*-acetylglucosaminyl lysozyme, should be ruled out.

Lowe *et al.*[217] further studied the hydrolysis of aryl β-*N*,*N'*-diacetylchitobioside and *S*-phenyl β-*N*,*N'*-diacetylchitobioside, and discussed the possibility that the carbonium ion is stabilized in a pyranose ring with a half-chair conformation (Eq. (33)).

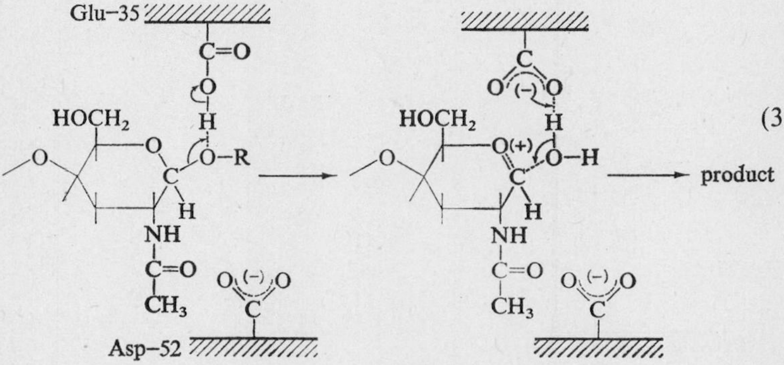

(33)

Raftery and Rand-Meir[152] found that the lysozyme-catalyzed transglycosylation of *N,N'*-diacetylchitobiose in the presence of methanol gave only β-anomer of methyl *N*-acetylglucosaminide. This result suggested that the hydrolysis of *N,N'*-diacetylchitobiose did not take place through a simple nucleophilic displacement. They attempted to utilize transglycosylation in elucidating the mechanism of lysozyme-catalyzed hydrolysis of β-1,4-*N*-acetylglucosaminide linkages. Oligosaccharides of *N*-acetylglucosamine were incubated with *p*-nitrophenyl β-*N*-acetylglucosaminide, *p*-nitrophenyl β-glucoside or *p*-nitrophenyl β-2-deoxyglucopyranoside in the presence of lysozyme, and β-anomers of *p*-nitrophenyl glycosides attached to split oligosaccharides of *N*-acetylglucosamine were enzymatically synthesized. For example, *p*-nitrophenyl β-G–(NAG)$_3$ and *p*-nitrophenyl β-2-deoxyglucose–(NAG)$_3$, etc. were isolated from the incubation mixture. These glucosides and glucosaminides released *p*-nitrophenol on catalytic hydrolysis with lysozyme. *p*-Nitrophenyl glycosides were not hydrolyzed by lysozyme unless oligosaccharides of *N*-acetylglucosamine were present in the solution. Thus, *p*-nitrophenyl β-*N*-acetylglucosaminide can be hydrolyzed through transglycosylation accompanied by cleavage of the β-1,4-*N*-acetylglucosaminide linkage (catalyzed by lysozyme), releasing *p*-nitrophenol which can be readily detected by spectrophotometry. The rate of hydrolysis of *p*-nitrophenyl β-glycosides of *N*-acetylglucosamine, D-glucose and 2-deoxy-D-glucose by lysozyme showed a ratio of 2 : 1 : 16. This indicates that for the hydrolysis of the β-1,4-glycoside linkage, the 2-acetamide group need not exist in the sugar residue which goes to the reducing end. However, it does not necessarily mean that for the oligo-

(a) (34)

Glu–35 ... Asp–52 → α-anomer

(b)

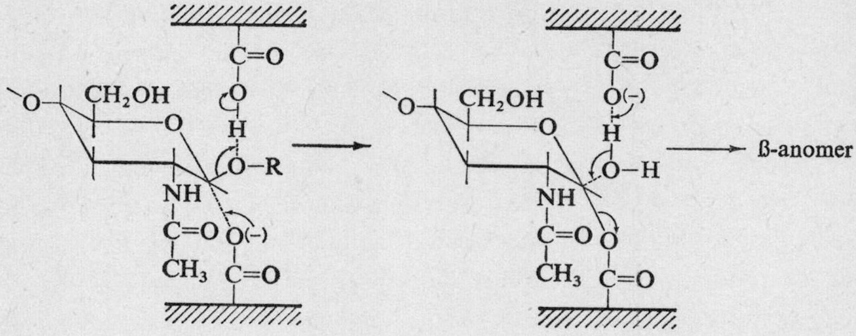

→ ß-anomer

(c)

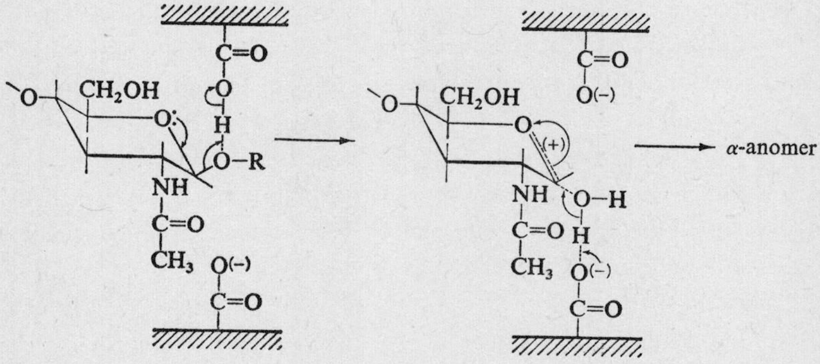

→ α-anomer

(d)

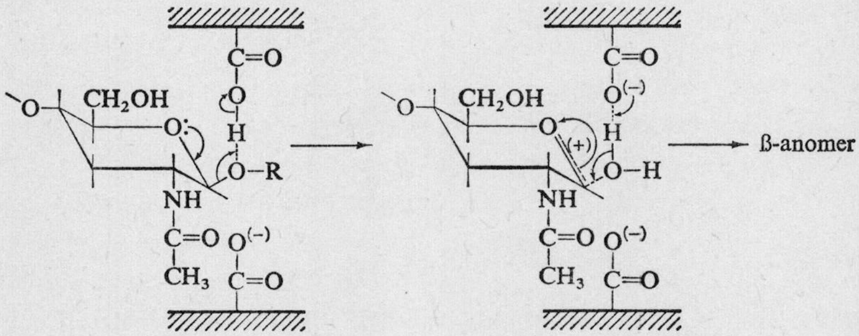

→ ß-anomer

(e)

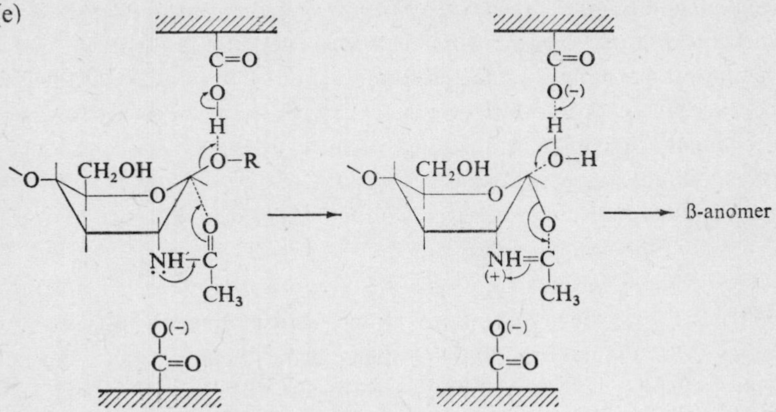

$\rightarrow$ ß-anomer

(f)

$\rightarrow$ ß-anomer

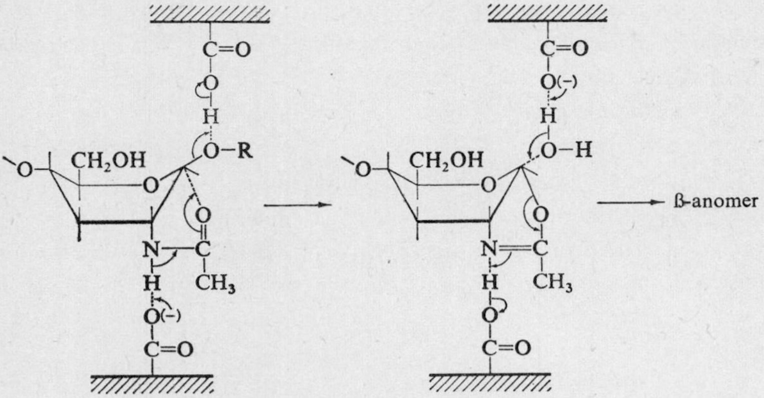

saccharides of *N*-acetylglucosamine, the 2-acetamide group does not accelerate the rate of hydrolysis. Raftery and Rand-Meir[152] discussed all the possible mechanisms on the basis of their experimental results. The mechanisms are illustrated is Eq. (34).

The groups —COOH (above the substrate) and —COO$^-$ (below the substrate) represent the carboxyl group of Glu-35 and the carboxylate anion of Asp-52, respectively. The mechanism (a) may be precluded because this mechanism gives α-anomer as the product. As described above, there is retention of anomeric structure in the hydrolysis of *N*-

acetylglucosaminide linkage by lysozyme. Mechanism (c) can also be ruled out for the same reason. Mechanism (e), which includes simple anchimeric assistance by the 2-acetamide group is not indispensable for the catalytic hydrolysis of glycoside linkage by lysozyme. However, mechanism (f) cannot be neglected because, as Piszkiewicz and Bruice have pointed out, general base catalysis by the carboxylate anion of Asp-52 toward the anchimeric assistance of the 2-acetamide group seems to occur in the real hydrolysis.

From the above considerations, it is obvious that the mechanism for hydrolysis by lysozyme should contain a carbonium ion intermediate giving the β-anomer, or a double displacement-like process. The mechanisms satisfying these conditions are (b) and (d). In the former, the carbonium ion is stabilized by formation of a pyranoxonium ion and by interaction with the β-carboxylate anion of Asp-52 in the form of an ion pair. Since the carbonium ion forms a glucosaminyl compound with β-carboxylate anion of Asp-52 in mechanism (d), this should be excluded as a possibility. Thus, it is presumed that (b) is the most probable of the possible mechanisms.

The 2-acetamide group of the sugar residue at subsite D is not indispensable for hydrolysis of the substrate. However, deacetylation of chitin reduced its digestibility by lysozyme, though the ability to form the lysozyme-substrate complex remained unchanged up to 50% deacetylation. The decrease in digestibility may be caused by Coulombic interaction between the protonated amino group at the C_2 position of

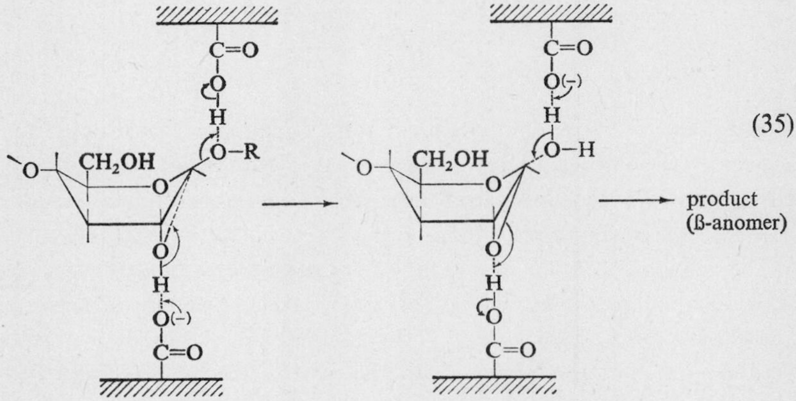

(35)

product (ß-anomer)

the residue at subsite D and the β-carboxylate anion of Asp-52.[147b] It is likely that Raftery's mechanism (b) may contain some ambiguity because it does not explain the exact role of the β-carboxylate anion of Asp-52. For the hydrolysis of *p*-nitrophenyl β-D-glucopyranoside, the following pathway (Eq. (35)) was proposed.

E. Hydrolysis of Oligosaccharide and Transglycosylation

Rupley and Gates[218] and Rupley[47] studied in detail transglycosylation between oligosaccharide of *N*-acetylglucosamine and ^{14}C-labeled *N*-acetylglucosamine in the presence of lysozyme. They found that radioactive trisaccharide, tetrasaccharide, etc. were formed in a mixture containing 0.4 M *N,N'*-diacetylchitobiose, 0.65 M radioactive *N*-acetylglucosamine and 3.6% lysozyme at pH 5.0 incubated for 20 days at 40°C. This result clearly indicates that the oligosaccharides were formed by transglycosylation. Summarized modes of transglycosylation with various oligosaccharides are shown in Table 23.[218] The more detailed mechanism with tetrasaccharide and *N*-acetylglucosamine is shown in

TABLE 23
Relative Rates of Hydrolysis and Transglycosylation for Various Oligosaccharides[218]

Substrate	Transglycosylation		Hydrolysis	
	bond	relative rate	bond	relative rate
NAG—NAG—NAG (bonds 1, 2)	1	1		
	2	0.2		
NAG—NAG—NAG—NAG (bonds 1, 2, 3)	1	1	1+3	1
	2	0.3		
	3	1.6	2	0.2
NAG—NAG—NAG—NAG—NAG (bonds 1, 2, 3, 4)	1	0.1[†1] 1.9[†2]		
	2	0.1 0.3	1+4	1
	3	0.8 2.2		
	4	1 1	2+3	0.5
NAG—NAG—NAG—NAG—NAG—NAG (bonds 1, 2, 3, 4, 5)	1	0.02		
	2	0	1+5	—
	3	0.01	2+4	1
	4	1	3	—
	5	0		

†1 Substrate concentration, 10^{-3} M.
†2 Substrate concentration, 10^{-2} M.

TABLE 24

Transglycosylation with Tetrasaccharide and *N*-Acetylglucosamine[47]

$$\mathrm{NAG_1{-}NAG_2{-}NAG_3{-}NAG_4 + NAG^*}$$
$$\rightarrow \begin{cases} \mathrm{NAG_1{-}NAG^* + NAG_2{-}NAG_3{-}NAG_4} \quad \ldots\ldots \text{(A)} \\ \mathrm{NAG_1{-}NAG_2{-}NAG^* + NAG_3{-}NAG_4} \quad \ldots\ldots \text{(B)} \\ \mathrm{NAG_1{-}NAG_2{-}NAG_3{-}NAG^* + NAG_4} \quad \ldots\ldots \text{(C)} \end{cases}$$

Table 24.[47] Radioactive *N*-acetylglucosamine always appeared in the reducing end. The reactions (A), (B) and (C) took place at the rate ratio of 1 : 0.3 : 1.6.

The mode of transglycosylation with pentasaccharide and hexasaccharide of *N*-acetylglucosamine depended upon the concentration of incubation mixture. In particular, abnormal relative rates were observed in the case of the pentasaccharide. When 10^{-3} M pentasaccharide was used, the relative rates of transglycosylation at bonds 1, 2, 3 and 4 showed the ratio of 0.1 : 0.1 : 0.8 : 1.0. For 10^{-2} M pentasaccharide, the ratio changed to 1.9 : 0.3 : 2.2 : 1.0. This change seemed to be caused by the formation of a 1 : 2 type of lysozyme-substrate complex at higher concentrations of substrate. The enzyme may bind two molecules of the substrate at the cleft to form the complex. As a result, the probability of bond 1 existing in the region between subsites D and E might increase at a high pentasaccharide concentration, bringing about a remarkable increase in the rate of hydrolysis and transglycosylation at bond 1.

The position of the cleaved bond and the rate of hydrolysis under the conditions where 1 : 1 type complex was formed are summarized in Table 25.[47]

Six *N*-acetylglycosamine residues can bind in the cleft of the lysozyme molecule. Since it is probable that the hexasaccharide just occupies all the subsites in the cleft, bond 4 is assumed to be in the region between subsites D and E. Therefore, bond 4 might be rapidly cleaved, accompanied by concomitant transglycosylation as well as the hydrolysis. A similar explanation can be applied in the case of bond 4 of the pentasaccharide. One problem remaining is that concerning the cleavage of bond 3 of the pentasaccharide, and the positions of hydrolyzable bonds of the tetrasaccharide and the trisaccharide. When the 1 : 1 type of complex is formed, the tetrasaccharide and trisaccharide may bind to the left of the lysozyme molecule without crossing subsites D and E.

TABLE 25
Position of Bond Cleaved and Rate of Hydrolysis with
1:1 Type Complex Formation[47]

Position of bond cleaved	Relative rate	Reaction order
NAG$_1$–NAG$_2$	0.003	1
NAG$_1$ ↓ NAG$_2$ ⇣ NAG$_3$	1	1
NAG$_1$ ↓ NAG$_2$ ⇣ NAG$_3$ ↓ NAG$_4$	8	1
NAG$_1$–NAG$_2$–NAG$_3$ ↓ NAG$_4$ ↓ NAG$_5$	4×10^3	0
NAG$_1$–NAG$_2$–NAG$_3$–NAG$_4$ ↓ NAG$_5$–NAG$_6$	3×10^4	0

This type of complex is a "nonproductive complex". In fact, these oligosaccharides can be hydrolyzed, though the rate of hydrolysis is very low. When the formation of 1 : 2 type of complexes is assumed, some of these problems may be solved rationally. In the case of the trisaccharide, one molecule of the substrate may occupy subsites A, B and C, and the other molecule may occupy subsites D, E and F. Consequently, bond 1 of the second molecule will be split catalytically because it is located in the position between subsites D and E. This is also the case for bond cleavage in the tetrasaccharide molecule.

However, the hydrolytic cleavage of bond 2 of the trisaccharide, bonds 2 and 3 of the tetrasaccharide, and bond 3 of the pentasaccharide cannot be explained by assuming the formation of 1 : 2 type complexes. They may be explained by assuming that the various oligosaccharides are able to bind at various positions in the cleft of the lysozyme molecule in solution.

Pollock *et al.*[201] studied transglycosylation using tritium-labeled cell-wall tetrasaccharide, ^{3}H–NAG$_1$–NAM$_2$–NAG$_3$–NAM$_4$. This oligosaccharide at 7.5×10^{-3} M was incubated at pH 5.25, 37°C with 0.025% lysozyme and two kinds of acceptors at the same concentration. The products were fractionated by paper chromatography (electrophoresis). In this experiment, it was found that the cell-wall tetrasaccharide was hydrolyzed into disaccharide, NAG$_1$–NAM$_2$ and NAG$_3$–NAM$_4$, and acceptors were transferred to the NAG$_1$–NAM$_2$ fragment (Table 26).[201] The data indicate that *N*-acetylglucosamine and disaccharides are much better acceptors in the transglycosylation than glucose and monosac-

TABLE 26

Rates of Transglycosylation using Cell-wall Tetrasaccharide[201]

Acceptor 1	Acceptor 2	Concentration (M)	Ratio of rates
NAG–NAG	NAG	1.5×10^{-3}	2:1
G–G	G	7.5×10^{-2}	2.8:1
NAG–NAG	G	3.75×10^{-2}	9:1
G–G	NAG	3.5×10^{-2}	0.67:1
NAG	G	2.5×10^{-2}	4.5:1
G	X	1.25×10^{-1}	1:1

charide. This may mean that the 2-acetamide group enhances the affinity of the sugar residue for subsite E of the cleft. Since the ratio of the ability of glucose as an acceptor to that of xylose was 1:1, it is obvious that there is no effect of the $-CH_2OH$ group at the C-5 position on the binding of the sugar residue at subsite E. It was found that the ability of *N*-acetylglucosamine as acceptor was 4.5 times greater than that of glucose. This result was further confirmed by a comparison of both *N,N'*-diacetylchitobiose–*N*-acetylglucosamine systems and *N,N'*-diacetylchitobiose–glucose systems, where the two acceptors were added to the same incubation mixture and the efficiency of both sugars as acceptors compared.

Recently, Rupley *et al.*[219] studied transglycosylation between oligosaccharides and various alcohols, and concluded that Glu-35 acts as a general acid catalyst in the hydrolysis of β-1,4-*N*-acetylglucosaminide linkages. Hayashi *et al.* (*unpublished*) measured the amount of released *p*-nitrophenol in incubation mixtures of lysozyme, glycol chitin and *p*-nitrophenol β-*N*-acetylglucosaminide, in comparison with the increase in reducing power. The presence of *p*-nitrophenyl β-*N*-acetylglucosaminide to some extent inhibited the increase in reducing power resulting from hydrolysis of the substrate, glycol chitin; the increase ceased after about 6 hr (see Fig. 58). In contrast to the change in reducing power, the release of *p*-nitrophenol proceeded linearly for a long time; for instance, after 48 hr the rate still remained virtually unchanged. The release of *p*-nitrophenol by hydrolysis of the bond between the *N*-acetylglucosamine residue and the *p*-nitrophenol moiety was also due to the presence of oligosaccharide that was isolated from the complete digest of glycol chitin by lysozyme. Thus, in the initial stages of the hydrolysis of glycol

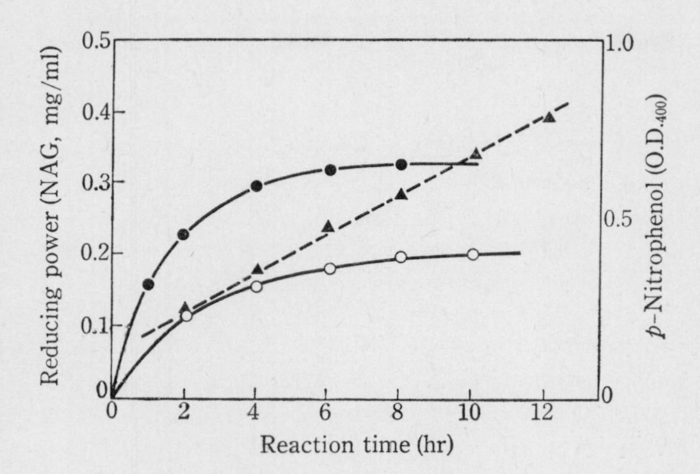

Fig. 58. Relation between reducing power and the release of *p*-nitrophenol at pH 5.5 and 40 °C.

	Lysozyme (%)	Glycol chitin (%)	NP-β-NAG (%)
●, reducing power	0.02	0.50	0
○, reducing power	0.02	0.50	0.25
▲, *p*-nitrophenol	0.28	0.11	0.50

chitin by lysozyme, the cleavage of β-1,4-*N*-acetylglucosaminide linkage is accompanied by transglycosylation at low efficiency. At the stage where no increase in reducing power was observed, the cleavage of *N*-acetylglucosaminide linkage seems to result in transglycosylation with almost 100% efficiency. A more detailed analysis of this type of phenomenon will surely offer further clues for understanding the precise mechanisms involved in lysozyme-catalyzed hydrolysis.

F. Problems Related to the Stabilization of Carbonium Ion Intermediates

At the present time, the most important problem to be solved concerning the mechanism of lysozyme action appears to be that of the stabilization of the carbonium ion intermediate. Possible structures for the carbonium ion intermediate are as follows.

(1) Pyranoxonium ion (half-chair) conformation)

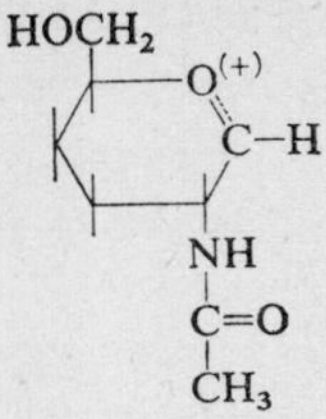

(2) Protonated oxazoline

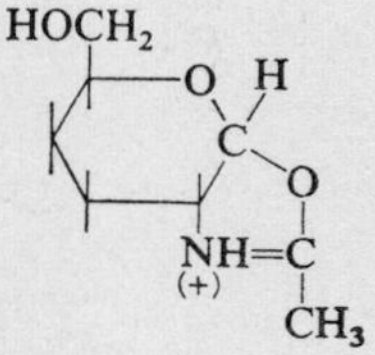

(3) Oxazoline

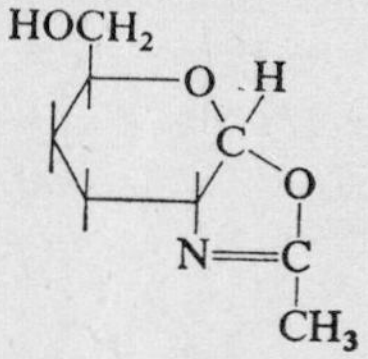

(4) Glucosaminyl enzyme

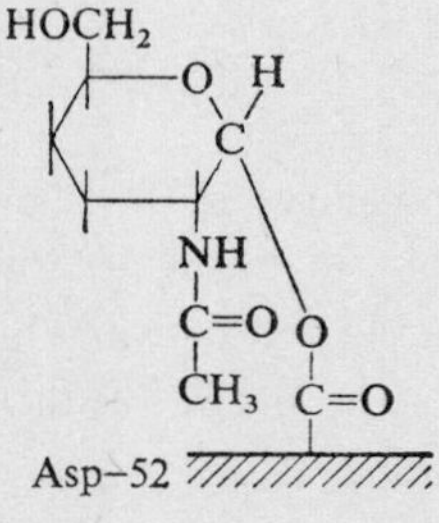

Asp–52

(5)

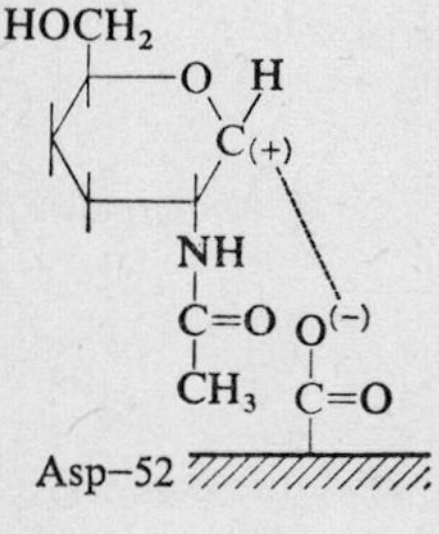

Asp–52

(6)

(hydrophobic region)

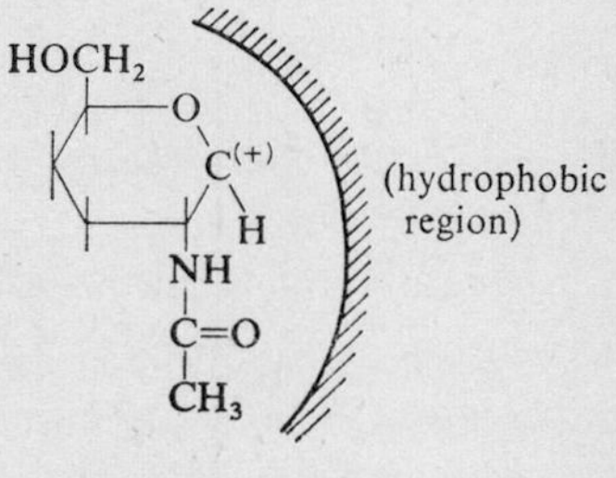

Considerable, detailed criticism has been made of some of these possible structures and some have been precluded by various observations. However, it is still necessary to obtain direct and exact evidence for judging the real structure of the intermediate in the lysozyme-catalyzed reaction.

The major part of the mechanism of lysozyme-catalyzed reactions has been clearly elucidated. The remaining problems, such as the real structure of the intermediate and more detailed characterization of the catalytic groups of the lysozyme molecule, may not be solved by physicochemical or by organic-chemical techniques alone. It is clear that only well-designed and integrated physical/organic-chemical studies will yield information suitable for elucidating the mechanism of enzymatic reactions.

References

1. E. P. Abraham and R. Robinson, *Nature*, **140**, 24 (1937).
2. L. R. Berger and R. S. Weiser, *Biochim. Biophys. Acta*, **26**, 517 (1957).
3. M. R. J. Salton and J. M. Ghuysen, *ibid.*, **36**, 552 (1959).
4. J. Jolles, J. Jauregui-Adell, I. Bernier and P. Jolles, *ibid.*, **78**, 668 (1963).
5. R. E. Canfield, *J. Biol. Chem.*, **238**, 2698 (1963).
6. C. C. F. Blake, G. A. Mair, A. C. T. North, D. C. Phillips and V. R. Sarma, *Proc. Roy. Soc.* (*London*), **B167**, 365 (1967).
7. C. C. F. Blake, L. N. Johnson, G. A. Mair, A. C. T. North, D. C. Phillips and V. R. Sarma, *ibid.*, **B167**, 378 (1967).
8. M. Imanishi, S. Shinka, N. Miyagawa and T. Amano, *Biken's J.*, **9**, 107 (1966).
9. J. Jolles, J. Hermann, B. Nieman and P. Jolles, *Eur. J. Biochem.*, **1**, 334 (1967).
10. A. C. Dianoux and P. Jolles, *Biochim. Biophys. Acta*, **133**, 472 (1967).
11. R. E. Canfield and S. McMurry, *Biochem. Biophys. Res. Commun.*, **26**, 38 (1967).
12. M. Kaneda, T. Kato, N. Tominaga, K. Chitani and K. Narita, *J. Biochem.* (*Tokyo*), **66**, 747 (1969).
13. N. Arnheim, E. M. Prager and A. C. Wilson, *J. Biol. Chem.*, **244**, 2085 (1969).
14. J. N. LaRue and J. C. Speck Jr., *Federation Proc.*, **28**, 662 (1969).
15. J. Hermann and J. Jolles, *Biochim. Biophys. Acta*, **200**, 178 (1970).
16. R. E. Canfield, *Brookhaven Symp. Biol.*, **21**, 136 (1969).
17. J. Jolles and P. Jolles, *Helv. Chim. Acta*, **52**, 2671 (1969).
18. D. C. Phillips, *Sci. Am.*, **215**, 78 (1966).

19. N. Greenfield, B. Davidson and G. D. Fasman, *Biochemistry*, **6**, 1630 (1967).
20. N. Greenfield and G. D. Fasman, *ibid.*, **8**, 4108 (1969).
21. B. Jirgensons, *J. Biol. Chem.*, **238**, 2716 (1963).
22. Y. Tomimatsu and W. Gaffield, *Biopolymers*, **3**, 509 (1965).
23. K. Ikeda, K. Hamaguchi, M. Imanishi and T. Amano, *J. Biochem.* (*Tokyo*), **62**, 315 (1967).
24. K. Hamaguchi and A. Kurono, *ibid.*, **54**, 111 (1963).
25. T. T. Herskovitz and L. Mescanti, *J. Biol. Chem.*, **240**, 639 (1965).
26. A. N. Glazer and N. S. Simmons, *J. Am. Chem. Soc.*, **87**, 2287 (1965).
27. N. Greenfield, B. Davidson and G. D. Fasman, *Biochemistry*, **6**, 1630 (1967).
28. M. J. Kronman, *Biochem. Biophys. Res. Commun.*, **33**, 535 (1968).
29. K. Hamaguchi, *J. Biochem.* (*Tokyo*), **56**, 441 (1964).
30. T. Miyazawa, *Polyamino Acids, Polypeptides and Proteins* (ed. M. A. Stahmann), p. 201, University of Wisconsin Press, 1962.
31. T. Miyazawa, K. Fukushima, S. Sugano and Y. Masuda, *Conformation of Biopolymers* (ed. G. N. Ramachandran), vol. 2, p. 557, Academic Press, 1967.
32. K. Hamaguchi and K. Imahori, *J. Biochem.* (*Tokyo*), **55**, 388 (1964).
33. C. Tanford and M. L. Wagner, *J. Am. Chem. Soc.*, **76**, 3331 (1954).
34. J. W. Donovan, M. Laskowski Jr. and H. A. Scheraga, *J. Mol. Biol.*, **1**, 293 (1959).
35. J. W. Donovan, M. Laskowski Jr. and H. A. Scheraga, *J. Am. Chem. Soc.*, **82**, 2154 (1960).
36. R. Sakakibara and K. Hamaguchi, *J. Biochem.* (*Tokyo*), **64**, 613 (1968).
37. A. J. Sophianopoulos and E. A. Sasse, *J. Biol. Chem.*, **240**, PC1864 (1965).
38. C. Tanford, *Advan. Protein Chem.*, **17**, 69 (1962).
39. A. J. Sophianopoulos, C. K. Rhodes, D. N. Holcomb and K. E. Van Holde, *J. Biol. Chem.*, **237**, 1107 (1962).
40. K. Ogasahara and K. Hamaguchi, *J. Biochem.* (*Tokyo*), **61**, 199 (1967).
41. J. W. Donovan, M. Laskowski Jr. and H. A. Scheraga, *J. Am. Chem. Soc.*, **83**, 2686 (1961).
42. C. Fromageot and G. Schneck, *Biochim. Biophys. Acta*, **6**, 113 (1950).
43. Y. Inada, *J. Biochem.* (*Tokyo*), **49**, 217 (1961).
44. T. Tojo, K. Hamaguchi, M. Imanishi and T. Amano, *ibid.*, **60**, 538 (1966).
45. K. Hayashi, T. Shimoda, K. Yamada, A. Kumai and M. Funatsu, *ibid.*, **64**, 239 (1968).
46. K. Hayashi, T. Shimoda, T. Imoto and M. Funatsu, *ibid.*, **64**, 365 (1968).
47. J. A. Rupley, *Proc. Roy. Soc.* (*London*), **B167**, 416 (1967).
48. S. S. Lehrer and G. D. Fasman, *Biochem. Biophys. Res. Commun.*, **23**, 133 (1966).

49. M. A. Raftery, F. W. Dahlquist, S. I. Chan and S. M. Parsons, *J. Biol. Chem.*, **243**, 4175 (1968).
50. S. S. Lehrer and G. D. Fasman, *ibid.*, **242**, 4644 (1967).
51. J. A. Rupley, L. Butler, M. Gerring, F. J. Hartdegen and R. Pecararo, *Proc. Natl. Acad. Sci. U.S.*, **57**, 1088 (1967).
52. F. W. Dahlquist, L. Lao and M. A. Raftery, *ibid.*, **56**, 26 (1966).
53. F. W. Dahlquist and M. A. Raftery, *Biochemistry*, **8**, 713 (1969).
54. A. J. Sophianopoulos and B. J. Weiss, *ibid.*, **3**, 1920 (1964).
55. K. C. Aune and C. Tanford, *ibid.*, **8**, 4579 (1969).
56. D. G. Hoare and D. E. Koshland Jr., *J. Am. Chem. Soc.*, **88**, 2057 (1966).
57. H. Horinishi, K. Nakaya, A. Tani and K. Shibata, *J. Biochem.* (*Tokyo*), **63**, 41 (1968).
58. A. Matsushima, K. Sakurai, M. Nomoto, Y. Inada and K. Shibata, *ibid.*, **61**, 328 (1967).
59. A. Matsushima, K. Sakurai, M. Nomoto, Y. Inada and K. Shibata, *ibid.*, **64**, 507 (1968).
60. K. Nakaya, H. Horinishi and K. Shibata, *ibid.*, **61**, 337 (1967).
61. K. Nakaya, H. Horinishi and K. Shibata, *ibid.*, **61**, 345 (1967).
62. D. J. Herzig, A. W. Rees and R. A. Day. *Biopolymers*, **2**, 349 (1964).
63. C. B. Hiremath and R. A. Day, *J. Am. Chem. Soc.*, **86**, 5027 (1964).
64. P. S. Marfey, M. Uziel and J. Little, *J. Biol. Chem.*, **240**, 3270 (1965).
65. D. H. Meadows, J. L. Markley, J. S. Cohen and O. Jardetzky. *Proc. Natl. Acad. Sci. U.S.*, **58**, 1307 (1967).
66. H. Horinishi, Y. Hachimori, K. Kurihara and K. Shibata, *Biochim. Biophys. Acta*, **86**, 477 (1964).
67. K. Kurihara, H. Horinishi and K. Shibata, *ibid.*, **74**, 678 (1963).
68. I. Covelli and J. Wolff, *Biochemistry*, **5**, 860 (1966).
69. J. Wolff and I. Covelli, *ibid.*, **5**, 867 (1966).
70. M. Z. Atassi and A. F. S. A. Habeeb, *ibid.*, **8**, 1385 (1969).
71. C. C. F. Blake, *New Scientist*, **29**, 333 (1966).
72. K. Ikeda and K. Hamaguchi, *J. Biochem.* (*Tokyo*), **66**, 513 (1969).
73. K. Hamaguchi, *ibid.*, **55**, 333 (1964).
74. K. Hamaguchi and A. Kurono, *ibid.*, **54**, 497 (1963).
75. K. Hamaguchi, A. Kurono and S. Goto, *ibid.*, **54**, 259 (1963).
76. K. Hamaguchi, K. Hayashi, T. Imoto and M. Funatsu, *ibid.*, **55**, 24 (1964).
77. A. Kurono and K. Hamaguchi, *ibid.*, **56**, 432 (1964).
78. C. C. Bigelow and I. I. Geschwind, *Compt. Rend. Trav. Lab. Carlsberg*, **31**, 283 (1960).
79. C. C. Bigelow and I. I. Geschwind, *ibid.*, **31**, 305 (1960).
80. M. J. Kronman, L. Cerankowski and L. G. Holmes, *Biochemistry*, **4**, 518 (1965).
81. J. W. Donovan, *ibid.*, **3**, 67 (1964).
82. H. L. Oppenheimer, J. Mercouroff and G. P. Hess, *Biochim. Biophys.*

Acta, **71**, 78 (1963).
83. C. J. Martin and G. M. Bhatnagar, *Biochemistry*, **5**, 1230 (1966).
84. E. J. Williams, T. T. Herskovits and M. Laskowski Jr., *J. Biol. Chem.*, **240**, 3574 (1965).
85. E. J. Williams and M. Laskowski Jr., *ibid.*, **240**, 3580 (1965).
86. K. Hayashi, T. Imoto and M. Funatsu, *J. Biochem.* (*Tokyo*), **54**, 381 (1963).
87. K. Hayashi, T. Imoto, G. Funatsu and M. Funatsu, *ibid.*, **58**, 227 (1965).
88. M. J. Kronman, F. M. Robbins and R. E. Andreotti, *Biochim. Biophys. Acta*, **143**, 462 (1967).
89. Y. Hachimori, H. Horinishi, K. Kurihara and K. Shibata, *ibid.*, **93**, 346 (1964).
90. F. J. Hartdegen and J. A. Rupley, *J. Am. Chem. Soc.*, **89**, 1743 (1967).
91. C. C. F. Blake, *Proc. Roy. Soc.* (*London*), **B167**, 435 (1967).
92. A. Previero, M. A. Coletti-Previero and P. Jolles, *J. Mol. Biol.*, **24**, 261 (1961).
93. T. A. Bewley and C. H. Li, *Nature*, **206**, 264 (1965).
94. A. N. Glazer and N. S. Simmons, *J. Am. Chem. Soc.*, **88**, 2335 (1966).
95. K. Ikeda and K. Hamaguchi, *J. Biochem.* (*Tokyo*), **66**, 513 (1969).
96. J. E. Churchich, *Biochim. Biophys. Acta*, **92**, 194 (1964).
97. J. E. Churchich, *ibid.*, **120**, 406 (1966).
98. G. Jori, G. Galiazzo, A. Marzotto and E. Scoffone, *J. Biol. Chem.*, **243**, 4272 (1968).
99. K. Hamaguchi, *Bull. Chem. Soc. Japan*, **31**, 123 (1958).
100. R. Cecil and R. G. Wake, *Biochem. J.*, **82**, 401 (1962).
101. P. Azari, *Arch. Biochem. Biophys.*, **115**, 230 (1966).
102. T. A. Bewley and C. H. Li, *Intern. J. Protein Res.*, **1**, 117 (1969).
103. K. Hamaguchi, *Nippon Kagaku Zasshi* (Japanese), **87**, 893 (1966).
104. S. Beychok, *Proc. Natl. Acad. Sci. U.S.*, **53**, 99 (1965).
105. J. R. Colvin, *Can. J. Chem.*, **30**, 831 (1952).
106. L. R. Wetter and H. F. Deutsch., *J. Biol. Chem.*, **192**, 237 (1951).
107. V. Luzzati, J. Witz and A. Nicholaieff, *J. Mol. Biol.*, **3**, 367 (1961).
108. L. K. Steinrauf, *Acta Cryst.*, **12**, 77 (1959).
109. A. N. Glazer, *Australian J. Chem.*, **12**, 304 (1959).
110. P. A. Charlwood, *J. Am. Chem. Soc.*, **79**, 776 (1967).
111. T. L. McMeekin and K. Marshall, *Science*, **116**, 142 (1952).
112. C. C. F. Blake, D. F. Koenig, G. A. Mair, A. C. T. North, D. C. Phillips and V. R. Sarma, *Nature*, **206**, 757 (1965).
113. G. Alderton, W. H. Ward and H. L. Fevold, *J. Biol. Chem.*, **157**, 43 (1945).
114. B. Jirgensons, *Arch. Biochem. Biophys.*, **39**, 261 (1952); **41**, 33 (1952).
115. J. Leonis, *ibid.*, **65**, 182 (1956).
116. M. Halwer, G. C. Nutting and B. A. Prince, *J. Am. Chem. Soc.*, **73**, 2786 (1951).

117. R. F. Steiner, *Arch. Biochem. Biophys.*, **47**, 56 (1953).
118. K. Imai, T. Takagi and T. Isemura, *J. Biochem.* (*Tokyo*), **53**, 1 (1963).
119. A. J. Sophianopoulos and K. E. Van Holde, *J. Biol. Chem.*, **236**, PC82 (1961).
120. A. J. Sophianopoulos and K. E. Van Holde, *ibid.*, **239**, 2516 (1964).
121. E. R. Bruzzesi, E. Chiancone and E. A. Antonini, *Biochemistry*, **4**, 1796 (1965).
122. A. J. Sophianopoulos, *J. Biol. Chem.*, **244**, 3188 (1966).
123. L. G. Butler and J. A. Rupley, *ibid.*, **242**, 1077 (1967).
124. U. Zehavi and A. Lustig, *Biochim. Biophys. Acta*, **194**, 532 (1969).
125. K. Hamaguchi and H. Sakai, *J. Biochem.* (*Tokyo*), **57**, 721 (1965).
126. K. C. Aune, A. Salahuddin, M. H. Zarlengo and C. Tanford, *J. Biol. Chem.*, **242**, 4486 (1967).
127. K. Hamaguchi and K. Rokkaku, *J. Biochem.* (*Tokyo*), **48**, 358 (1960).
128. K. Tanaka and K. Hamaguchi, *unpublished.*
129. C. Tanford, *Advan. Protein Chem.*, **23**, 122 (1968).
130. C. Tanford, K. Kawahara, S. Lapanje, T. M. Hooker Jr., M. H. Zarlengo, A. Salahuddin, K. C. Aune and T. Takagi, *J. Am. Chem. Soc.*, **89**, 5023 (1967).
131. C. Tanford, R. H. Pain and N. S. Otchin, *J. Mol. Biol.*, **15**, 489 (1966).
132. K. C. Aune and C. Tanford, *Biochemistry*, **8**, 4586 (1969).
133. C. Tanford and K. C. Aune, *ibid.*, **9**, 206 (1970).
134. N. Shimaki, K. Ikeda and K. Hamaguchi, *J. Biochem.* (*Tokyo*), in press.
135. V. S. Ananthanarayanan and C. C. Bigelow, *Biochemistry*, **8**, 3717 (1969).
136. V. S. Ananthanarayanan and C. C. Bigelow, *ibid.*, **8**, 3723 (1969).
137. K. Hamaguchi, *J. Biochem.* (*Tokyo*), **42**, 449 (1955).
138. K. Hamaguchi, *ibid.*, **42**, 705 (1955).
139. K. Hamaguchi, *ibid.*, **43**, 83 (1956).
140. K. Hamaguchi, *ibid.*, **43**, 355 (1956).
141. T. Yamashita and H. B. Bull, *J. Colloid Interface Sci.*, **24**, 310 (1967).
142. R. Repaske, *Biochim. Biophys. Acta*, **22**, 189 (1956).
143. K. Meyer, R. Thompson, J. W. Palmer and D. Khorazo, *J. Biol. Chem.*. **113**, 303 (1936).
144. M. R. J. Salton, *Nature*, **170**, 746 (1952).
145. M. R. J. Salton and R. W. Horme, *Biochim. Biophys. Acta*, **7**, 177 (1951).
146. K. Hamaguchi and M. Funatsu, *J. Biochem.* (*Tokyo*), **46** 1659 (1959).
147. *a*: Y. Matsushima, S. Hara, T. Miyahara and Y. Unemura, *Seikagaku* (Japanese), **37**, 616 (1965); *b*: K. Hayashi, N. Fujimoto, M. Kugimiya and M. Funatsu, *J. Biochem.* (*Tokyo*), **65**, 401 (1969).
148. P. Jolles, *Angew. Chem. Intern. Ed. Engl.*, **3**, 28 (1964).
149. M. R. J. Salton, *Ann. Rev. Biochem.*, **34**, 143 (1965).
150. M. Leyh-Bouille, J. M. Ghuysen, D. J. Tipper and J. L. Strominger,

Biochemistry, **5**, 3079 (1966).
151. T. Osawa, *Carbohydrate Res.*, **1**, 435 (1966).
152. M. A. Raftery and T. Rand-Meir, *Biochemistry*, **7**, 3281 (1968).
153. *a*: D. Shugar, *Biochim. Biophys. Acta*, **8**, 302 (1952); *b*: P. Jolles and J. Fromageot, *ibid.*, **11**, 95 (1953); *c*: M. H. Richmond, *ibid.*, **33**, 78 (1959).
154. T. Viswanatha and W. B. Lawson, *Arch. Biochem. Biophys.*, **93**, 130 (1961).
155. K. Hayashi, N. Yamasaki and M. Funatsu, *Agr. Biol. Chem.* (*Tokyo*), **28**, 517 (1964).
156. A. N. Smolelis and S. E. Hartsell, *J. Bacteriol.*, **63**, 665 (1954).
157. F. H. Wilcox and L. J. Daniel, *Arch. Biochem. Biophys.*, **52**, 305 (1954).
158. N. Yamasaki, K. Hayashi and M. Funatsu, *Agr. Biol. Chem.* (*Tokyo*), **32**, 64 (1968).
159. M. R. J. Salton, *J. Gen. Microbiol.*, **9**, 512 (1953).
160. M. R. J. Salton, *Biochim. Biophys. Acta*, **22**, 495 (1956).
161. R. W. Jeanloz, N. Sharon and H. M. Flowers, *Biochem. Biophys. Res. Commun.*, **12**, 20 (1963).
162. K. Hamaguchi, K. Rokkaku, M. Funatsu and K. Hayashi, *J. Biochem.* (*Tokyo*), **48**, 352 (1960).
163. K. Hayashi, T. Imoto and M. Funatsu, *ibid.*, **54**, 381 (1963).
164. K. Hayashi, M. Kugimiya and M. Funatsu, *ibid.*, **64**, 93 (1968).
165. T. Imoto, Y. Doi, K. Hayashi and M. Funatsu, *ibid.*, **65**, 667 (1969).
166. N. A. Kravchenko, *Proc. Roy. Soc.* (*London*), **B167**, 429 (1967).
167. P. Jolles, *ibid.*, **B167**, 350 (1967).
168. T. Isemura, T. Takagi, Y. Maeda and K. Imai, *Biochem. Biophys. Res. Commun.*, **5**, 373 (1961).
169. F. H. White, *Federation Proc.*, **21**, 233 (1962).
170. N. A. Kravchenko, G. V. Kleopina and E. D. Kaverzneva, *Biochim. Biophys. Acta*, **92**, 412 (1963).
171. J. Jolles, G. Spotorno and P. Jolles, *Nature*, **208**, 1204 (1965).
172. H. Fraenkel-Conrat, *Arch. Biochem. Biophys.*, **27**, 109 (1950).
173. C. Wauters and J. Leonis, *Proc. 2nd Chromatog. Symp.*, p. 125, 1962.
174. N. A. Kravchenko, G. V. Kleopina and E. D. Kaverzneva, *Biochim. Biophys. Acta*, **59**, 507 (1962).
175. N. Yamasaki, K. Hayashi and M. Funatsu, *Agr. Biol. Chem.* (*Tokyo*), **32**, 55 (1968).
176. S. M. Parsons, L. Jao, F. W. Dahlquist, C. L. Borders Jr., T. Groff, J. Racs and M. A. Raftery, *Biochemistry*, **8**, 700 (1969).
177. A. McCoubrey and M. H. Smith, *Biochem. Pharmacol.*, **15**, 1623 (1966).
178. I. I. Geschwind and C. H. Li, *Biochim. Biophys. Acta*, **25**, 171 (1957).
179. J. Jauregui-Adell and P. Jolles, *Bull. Soc. Chim. Biol.*, **46**, 141 (1964).
180. L. Weil, A. R. Buchert and J. Maher, *Arch. Biochem. Biophys.*, **40**, 245 (1952).

181. T. Murachi, T. Miyake and N. Yamasuki, *Seikagaku* (Japanese), **35**, 587 (1963).
182. I. Bernier and P. Jolles, *Compt. Rend.*, **253**, 745 (1961).
183. G. T. S. Rao and L. K. Ramachandran, *Biochim. Biophys. Acta*, **59**, 507 (1962).
184. T. Takahashi, K. Hamaguchi, K. Hayashi, T. Imoto and M. Funatsu, *J. Biochem.* (*Tokyo*), **58**, 385 (1965).
185. F. J. Hartdegen and J. A. Rupley, *Biochim. Biophys. Acta*, **92**, 625 (1964).
186. J. A. Rupley and T. Imoto, in preparation.
187. J. C. Sheehan and G. P. Hess, *J. Am. Chem. Soc.*, **77**, 1067 (1955).
188. D. G. Hoare and D. E. Koshland Jr., *J. Biol. Chem.*, **242**, 2447 (1967).
189. K. Hayashi *et al., unpublished.*
190. T-Y. Lin and D. E. Koshland Jr., *J. Biol. Chem.*, **244**, 505 (1969).
191. S. M. Parsons and M. A. Raftery, *Biochemistry*, **8**, 4199 (1969).
192. M. Shinitzsky, V. Grisaro, D. M. Chipman and N. Sharon, *Arch. Biochem. Biophys.*, **115**, 232 (1966).
193. D. M. Chipman, V. Grisaro and N. Sharon, *J. Biol. Chem.*, **242**, 4388 (1967).
194. K. Hayashi *et al., unpublished.*
195. B. J. Adkins and J. T. Yang, *Biochemistry*, **7**, 266 (1968).
196. E. W. Thomas, *Biochem. Biophys. Res. Commun.*, **24**, 611 (1966).
197. T. Imoto, K. Hayashi and M. Funatsu, *J. Biochem.* (*Tokyo*), **64**, 387 (1968).
198. J. P. Loquet, J. Saint-Blancard and P. Jolles, *Biochim. Biophys. Acta*, **167**, 150 (1968).
199. K. Hayashi, M. Kugimiya, T. Imoto, M. Funatsu and C. C. Bigelow, *Biochemistry*, **7**, 1461 (1968).
200. K. Hayashi, M. Kugimiya, T. Imoto, M. Funatsu and C. C. Bigelow, *ibid.*, **7**, 1467 (1968).
201. J. J. Pollock, D. M. Chipman and N. Sharon, *Biochem. Biophys. Res. Commun.*, **28**, 779 (1967).
202. B. B. Sykes, *Biochemistry*, **8**, 1110 (1969).
203. C. C. McDonald and W. D. Phillips, *J. Am. Chem. Soc.*, **89**, 6332 (1967).
204. H. Sternlicht and D. Wilson, *Biochemistry*, **6**, 2881 (1967).
205. C. Armour, C. A. Bunton, S. Patai, L. H. Selman and C. A. Vernon, *J. Chem. Soc.*, **1961**, 142.
206. C. A. Bunton, D. R. Llewellyn, K. G. Oldham and C. A. Vernon, *ibid.*, **1958**, 3588.
207. C. A. Vernon, *Proc. Roy. Soc.* (*London*), **B167**, 389 (1967).
208. D. E. Koshland Jr., *The Mechanism of Enzyme Action*, p. 608, John Hopkins Press, 1954.
209. D. E. Koshland Jr., *Discussions Faraday Soc.*, **20**, 142 (1956).
210. F. C. Mayer and J. Larner, *J. Am. Chem. Soc.*, **81**, 188 (1959).

211. M. Halpern and J. Leibowitz, *Biochim. Biophys. Acta*, **36**, 29 (1959).
212. F. W. Dahlquist, C. L. Borders Jr., G. Jacobson and M. A. Raftery, *Biochemistry*, **8**, 694 (1969).
213. D. Piszkiewicz and T. C. Bruice, *J. Am. Chem. Soc.*, **89**, 6237 (1967).
214. D. Piszkiewicz and T. C. Bruice, *ibid.*, **90**, 2156 (1968).
215. D. Piszkiewicz and T. C. Bruice, *ibid.*, **90**, 5844 (1968).
216. G. Lowe, *Proc. Roy. Soc.* (*London*), **B167**, 431 (1967).
217. G. Lowe, G. Sheppard, M. L. Sinnott and A. Williams, *Biochem. J.*, **104**, 893 (1967).
218. J. A. Rupley and V. Gates, *Proc. Natl. Acad. Sci. U.S.*, **57**, 496 (1967).
219. J. A. Rupley, V. Gates and R. Bilbrey, *J. Am. Chem. Soc.*, **90**, 5633 (1968).

T2 and T4 Phage Lysozyme and λ Phage Endolysin†

Akira TSUGITA and Yoshiko IKEYA-OCADA*
Laboratory of Molecular Genetics, Medical School, Osaka University
Torii Hall, Nakanoshima, Kita-ku, Osaka, Japan

**Present address: Department of Biology, Medical School, University of North Carolina*
Chapel Hill, North Carolina, U.S.A.

† This paper is dedicated to the late Professor Wendel Stanley, who stimulated the author(A.T.)'s interest in molecular biology, and constantly encouraged and guided his studies.

1. Introduction

When bacteriophages infect a host bacterial cell, replication of phage particles soon follows. After maturation, phage particles are released from the host cell by lysis of the bacterial cell-wall. The initial invasion of the phages into the host cell is not yet fully understood, but the last step involving release of the phages is known to involve, at least partially, the action of phage-induced enzymes.

There is a good deal of genetic information available about both the T-series and λ phages as well as the host *Escherichia coli*, and in addition the lytic activity is readily measurable. Consequently, research on the relationship between genes and their expressing process or product proteins appeared timely, and has been carried out using a T4 phage lysozyme system.

T4 phage lysozyme was selected for illustration of the genetic code because it is an accessible protein of low molecular weight and relatively easy to isolate. The research has been concerned with a comparison of the base sequences in wild-type lysozyme with that of pseudo-wild mutants carrying certain pairs of frameshift mutations in the lysozyme gene. The genetic message is translated sequentially by triplets (frames), starting from a defined point. Deletions or insertions of one or more bases into the gene cause a shift of the reading frame of the message.[1] If deletion of a base is followed by addition of another base, as can result from use of the proflavin mutagen, the reading frame (i.e. the amino acid sequence) may be altered but only in the region of the mutations.

Using codons proposed from *in vitro* studies, we were able to assign a sequence of bases that would code both for the wild type amino acid sequences and for the altered sequence of amino acids in the mutant strains. These codons were also found to be used *in vivo* and, in addition, some of the codons have been assigned.[2–11] These studies have contributed to an understanding of the directions of transcription and translation of the lysozyme gene, the mechanism of frameshift mutations, and the essential and non-essential sequences required for enzyme action.

Nonsense mutants of T4 lysozyme have been isolated. The alterations of amino acids by the suppressive host have provided further information about the essential amino acids in the sequence of T4 phage lysozyme. Studies of nonsense mutants in the presence of adequate host cells which result in insertion of original or altered amino acids, have also supplemented our knowledge of lysozyme production. From these researches it has been observed that a variation of suppression efficiencies which depends on the location of the nonsense code involves the effect of an adjacent codon on the nonsense codon, and this in turn may suggest a valid ending mechanism in protein synthesis, as discussed below.

2. History

Twort (1915)[12] first described a filtrable "lytic agent" which caused lysis of *Staphylococcus*. In 1917, d'Herelle,[13] who reported his independent discovery of the so-called "bacteriophage" and shared with Twort credit for the discovery of phages, carried out much of the basic research on these agents including their lytic activity. He first suggested in 1926[14] that "a lytic enzyme" was secreted by phages, and Bronfenbrenner and Muckenfuss in 1928[15] claimed that "the ferment-like substance responsible for the lysis was different from the bacteriophage" using the same system. Steric, in 1929,[16] and Schuurman, in 1936,[17] both using virulent *coli* phages reported evidence in support of this view, clearly indicating that an enzyme which diffuses from the plaque into the surrounding agar and lyses heat-killed bacteria and also causes a surface change in living bacteria, is produced by the phage-bacteria system. Wollman and Wollman, in 1933 and 1934,[18,19] presented evidence that the enzyme is produced by the phage rather than by the bacterium.

In 1939, Pirie,[20] using a virulent phage WLL on *Bacterium coli* 14, observed an increase in reducing sugars in the medium during lysis of the bacteria. He also isolated a similar enzyme from uninfected *coli* cells which was clearly distinguished by its reactivity to a chemical inhibitor, iodoacetate. He extended this idea[20] by suggesting that an enzyme, not of bacterial origin, was produced during lysis of the bacteria by phages and that it acts on the bacterial cell-wall, and thus increased the reducing sugar concentration in a manner similar to that of egg-white lysozyme.

These extensive studies were sufficient to establish the now current concept of a phage lysozyme. Several phages of *E. coli* were subsequently examined in detail, and considerable information has been accumulated about the T-series phages including useful genetic details.

Barrington and Kozloff[21,22] showed that incubation of several T-series phages with *E. coli* B cell-walls caused the release of as much as 15% of the total nitrogen from the host cell-walls and suggested that the activity was located in the virus tail. There was some confusion however concerning the mechanisms of invasion and lysis in the phage-host interaction. Kozloff and Lute[23] found an ATP-dependent contractile protein in the tail part of the phage which was shown to be essential for successful viral invasion. In this case it would appear that the invasion mechanism of the phage is differentiated from the lysis mechanism of the host cell. Brown[24] found that phage free T6 lysates are able to lyse chloroform-treated bacteria. Koch and Weidel[25] showed that the active part of the T2 phage tail is able to repeat its action on bacterial membranes and suggested that this action is enzymatic in nature, so confirming the results obtained by Barrington and Kozloff.[21,22] Further, Koch and Jordan[26] demonstrated that there were no differences in the composition of the split product released by phage free lysates when compared with that released by intact phages or phage ghosts. Of special importance, these authors also found that the enzymatically active material was a low molecular weight protein and suggested that the host range specificity, in addition to the phage adsorption and invasion mechanism, were not connected to this enzyme action.

The enzyme has now been characterized and shown to be able to split off a component from isolated bacterial cell-walls.[27] Chemical analysis of the material released was made by Koch;[28] it was found to consist of glucosamine, muramic acid, glutamic acid, diaminopimelic acid and alanine, as part of a homogeneous compound of molecular weight about 10,000.[29] Increase in the reducing sugar concentration during the enzyme reaction again confirmed Pirie's original observation. A partial purification of the enzyme was carried out and the preparation shown to lyse *Micrococcus lysodeikticus*, which is commonly used as a substrate for, and also to react competitively with egg-white lysozyme. The lytic enzyme was named phage "lysozyme".[30]

These workers were also able to separate the lysozyme activity from the killing activity of the phage. Substantiated evidence for the specificity of T2 phage lytic enzyme as lysozyme has been obtained by Weidel's group[31–33] in parallel with work involving the purification and crystallization of the T2 enzyme.[34,35]

It had been reported earlier that lysozyme purified from particles (tail) of phage T2 differed from that found free in the lysate, but more exhaustive studies have shown that this is not so.[35] Streisinger (1961) discussed the structure gene of lysozyme in the T4 phage genome, designated by the letter "*e*".

The T4 phage lysozyme is a representative enzyme protein known to be synthesized as a late protein. After phage infection early proteins are first synthesized, followed by late proteins. Thus, lysozyme activity has been employed to study the switching mechanism from early to late protein groups in connection with the synthesis of *m*RNA and the 'trigger' role of RNA polymerase.[36–38] The *m*RNA specific for the lysozyme gene, demonstrable by a hybridization test, was observed in a very early period and during the late period after infection by the phage.[39,40] The very early *m*RNA of the lysozyme gene is not translated to the protein, but the late *m*RNA can be translated into active enzyme.[41,42]

Salser *et al.*[43,44] reported the results of a successful experiment involving *in vitro* synthesis of lysozyme protein under the control of *m*RNA, and this was confirmed by using *m*RNA from the suppressive hosts. This is one of the few cases in which *de novo* synthesis of a native protein with biological activity has been carried out in an *in vitro* system. The above observations in which *m*RNA was used for translation of lysozyme protein have been confirmed. DNA-dependent *in vitro* synthesis was also reported recently by Schweiger and Gold, and together with the Salser group's use of an *m*RNA-dependent system, much general information about protein synthetic pathways has emerged.[45]

It has been known for some time that lytic activity appears during the development of the lysogenic bacteriophage. Such activity cannot be detected in the host cell prior to infection by the phage or in the cell made lysogenic for the phage prior to induction.

Jacob *et al.*[46] observed lytic activity after induction of a λ phage using *E. coli* K12 as host cell and termed this activity "endolysin". They partially purified the enzyme[47,48] and examined the specificity, which still remains ambiguous. The structural gene of this lytic enzyme, *R*, is located near one end of the linear DNA isolated from mature λ phage. The location of *R* as the terminal gene on the right side of the genetic map of the vegetative λ[49,50] has been established by determining the gene content of fragments of λ DNA which vary in size but have in common the right end of the intact λ DNA molecule.[51] A map of the *R* cistron has also been deduced from a set of temperature sensitive and nonsense mutants.[52-54] Black and Hogness[55,56] studied the purification, specificity, molecular weight and preliminary protein structure of this endolysin. We[57] purified the λ lytic enzyme and subjected it to sequence studies. No homology in the primary sequence was found between the λ lytic enzyme and T4 phage lysozyme.[57]

3. Roles of Lysozyme and Other Lytic Agents in the Life Cycle of T4 Phage in the Host Cell

Bacteriophage infection causes lysis of the bacterial cell and in the early days of phage research it was commonly supposed that lysozyme was the only active agent for this lytic process. During the life cycle of the infecting phage, we may consider the lytic action to be divided into two steps: (i) penetration of phage into the host cell, and (ii) lysis of the host cell and multiplication of the daughter phages. The lysozyme plays a role on behalf of the phage only when it lyses the host cells after reproduction of the phages.[58-60] In this section, the two (penetration and lysis) mechanisms will be discussed in the light of current knowledge and will include discussion of some hypothetical processes.

3.1. Host Cell Recognition and Absorbance

The initial step prior to phage interaction with the host cell may be a process of recognition and absorbance. The recognition site is considered to be a lipopolysaccharide component of the bacterial cell-wall.[61,62]

The host cell recognition was found to be dependent on the tail fiber component of the phage particle.[60] The T4 phage particle has six equally shaped tail fiber rods (20 Å × 800 Å),[63] and each fiber appears to consist of two morphologically almost identical pieces (20 Å × 400 Å) which however differ from each other in chemical composition.[64,65] The distal half of the fiber is controlled by genes 35, 36, 37 and 38, and the proximal part by gene 34.[66,67] Gene 57 is one of the tail fiber genes but its role is not yet known. Furthermore the distal part of the tail fiber has been isolated and found to be responsible for the host cell recognition.[65,68] In the distal part the role of gene 35 product and gene 36 product has to some extent become clear and one of the remaining genes, 37 or 38, is now suspected to be responsible for the host recognition.[65]

The components of the tail's short pins which are attached to the base plate were found to be responsible for adsorption followed by fiber recognition.[60] However, a fiberless particle has decreased infectivity, suggesting that a fiber component is possibly involved in this adsorption step.[69] It was also found that urea treated phage in which the base plate is observed to be degraded is still able to infect the spheroplast of the host cell.[71]

Weidel and Kellenberger[62,72] found that the receptors for T2 and T1 phages are lipoproteins which are soluble in 90% phenol, whilst receptors for T3 and T7 as well as for T4 phages are lipopolysaccharide layers in the cell-wall which can be solubilized by treatment with water. It is interesting to note that a lysate of tail fiber deficient phages does not exhibit "lysis inhibition"; in other words, bacteria infected with phage and later superinfected with the same phage, lyse later than bacteria that have been infected only once.

3.2. Penetration of Phage DNA into the Host Cell

A successful infection involves host cell penetration by phage DNA. During this process the tail fibers bend and the "tail sheath" contracts, giving the "base plate" access to the bacterial surface. By this time the tail pin has altered to a "short tail fiber" allowing the phage more access to the bacterial cell-wall. The "tail core" penetrates the cell-wall (about 120 Å thick) but not the cytoplasmic membrane, and injects phage DNA.[60] A contracting enzyme was observed in association with the "tail" of T2 phage[73] which was found to be ambiguous. Mutants defective in genes 11 and 12 are capable of being adsorbed on bacteria but are unable to kill the host presumably because they lack the DNA injection mechanism.[74]

Phage lysozyme has been reported to be associated with phage particles of the related phage T2[30] and it seems plausible that digestion of the bacterial cell-wall with this phage associated lysozyme is a crucial step in the penetration process. However, Emrich and Streisinger,[58] after comparing a lysozyme deficient mutant (*e*) with the wild-type strain, questioned that phage lysozyme is necessary for phage infection. They found that even this mutant phage produces an identical time course for the appearance of intracellular phages in bacteria, suggesting that the lysozyme has no role in the penetration step. A test of "lysis from without"[75] measures the lytic activity from outside the cell. Bacteria infected with a very high multiplicity of phage lyse immediately after infection. This is called "lysis from without". The *e* mutant phage causes "lysis from without" equally as well as the wild-type phage.

The amount of lysozyme associated with the purified wild-type phage particle was calculated as 0.5[58] or 0.01[76] molecule per phage particle. Thus it can be concluded that no lysozyme activity is associated with the *e* mutant phage particle. However, after disruption freezing and thawing of the purified e^+ phage, a "new" lytic activity appears which cannot be eliminated by treatment with anti-phage lysozyme serum. Although this new lytic activity was found to be firmly associated with phage particles, its level is extremely low compared with the level of the wild-type phage lysate. Further studies are required to reveal whether the "new" lytic activity is associated with one of the known structures of the phage, or whether it resembles lysozyme with respect to enzymatic specificity. However, the "new" lytic enzyme seems to act as an essential tool during the invasion process.

3.3. Changes in Host Metabolism and in Cell-wall or Cell-membrane

Puck and Lee[77,78] found that after phage infection the host cell-wall permeability increases over a short period. Just after infection, most of the host dependent protein synthesis is stopped immediately, and the synthesis of "very early" or "early" proteins directed by the phage genome is initiated. By employing a trigger mechanism (due to the phage specific δ factor of RNA polymerase) the late phage proteins appear at a certain time after infection.

Oxygen uptake continues until cell lysis (about 30 min after infection), which is independent of the presence of lysozyme and the cessation of metabolism, has occurred.[79] A bacterium infected with phage, and later superinfected with the same phage, lyses later than a bacterium infected once. This phenomenon is called "lysis inhibition".[80] The reason for lysis inhibition is not known and some mutants such as spackle (*s*)[81] and rapid lysis (*r*)[82] do not exhibit lysis inhibition, whilst others such as the *e* mutant do. During this delay the oxygen metabolism continues normally.[79] A metabolic inhibitor such as KCN induces a premature lysis,[83] and the cessation of metabolism is followed by disappearance of the cytoplasmic membrane and corresponding increase in permeability.

The mutant called spackle (*s*) has been described.[81] The *s* mutant phage does not induce lysis inhibition and furthermore infected bacteria are more sensitive to "lysis from without", when compared with bacteria infected with r_{I} and r_{II} mutants which develop normal resistance to "lysis from without". Dominancy tests showed that s^{+} is always dominant to *s*. These results indicate that the *s* mutant permits lysis of the phage infected host even in the absence of lysozyme and affects a phage directed synthesis of bacterial cell-wall components.[81] Although the role of the *r* gene is still ambiguous, the *r* gene product seems to connect with cell-membrane or cell-wall components, to a certain extent, with rapid cessation of infected cell metabolism and rapid lysis of the host cell.[84] Another gene, the acridine sensitive gene, was also reported to affect the cell permeability.[85] The genetic mapping for these genes which differ from the *e* gene, is shown in Fig. 1. Following infection, the host cell immediately stops its own activity whilst early functions of the phage metabolism, including the *r* function, *s* function and possibly other membrane or wall synthesizing actions, replace the host synthesizing activity and other functions under the direction of the original host. Thus, during that period the phage directed metabolism continues within the closed bacterial cell. At the end of the latent period, cessation of metabolism including damage to the cytoplasmic membrane and cell-wall (i.e. breakdown of the permeability barrier) takes place.

Even the empty protein "ghost" of the T-even phage can inhibit some host macromolecular synthesis and cause cell death.[86,87] The ghost infection completely inhibits synthesis of DNA, RNA, and protein within 2 min, and causes a sudden increase in host cell permeability,[88] and the ghost infected bacteria prevent superinfected phages from multiplying.[86,89] Using *e* mutant phage, the effect mediated by the ghost was shown not to involve phage lysozyme activity.[90] Thus the mech-

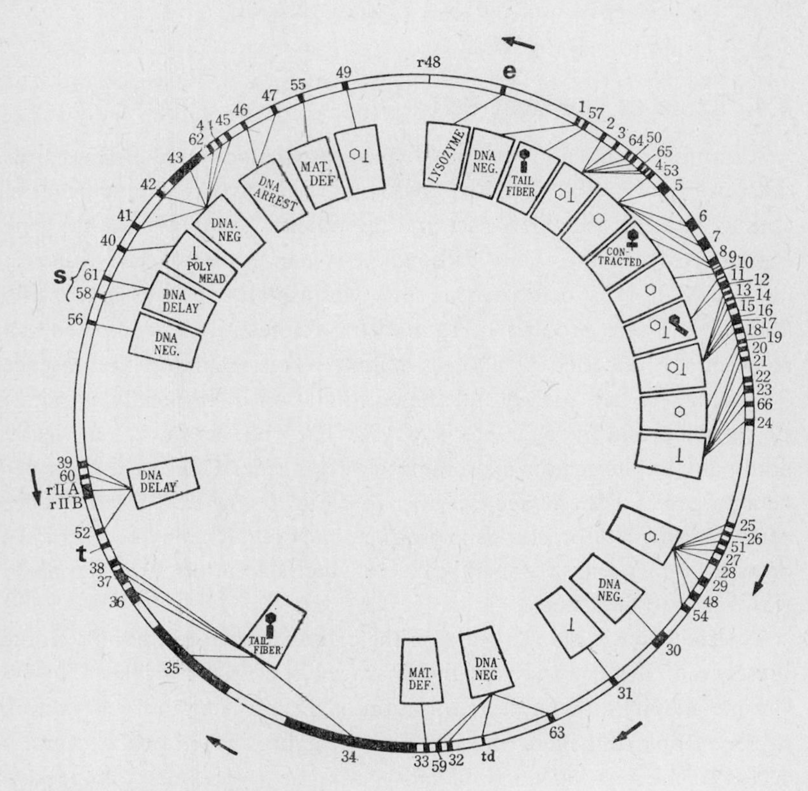

Fig. 1. Genetic map of phage T4.

anism of ghost infection is quite similar to that of the phage infection process except that the phage carries its own syntheses of cell-wall components which maintain the permeability characteristic of uninfected bacteria and a similar rate of oxygen uptake. The former difference has not been studied in detail but the latter has been shown to be related to the switching from bacterial to phage directed metabolism. The extent of switching depends on the kind of phage. The immediate permeability change following ghost infection is caused by an unknown mechanism which does not require protein synthesis[90b] and does not cause any detectable soluble cell-wall components to be released. A kind of conformational change in cell-wall or cell-membrane is postulated for this phenomenon which successively causes the cessation of metabolism, including that of the DNA–cell-membrane moiety in the bacteria.

3.4. Lysis of the Host Cell

Although we do not yet know what triggers lysis at the latent lysis time, two lytic enzyme activities are known to be involved in the process. One we call the gene *t* product and the other is the conventional gene *e* product, lysozyme. The "Tithonus" (*t*) mutant which constitutes a new class of lysis defective mutants was isolated by Josslin.[59] The functional *t* gene product is required for the cessation of infected cell metabolism. Infection by the $e^{+}t$ mutant phage results in the absence of lysis at the late period and the continuation of intracellular phage synthesis beyond the characteristic lysis. The *t* phenotype is genetically dominant and its genetic mapping is between genes 52 and 38, the latter being a gene for the phage tail fiber (see Fig. 1). Several possible roles can be considered for the *t* gene product, one of which may be to modify or break the cytoplasmic membrane, in which one of the free fatty acids was found to be affected.[59]

The lysozyme is able to react on the cell-wall from inside only in the presence of the *t* product at the characteristic time for lysis. The lysozyme activity also requires cessation of metabolism and degradation of the cytoplasmic membrane by the *t* gene product in order to attain access to the cell-wall. The phenotype of the *s* gene may compensate this lysozyme activity. From the substrate affinity it is concluded that the active site of the lysozyme must attack the cell-wall but not the cytoplasmic membrane.

The λ *s* gene is also required for cessation of phage infected cell metabolism just as is that of the T4 *t* gene.[91] Analogous considerations may be extended to the smaller *coli* phages, ϕx174,[92] S13[93] and R17.[94] The two former DNA phages have a gene, the function of which seems analogous to that of the *t* gene, and the coat protein of RNA phage R17 appears to be the *t* gene function. However, no phage induced lysozyme activity was found in ϕx174 infected cell lysate,[95] nor was any lytic activity proven in the R17 coat protein.[94]

4. Genetics of the Lysozyme Gene of Phage T4

Streisinger *et al.*[96] reported a detection method for mutants of the lysozyme gene of phage T4 in 1961, and discovered mutations that affect both the lysozyme activity produced by phage T4 and the primary structure of the lysozyme molecule itself. The structural gene of lysozyme in T4 phage was designated by the letter "*e*".

Most lysozyme-negative phages fail to produce plaques when plated on the usual media, but do form plaques on special plates supplemented with egg-white lysozyme and citrate buffer, pH 8.0.[3]

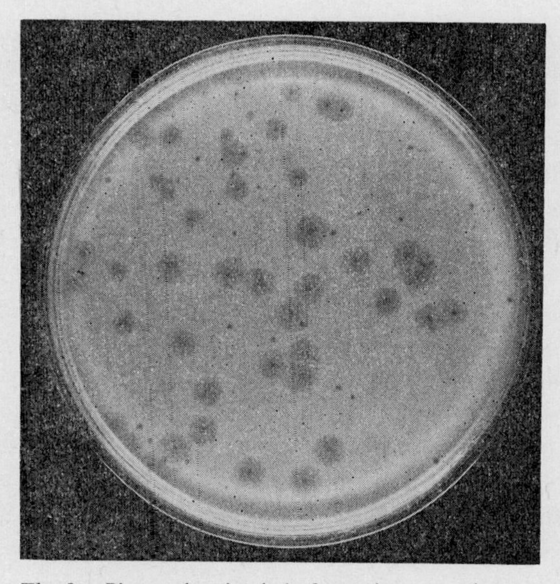

Fig. 2. Plaque showing halo formation around mutant phages.

The selection of mutant phages was aided by the nature of the halo formation around plaques after exposure of the plates to chloroform vapor. Plaques of wild type phage are surrounded by large halos whereas mutant plaques either lack a halo or are surrounded by a small halo[96] (see Fig. 2). The halos are due to the action of the phage lysozyme, which diffuses from the plaque and lyses the chloroform killed bacteria which surround the plaque.

The relative position of the *e* gene in a circular linkage map of phage T4 was established by three-factor crosses and spot crosses against deletion mutants, as used for many other genes of phage T4 (Streisinger, *personal communication*). Fig. 1 (above) shows the position of the *e* gene with respect to certain of the other genes.[97] Details of the genetic mapping within the *e* gene have also been accomplished.

4.1. Isolation of the Mutant Phage: Frameshift Mutants

Genetic analysis of acridine induced mutations by Crick *et al.*[1] has shown that mutations of this kind can be classified into one of two different sets. When combined as a double mutant, mutations of the same set produce the defective phenotype as either of the single mutants, whereas two mutations belonging to different sets often suppress each other, resulting in a wild or pseudo-wild phenotype.

A previous proposal by Brenner *et al.*[98] that acridines (proflavin) induced mutations by causing deletion or insertion of a single nucleotide in DNA, suggested the identification of the two sets of mutations with in one case deletions, and in the other case insertions of single nucleotides. On the basis of this interpretation the ability of mutations of the opposite set to suppress each other could be explained by assuming that the nucleotide sequence of a cistron is translated into its protein product by the sequential reading of triplets of bases (in *m*RNA) from a fixed starting point which corresponds to the beginning of the cistron. The genetic message was thus imagined to be separated into groups of three bases by a "reading frame" which is set in register at the beginning of the message. The addition or deletion of a base would thus cause a shift in the reading frame and would result in a grossly different translation of the message beyond the site of the mutation.

Fig. 3 shows a schematic illustration of a mutant carrying double frameshift mutations. On the other hand, if a deletion were followed by an insertion (or vice versa), as would be the case for a double mutant composed of two acridine mutations in different sets, the reading frame, and therefore the amino acid sequence of the protein product, would be altered only in the region between the mutations. If the mutations are

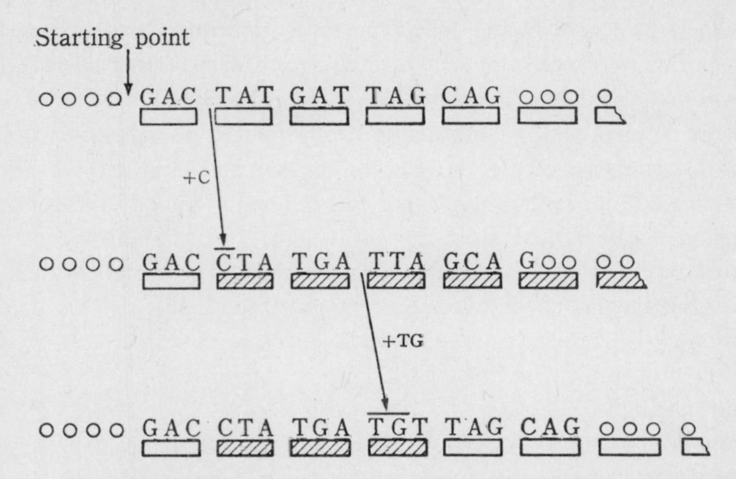

Fig. 3. Schematic illustration of frameshift mutation. This figure shows how the insertion of a single base (C) followed by insertion of two bases (TG) changes the reading frame in a DNA. Each triplet corresponds to a definite amino acid. The shaded parts indicate amino acid sequences which differ from the ones which would be obtained if no insertion occurred.

close to one another, and if the altered region does not cover an essential part of the enzyme protein, the protein product of the cistron might in some cases be active.

Isolation of pseudo-wild mutant strains carrying double (or triple) frameshifts were carried out in the following steps.

(i) *Isolation of a single frameshift mutant.* The bacteria infected with the e^+ phage in the presence of proflavin were cultivated for one generation of the phage life cycle and the proflavin removed to allow maturation of the phages. When the lysate was plated on media supplemented with egg-white lysozyme together with citrate buffer, pH 8.0 ("citrate plate"), the e^- mutants were selected as pin-point plaques whilst the e^+ strain showed large and clear plaques. The selected e^- mutant did not produce any plaques on the normal plate, whereas the e^+ mutant showed clear plaques having large halos after exposure to chloroform vapor.[6,81)]

(ii) *Isolation of pseudo-wild mutants carrying double* (*or triple*) *frameshifts.* This was carried out by either pair-wise crosses of the above two independent single mutant strains (recombination), or by spontaneous or proflavin induced reversion of the single mutant strains (reversion).[5,9,11)]

The selection of pseudo-wild mutants is performed by observing the type of halo formed around plaques after exposure to chloroform when pseudo-wild mutant strains show smaller halos than those of the e^+ strain.[96] In order to confirm the newly formed pseudo-wild mutant strains, pair-wise crosses between the mutant strains and the e^+ strain are carried out. In the case of the recombinants, the crosses yield two original single mutant strains, whilst the crosses for the revertants yield two types of lysozyme negative (*e*) mutants, one of which is identical with the original single mutant strain and the other is a single mutant strain newly induced by the reversion process. The back-crosses between this and the original single mutant should result in the pseudo-wild revertant. To date, 26 pseudo-wild strains including 25 double frameshift mutants and one triple mutant have been isolated, and of these, 15 strains have had their amino acid sequences determined.

4.2. Nonsense Mutants

Protein biosynthesis is a sequential process during which a peptide chain grows unidirectionally, by increments of one amino acid, from the amino-terminal towards the carboxyl-terminal residue.[99–101] Accordingly, if a nonsense triplet is present at any of the positions in a peptide chain, a gap will appear in the chain and cause premature termination of chain growth. Three kinds of nonsense codons are known; UAG (amber), UAA (ochre) and UGA (opal). These nonsense triplets do not normally occur within the coding regions of *m*RNA, but they can be generated from certain codons by mutations. Amber mutations (UAG) are related to the tryptophan codon (UGG) and the glutamine codon (CAG); ochre mutants (UAA) are related only to the glutamine codon (CAA), and opal mutations (UGA) to the arginine (CGA) and tryptophan (UGG) codons by way of a single step transitional mutation.

Four amber and four ochre mutations were obtained for the lysozyme locus from T4 e^+ strains by treatment with a chemical mutagen, hydroxylamine. T4 lysozyme contains three tryptophan and five glutamine residues.[102,103] It is conceivable that three of the four amber mutations are derived from tryptophan residues, and this has already been implied directly in a structural study.[104,105] The residual amber and the four ochre mutations are supposed to be derived from the five glutamine residues in the e^+ gene.[106] (Two of the ochre mutations have been

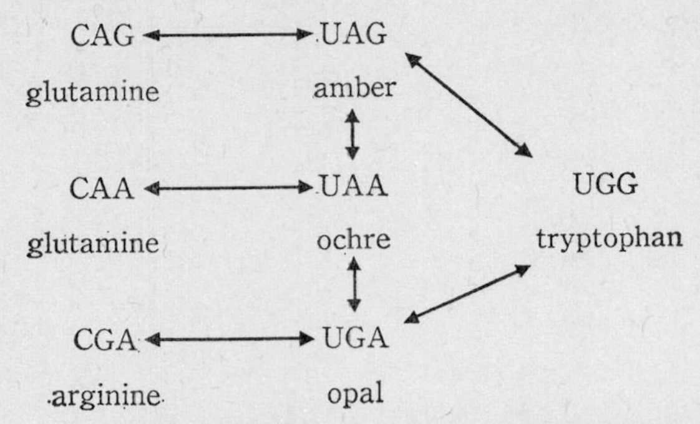

Fig. 4. Correlation between nonsense codons and the related amino acids by single step transition mutations.

proved to be mutated from the glutamine codon.[107]) Ochre mutants can be converted to amber and opal mutants by a single step transition as shown in Fig. 4. These nonsense mutants can easily be selected by plating them on the respective nonsense suppressive bacteria and nonsuppressive hosts. The nonsense mutants can grow only on their corresponding suppressive hosts, whereas the other wild and pseudo-wild mutants grow on either suppressive or nonsuppressive hosts.

4.3. Genetic Mapping of the Mutants

In order to locate the mutual positions of the various mutations, three factor crosses as well as spot test crosses were carried out. The latter tests are made against a set of deletion mutants in the *e* gene. These tests are rather simple but useful for the determination of the rough position of the mutations. On the other hand, the former tests are useful for determination of the mutual order of the mutations.[96])

When one considers three markers a, b, and c, closely linked in that order, and three crosses

$$\frac{a\quad b\quad\quad}{\quad\quad\quad c}\qquad \frac{a\quad\quad\quad}{\quad b\quad c}\qquad \frac{a\quad\quad c}{\quad b\quad}\ ,$$

wild-type recombinants will be produced in the first two crosses with relatively higher frequencies, and in the last cross with a relatively lower frequency, since one recombinational event is required for the former and two recombinational events for the latter. For instance, in the *e* gene the orders of the seven mutants were determined from the results of complementation tests as shown in the matrix of Fig. 5 and from

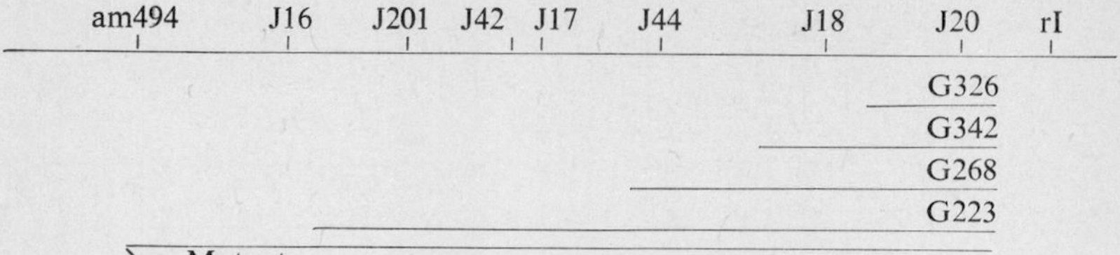

am494 J16 J201 J42 J17 J44 J18 J20 rI

G326
G342
G268
G223

Mutant / Delection	J16	J201	J42	J44	J18	J20
G326	+	+	+	+	+	−
G342	+	+	+	+	−	−
G268	+	+	+	−	−	−
G223	+	−	−	−	−	−

Fig. 5. Relative order of mutant sites in the *e* gene. In the matrix, + indicates plaques, due to recombination, in the spot test; − indicates no plaques due to recombination.

the factors presented in Table 1. In this table, plaques produced by e^+ (recombinant) phage among the progeny were scored as *am* or am^+. Since *am*A494 is known to be in gene 1 (outside the *e* gene), the ratio am:am^+ established the relative order of the markers in any one cross, and thus the orientation of those mutant sites within the *e* gene with respect to the genetic map is established. Cross 4 in Table 1 serves to determine the order of *e*J201 and *e*J42 relative to the other markers, which is not determined by the spot test (Fig. 5 matrix). However, the mutations *e*J42 and *e*J17 are closely linked and the order cannot be determined by the present genetic analysis without information on the protein sequence.

TABLE 1

The Order of *e* Mutants Relative to amA494

Cross	Among the progeny ame^+/ame^+ am^+e^+
(1) *e*J16 × *e*J42*am*A494	0.70
(2) *e*J18 × *e*J42*am*A494	0.44
(3) *e*J20 × *e*J42*am*A494	0.41
(4) *e*J201 × *e*J42*am*A494	0.73

E. coli B was grown to a concentration of 10^8/ml and concentrated to 1.5×10^9/ml. 0.5 ml of a mixture containing 7.5×10^9/ml of each of the parental phages was added to 0.5 ml of the concentrated bacteria. 3 min after infection the bacteria were diluted in broth and 19 min after infection were lysed by adding 1/10 vol chloroform and 10 μg/ml of egg-white lysozyme. The lysate was further diluted and plated.

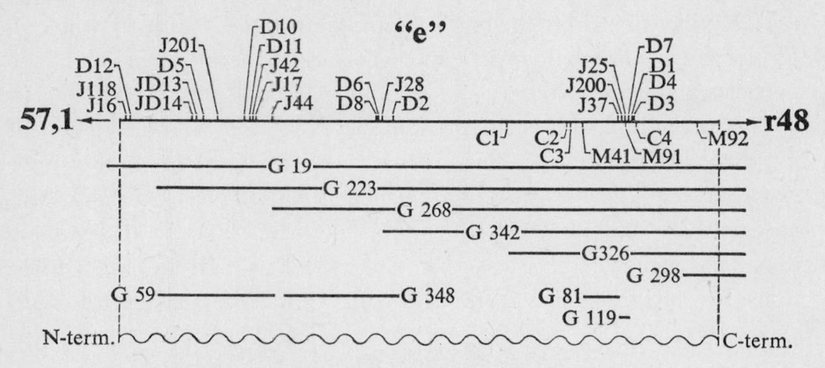

Fig. 6. Genetic map of the lysozyme gene (*e*) of phage T4.

The three factor crosses and/or the spot test were performed for the isolated individual mutants. Fig. 6 shows the detailed genetic map of the lysozyme locus of the T4 phage.

5. Biochemistry of T4 Phage Lysozyme

5.1. Purification[108]

Because T4 phage lysozyme is both a basic and a small protein, conventional methods designed for such types of proteins may be used for its purification. They include the use of weak acidic resins for chromatography and Sephadex for molecular sieving. In addition, it is essential to remove DNA and to protect SH-groups during the purification process.

5.1.1. Starting Lysate

Starting lysate was made in a high titer ($>10^{11}$ phages/ml). Bacterial strain B/1[2,108] was selected as the host bacterium because of its resistance to the strongly infectious T1 phage. Addition of chloroform was required for completion of the lysis. (Mark observed that the chloroform affects cell-membrane structure but not the cell-wall (K.K. Mark, *personal communication*).) The lysozyme was found largely in the supernatant of the lysate and appears not to be a structural component but rather to adsorb to the phage particles during certain culturing conditions. The lysate and associated phage particles, after removal of DNA (as described in the following section), were poured directly onto an ion exchange column in an attempt to dissociate the enzyme from phage particles and other proteins.

This procedure has so far proved successful for the purification of T4 phage lysozyme,[108] T2 phage lysozyme[109] and λ phage endolysin,[57] as well as the lytic enzyme of N20F′ phage. Although the binding complex between the lysozyme and the phage particle may be more stable than the isolated lysozyme, pH values below 5 cause substantial inactivation of isolated enzymes from T4,[108] T2[109] and the λ lytic enzyme.[57]

5.1.2. Removal of DNA

Lysate which contains DNA is a viscous solution and the acidic character of the DNA results in strong association with basic proteins. An efficient method of removing DNA is to add a basic dye, rivanol (6,9-diamino-2-ethoxyacridine), to the lysate until no further precipitation occurs. The precipitate, which contains DNA, cell debris, intact cells and acidic proteins, is removed by sedimentation.[108] The dye does not inhibit lysozyme activity to any noticeable extent and has been successfully used with T4 and T2 lysozymes and λ endolysin in our laboratory,[57,108,109] and with N20F′ in another laboratory.[110]

5.1.3. Ion-exchange Chromatography

Various weakly acidic resins with carboxyl functional groups were used for chromatographic separation of phage lysozymes, as for egg-white lysozyme. Amberlite CG-50 was chosen for T4 phage lysozyme because of its easy handling properties. The chromatographic conditions employed involved variation of pH from a low to a high value within the stability range of the enzyme (6.0–7.5), and variation of the ionic strength (0.1–0.5 M Na^+) by use of a gradient or in a stepwise manner.[108] When concentration of the enzyme was required, the weakly acidic resin column could be used by changing the ionic strength of the eluent in a stepwise fashion.[108] CM-cellulose was sometimes preferred to Amberlite CG-50.[54,57] A weakly basic resin is also suitable for purification of the enzyme in the final step. To obtain crystalline material Kretsinger[111] used DEAE Sephadex A50 for an additional purification of the purified T4 phage lysozyme which had been obtained by conventional methods.

5.1.4. Molecular Sieve

Since lysozyme is a rather small molecule, molecular sieving is one of the most efficient purification methods. Sephadex G75 or G100 has been used for T4 and T2 lysozymes,[108,109] λ endolysin[54,57] and N20F′ lytic enzyme.[110]

The ionic strength of the phosphate buffer eluent must be maintained above 0.2 M to avoid irreversible adsorption of the enzymes on the Sephadex. When the lysozyme protein is present in low concentration and the solution is of low ionic strength, such adsorption was observed to occur not only to Sephadex but also to glass surfaces, dialyzing bags, celite or cellulose powder, etc. The complete purification process, with purities and yields at each step, is summarized in Table 2.

TABLE 2
Recovery and Specific Activity at Each Step of Purification of T4 Phage Lysozyme

Purification procedure	Lysozyme activity Total (u)	Recovery (%)	Specific activity (u/A_{280})
Lysate	7.5×10^4		0.17
1. Rivanol treatment	7.5×10^4	100	—†
2. Initial IRC-50 column concentration	6.4×10^4	85.5	13.9
3. First column chromatography on IRC-50	6.0×10^4	80.0	120
4. Second column chromatography on IRC-50	4.4×10^4	58.0	200
5. First gel-filtration on Sephadex G-75	3.8×10^4	50.2	240
6. Second gel-filtration on Sephadex G-75	3.0×10^4	39.5	260

† Absorbance was not measured because Rivanol remaining in solution has a high absorbance at 280 nm.

5.1.5. Crystallization

Kretsinger further purified a T4 lysozyme preparation[108] by passing it through three Amberlite 1RC-50 columns and a DEAE Sephadex column. The protein was initially precipitated with 62–68% saturated ammonium sulfate, then redissolved at a lower salt concentration, and finally the salt concentration was increased to 0.5–1.0% by dialysis. Crystals were eventually obtained at a protein concentration of 3 mg/

ml, the largest being a needle about 0.4 mm long and 0.05 mm thick.[111] Katz[35] also crystallized T2 lysozyme from a solution containing 0.15 M KCl and 0.02 M sodium bicarbonate (pH 10.1).

5.2. Physicochemical Properties

5.2.1. Homogeneity

The homogeneity of the purified T4 enzyme was examined by sedimentation analysis. The enzyme appeared homogeneous as evidenced by a single boundary observed with Schlieren optics. The homogeneity in molecular size was also shown by gel-filtration on Sephadex G-75. The enzyme was examined for its ionic character by Tiselius type electrophoresis using Schlieren optics.[108] Polyacrylamide gel-electrophoresis in the presence and absence of 5% SDS and 8 M urea revealed only a single band (*unpublished data*). The homogeneity of the material was further confirmed by amino acid analysis, in which the molecular ratios of each amino acid appeared as reasonable integer ratios.[108] The purified T2 lysozyme[34,109] and the λ endolysin[54,57] were also found to be homogeneous to about the same extent.

5.2.2. Molecular Weight

The sedimentation coefficient of the enzyme ($s_{20,w}$) was estimated to be 1.9 S. The molecular weight of the enzyme was determined by the Archibald sedimentation equilibrium method, assuming that the partial specific volume of the enzyme derived from its amino acid composition was 0.741. The value thus calculated was $19{,}000 \pm 1000$.[108] From the amino acid sequence the molecular weight was calculated to be 18,720,[102] which is compatible with the above value.

From the sedimentation coefficient and the diffusion constant of T2 enzyme roughly purified by Katz and Weidel,[34] the molecular weight was estimated to be between 21,000 and 15,200 ($\pm 10\%$). A crystalline enzyme preparation was later studied by sedimentation analysis and the molecular weight of the free lysozyme and the phage bound enzyme, respectively, were estimated to be 14,300 and 14,500 at pH 7.0 and 13,850 at pH 5.1.[35] The molecular weight determined from the amino acid composition and sequential analysis was 18,607,[109] which is about 4000 greater than the value reported by Katz.[35]

The sedimentation coefficient ($s_{20,w}$) of λ endolysin is 2.06 S. The diffusion coefficient ($D_{20,w}$) was measured by the spreading of the boundary formed in a synthetic boundary cell as a function of time and found to be 10.2–10.5 × 10^{-7} cm^2/sec. Combining these results gave a molecular weight for λ endolysin of 1.78 × 10^4.[55] The partial specific volume computed from the amino acid composition is 0.732 cm^3/g. The molecular weight was also estimated from results of sedimentation equilibrium; the enzyme was sedimented at a relatively high speed and at lower speeds, and the values were calculated to be 1.75 × 10^4 and 1.82 × 10^4, respectively.[55]

A minimum molecular weight can be computed from the frequencies of the amino acid residues in pure proteins. Black and Hogness[55] calculated the minimum molecular weight of the λ phage endolysin to be about 17,873 from its amino acid composition, which is consistent with the value found independently by sedimentation analysis. From the amino acid composition derived by sequential analysis in our laboratory,[57] the minimum molecular weight is calculated to be 17,558.

5.2.3. Absorption Spectrum

The absorption coefficient of purified T4 lysozyme at 280 nm (1 cm light path) was estimated to be 1.28 per mg of enzyme per ml, and the ratio of the absorption at 280 nm to that at 260 nm was 1.92 (Fig. 7). This indicates that nucleic acid components are absent in this protein preparation.[108]

The purified T2 phage lysozyme shows almost the same ultraviolet spectrum and absorption spectrum as the T4 phage lysozyme.[109] The absorption coefficient of the purified λ endolysin at 280 nm was calculated to be 1.7–1.8 per mg of enzyme per ml, and the ratio of the absorption at 280 nm to that at 255 nm was 2.0.[57]

5.2.4. Stability of the Enzyme

The enzyme proved to be stable when incubated at 37°C for 20 hr in 0.1 M phosphate buffer, pH 6.0 to 6.5, but was inactivated by prolonged incubation at higher or lower pH values. Magnesium and sodium ions seemed to exert a stabilizing effect. Enzyme inactivation was observed at temperatures above 40°C, and 50% inactivation occurred at 53.5°C.[108] SH-reagents were found to be effective to a small extent in stabilizing the enzyme because of the presence of a single SH-group in this protein.[102] The effect was especially marked in some mutant strains,[10] where purification required the presence of a high concentration (0.15 M) of mercaptoethanol.

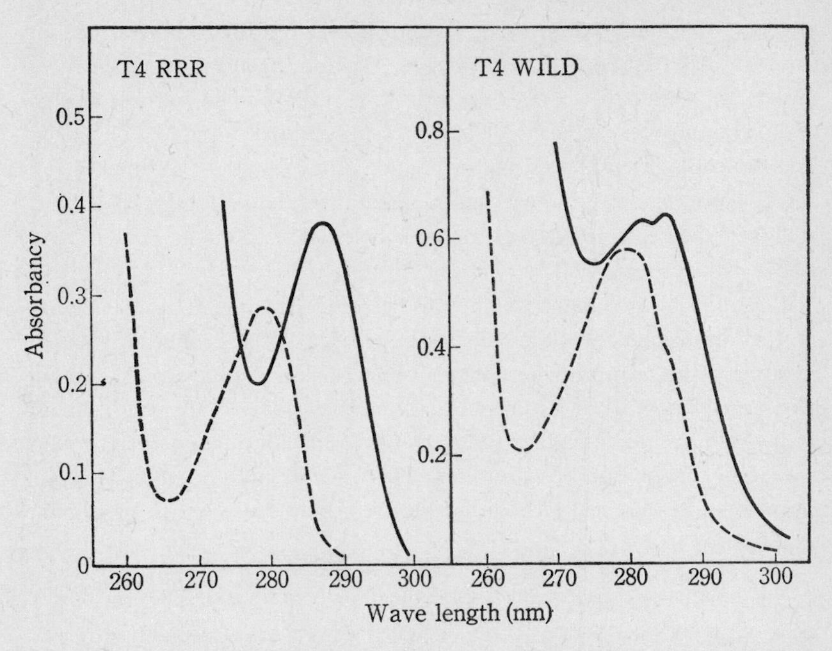

Fig. 7. Ultraviolet absorption spectra of wild-type T4 phage lysozyme and triple revertants (RRR) measured in 0.1 N NaOH (—) and at pH 7.0 (···).

The stability of the T2 lysozyme, as expected, is quite similar to that of the T4 phage lysozyme.[109]

λ Phage endolysin is stable between pH 4.0 and 8.0. Below pH 4.0 a sudden decrease of enzyme activity was observed, whilst above pH 8.0 the decrease of enzyme activity occurred gradually until pH 9.5. The lysozyme is denatured to a large extent in the presence of Fe^{2+} ion, but the addition of quinoline (10^{-3} M) can prevent this denaturation.[57] The presence of 10^{-3} M mercaptoethanol stabilizes the λ endolysin since the free sulfhydryl group of the cysteine residue in the protein is protected by this reagent.[48,54,57]

5.3. Chemical Properties

5.3.1. Amino Acid Composition

T4 phage lysozyme consists of 164 amino acid residues [Asp (11), Asn (11), Thr (11), Ser (6), Glu (7), Gln (6), Pro (3), Gly (11), Ala (15),

TABLE 3
Amino Acid Composition of T4, T2, λ and Egg-white Lysozymes

	T4		T2		λ	Egg-white	
Aspartic acid	11	22	11	21	19	8	21
Asparagine	11		10			13	
Threonine	11		10		6	7	
Serine	6		7		9	10	
Glutamic acid	7	13	7	13	16	2	5
Glutamine	6		6			3	
Proline	3		3		5	2	
Glycine	11		11		15	12	
Alanine	15		15		13	12	
Cys/2	2		2		1	8	
Valine	9		10		7	6	
Methionine	5		5		3	2	
Isoleucine	10		10		9	6	
Leucine	16		16		14	8	
Tyrosine	6		6		5	3	
Phenylalanine	5		5		5	3	
Tryptophan	3		3		3	6	
Lysine	13		13		12	6	
Histidine	1		1		3	1	
Arginine	13		13		12	11	
NH_2	17		16		13–14	16	
Total amino acid residues	164		164		157	129	

Cys/2 (2), Val (9), Met (5), Ile (10), Leu (16), Tyr (6), Phe (5), Trp (3), Lys (13), His (1), Arg (13)].[108] The total number of basic amino acids is 26, which exceeds the number of acidic amino acids by 8, and this presumably accounts for the basic character of the protein. One of the three tryptophan residues plays as important a role[112,113] as in the analogous egg-white lysozyme.[114–116] No common compositional features were otherwise observed between T4 phage lysozyme and egg-white lysozyme[117,118] (Table 3).

The T2 phage lysozyme consists of 164 amino acid residues [Asp (11), Asn (10), Thr (10), Ser (7), Glu (7), Gln (6), Pro (3), Gly (11), Ala (15), Cys/2 (2), Val (10), Met (5), Ile (10), Leu (16), Tyr (6), Phe (5), Trp (3),

Lys (13), His (1), Arg (13)].[109] This composition is almost identical to that of the T4 phage lysozyme, except for a decrease of one mole each of the asparagine and threonine residues and an increase of one mole in each of the serine and valine residues. No variation in the number of acidic and basic amino acids was observed in the T2 and T4 lysozymes, which is compatible with the fact that the same purification method can be applied to both proteins and that they are also chromatographically similar. The lack of variation of aromatic amino acids explains the almost identical absorption spectrum observed for both enzymes.

λ Phage endolysin consists of 157 amino acid residues [Asp + Asn (19), Thr (6), Ser (9), Glu + Gln (16), Pro (5), Gly (15), Ala (13), Cys/2 (1), Val (7), Met (3), Ile (9), Leu (14), Tyr (5), Phe (5), Trp (3), Lys (12), His (3), Arg (12)], as determined by a preliminary sequential study.[57] The results of direct amino acid analysis carried out independently[56] gave [Asp + Asn (19), Thr (6), *Ser* (*10*), *Glu* + *Gln* (*17*), Pro (5), Gly (15), Ala (13), Cys/2 (1), Val (7), Met (3), Ile (9), Leu (14), Tyr (5), Phe (5), Trp (3), Lys (12), His (3), Arg (12)]; *total 159*.

Although the phage strains and the preparation methods were different, the amino acid compositions were in good agreement, the only differences being italicized. The amide content was roughly estimated from the amount of ammonia liberated after alkali treatment, and found to correspond to about 15 residues.[57] From the sequential analysis,[57] of 19 moles of [Asp + Asn], 5 moles were found in the amidated form (Asn), 12 moles were identified as aspartic acid and the remaining 2 are still undetermined. Of the 16 moles of [Glu + Gln], 8 moles were identified as glutamine, 7 moles as glutamic acid and the remaining 1 is still uncertain. Thirteen moles out of a total of 15 moles of amide residues were confirmed. The number of strongly basic amino acid residues (Arg + Lys) amounts to 24 moles and acidic amino acid residues to approximately 20 moles, which indicates that λ phage endolysin is also a basic protein. However, in comparison to the T-even phage lysozymes, the numbers of basic (25 moles) and acidic (18 moles) amino acid residues suggest that the basic nature of these lysozymes may be stronger than that of λ endolysin, although it is a little dangerous to discuss this point without knowing the tertiary structure of these lytic enzymes. Some support for the expectation of a weaker basic character for λ endolysin was however provided by chromatography as well as electrophoresis. From a comparison of the composition of λ phage endolysin with that of T4 phage lysozyme on the basis of five functional groups (aliphatic, hydroxyl, aromatic, acidic, and strongly basic groups), Black and Hogness suggested that a similarity exists between the two lysozymes among these groups. However, little similarity in their sequences was found.[57] By reaction with 5,5'-dithiobis-(2-nitrobenzoic acid) one mole of cysteine was found to account for a free SH-group in λ endolysin.[54]

5.3.2. Primary Structure

The results of a preliminary amino acid sequence study of T4 phage lysozyme were published by Inouye and Tsugita[103] and the complete sequence in 1968.[102] The enzyme is a single polypeptide chain of 164 amino acid residues with an NH_2-terminal methionine residue and a COOH-terminal leucine residue.[103] The sequence was deduced from the sequential analysis of four kinds of partial hydrolyzates; tryptic,[119]

H–Met–Asn–Ile–Phe–Glu–Met–Leu–Arg–Ile–Asp–Glu–Gly–Leu–Arg–Leu–Lys– (5, 10, 15; T1, T2, T3)
Ile–Tyr–Lys–Asp–Thr–Glu–Gly–Tyr–Tyr–Thr–Ile–Gly–Ile–Gly–His–Leu– (20, 25, 30; T4, T5)
Leu–Thr–Lys–Ser–Pro–Ser–Leu–Asn–Ala–Ala–Lys–Ser–Glu–Leu–Asp–Lys–Ala– (35, 40, 45; T6, T7)
Ile–Gly–Arg–Asn–Cys–Asn–Gly–Val–Ile–Thr–Lys–Asp–Glu–Ala–Glu–Lys–Leu– (50, 55, 60, 65; T8, T9, T10)
Phe–Asn–Gln–Asp–Val–Asp–Ala–Ala–Val–Arg–Gly–Ile–Leu–Arg–Asn–Ala–Lys– (70, 75, 80; T11, T12, T13)
Leu–Lys–Pro–Val–Tyr–Asp–Ser–Leu–Asp–Ala–Val–Arg–Arg–Cys–Ala–Leu– (85, 90, 95; T14, T15)
Ile–Asn–Met–Val–Phe–Gln–Met–Gly–Glu–Thr–Gly–Val–Ala–Gly–Phe–Thr– (100, 105, 110, 115; T16)
Asn–Ser–Leu–Arg–Met–Leu–Gln–Gln–Lys–Arg–Trp–Asp–Glu–Ala–Ala–Val– (120, 125, 130; T17, T18, T19)
Asn–Leu–Ala–Lys–Ser–Arg–Trp–Tyr–Asn–Gln–Thr–Pro–Asn–Arg–Ala–Lys–Arg– (135, 140, 145; T20, T21, T22, T23)
Val–Ile–Thr–Thr–Phe–Arg–Thr–Gly–Thr–Trp–Asp–Ala–Tyr–Lys–Asn–Leu•OH (150, 155, 160; T24, T25, T26)

Fig. 8. The complete amino acid sequence of T4 phage lysozyme. Underlining indicates tryptic peptides, and the numbers correspond to those of the tryptic peptides from the amino terminal peptide.

chymotryptic[120] and peptic[121] digestions and dilute acid hydrolysis.[122] The protein was found to have two cysteine residues present as free sulfhydryl groups, in contrast to the four S-S bridges in egg-white lysozyme.[117] This was shown by titration with *p*-chloromercuribenzoate.[122] No common feature in the first order structure was found between T4 phage lysozyme and egg-white lysozyme,[117,118] although some structural similarity was discussed from the nature of the sequenced amino acids[113] (Fig. 8).

From the amino acid composition of T2 phage lysozyme, two changes only were expected from the sequence of T4 phage lysozyme. From a comparison of tryptic peptides of the two lysozymes, it was found however that there were three amino acid substitutions in which they differed. These were Asn_{40}/Ser, Ala_{41}/Val and Thr_{151}/Ala, from T4 lysozyme to T2 lysozyme, all of which can be explained by transitional mutations consisting of single base alterations. The sequence of T2 lysozyme is shown in Fig. 9.[123]

The NH_2-terminal sequence of λ endolysin was determined by the cyanate cleavage to be methionyl-valine.[55] The COOH-terminus was determined to be valine by hydrazinolysis,[55] since carboxypeptidase did not work on this protein.[55,57] The protein with one NH_2-terminal

5 10 15
H•Met–Asn–Ile–Phe–Glu–Met–Leu–Arg–Ile–Asp–Glu–Gly–Leu–Arg–Leu–Lys–
20 25 30
Ile–Tyr–Lys–Asp–Thr–Glu–Gly–Tyr–Tyr–Thr–Ile–Gly–Ile–Gly–His–Leu–Leu–Thr–
35 40 45 50
Lys–Ser–Pro–Ser–Leu–*Ser*–*Val*–Ala–Lys–Ser–Glu–Leu–Asp–Lys–Ala–Ile–Gly–
55 60 65
Arg–Asn–Cys–Asn–Gly–Val–Ile–Thr–Lys–Asp–Glu–Ala–Glu–Lys–Leu–Phe–
70 75 80
Asn–Gln–Asp–Val–Asp–Ala–Ala–Val–Arg–Gly–Ile–Leu–Arg–Asn–Ala–Lys–
85 90 95
Leu–Lys–Pro–Val–Tyr–Asp–Ser–Leu–Asp–Ala–Val–Arg–Arg–Cys–Ala–Leu–
100 105 110 115
Ile–Asn–Met–Val–Phe–Gln–Met–Gly–Glu–Thr–Gly–Val–Ala–Gly–Phe–Thr–
120 125 130
Asn–Ser–Leu–Arg–Met–Leu–Gln–Gln–Lys–Arg–Trp–Asp–Glu–Ala–Ala–Val–
135 140 145
Asn–Leu–Ala–Lys–Ser–Arg–Trp–Tyr–Asn–Gln–Thr–Pro–Asn–Arg–Ala–Lys–Arg–
150 155 160
Val–Ile–*Ala*–Thr–Phe–Arg–Thr–Gly–Thr–Trp–Asp–Ala–Tyr–Lys–Asn–Leu•OH

Fig. 9. The complete amino acid sequence of T2 phage lysozyme. Position which are different from the sequence of T4 phage lysozyme are shown in *italics*.

5 10 15
Met–Val–Glu–Ile–Asn–Asn–Gln–Arg–Lys–Ala–Phe–Leu–Asp–Met–Leu–Ala–
20 25 30
Trp–Ser–Glu–Gly–Thr–Asp–Asp–Gly–Arg–Gln–Lys–Thr–Arg–Asn–His–Gly–Tyr–
35 40 45 50
Asp–Val–Ile–Val–Gly–Gly–Glu–Leu–Phe–Thr–Asp–Tyr–Ser–Asp–His–Pro–Arg–
55 60 65
Lys–Leu–Val–Thr–Leu–Asn–Pro–Lys–Leu–Lys–Ser–Thr–Gly–Ala–Gly–Arg–
70 75 80
Tyr–Gln–Leu–Leu–Ser–Arg–Trp–Asp–Ala–Tyr–Arg–Lys–Gln–Leu–Gly–Leu–
85 90 95
Lys–Asp–Phe–Ser–Pro–Lys–Ser–Gln–Asp–Ala–Val–Ala–Leu–Gln–Gln–Ile–
100 105 110 115
Lys–Glu–Arg–Gly–Ala–Leu–Pro–Met–Ile–Asp–Arg–Gly–Asp–Ile–Arg–Gln–Ala–
120 125 130
Ile–Asp–Arg–Cys–Ser–Asn–Ile–Trp–Ala–Ser–Leu–Pro–Gly–Ala–Gly–Tyr–Gly–
135 140 145
Gln–Phe–Glu–His–Lys–Ala–Asp–Ser–Leu–Ile–Ala–Lys–Phe–Lys–Glu–Ala–Gly–
150 155
Gly–Thr–Val–Arg–Glu–Ile–Asp–Val

Fig. 10. Amino acid sequence of λ phage endolysin.

and two internal methionine residues was subjected to cyanogen bromide cleavage. In addition to one homoserine residue (derived from the NH_2-terminal methionine), two homoserine-containing peptides were isolated and analyzed for their amino acid compositions and terminal residues. From these results, together with the amino acid composition of λ phage endolysin, Black and Hogness[5] suggested the order of the three cyanogen bromide peptides as Met–Val–Glx——Met–Glx——Met–Ile——Val.
(Met 1, Val 2, Glx 3; Met 14, Glx 15; Met 115, Ile 116; Val 159)

Imada in our laboratory[57] has sequenced λ phage endolysin using the same (cyanogen bromide) method in addition to the conventional partial hydrolysis methods employing tryptic, chymotryptic, peptic and thermolytic (digestion with thermolysin) digestions. The sequence so far analyzed is given in Fig. 10. Although the compositions of the cyanogen bromide peptides are quite similar to those reported by Black and Hogness, minor but significant differences were seen.[55,57] Residue 15 is leucine instead of the glutamic acid or glutamine claimed by Black and Hogness. The NH_2-terminal dipeptide sequence and the COOH-terminus as well as the amino acid next to the third methionine are identical with each other. Further study is still required for completion of the sequence of λ phage endolysin including amide moieties.

5.3.3. Tertiary Structure

The tertiary structure of hen egg-white lysozyme has been clarified by Phillips' group employing X-ray crystallography,[114,115] but similar studies on phage lysozyme have hardly begun, although a trial result was reported by Kretsinger[111] who obtained crystals good enough for a preliminary X-ray crystallographic study. The spacing along the needle axis, which is a two-fold screw axis, is $a = 29.3$ Å, $b = 133$ Å. There is some uncertainty about the c axis because the macroscopic needles were frequently found to consist of finer needles which had grown together. The reflections were indexed with $c = 90$ Å, $\beta = 91.5°$. The volume of the unit cell was calculated to be 351,000 Å^3. Using 2.37 Å^3/dalton as a mean value, and the molecular weight, a figure of 8.0 molecules per unit cell (or four per crystallographic asymmetric unit in space group P_{21}) was calculated. Since there is no evidence to suggest that the lysozyme functions as an oligomer, it is rather surprising to find four molecules per asymmetric unit.

5.4. Catalytic Properties

5.4.1. Enzyme Assay

The activity of the phage lytic enzyme can be determined by various methods. These are mainly either turbidimetric or involve the estimation of solubilized components of the cell-wall. The choice of method depends to a large extent on the type of research being carried out and in particular the minimal amount of enzyme which has to be measured. Bacteriologists or microbial geneticists frequently use growing bacteria and plates for assay, whereas biochemists generally prefer to analyze the decrease in turbidity and solubilized components.

A. Turbidimetry of Growing Cells[123]

The harvested logarithmically growing cells of *E. coli* are suspended in a chloroform-saturated buffer with occasional shaking. The concentration of cells in the reaction mixture is adjusted to an absorption of 0.6 at 450 nm. The reaction proceeds in a linear fashion until the turbidity decreases to 65% of the initial value. One unit of enzyme activity is defined as the amount which catalyzes the reaction at a rate of decrease of optical density of 1.0/min[123] (also see Ref. 24, 124).

B. Turbidimetry of Dry Cells[81,108]

Logarithmically growing *E. coli* are chilled, collected, washed twice, then treated with chloroform-saturated buffer and gently mixed at room temperature for one hour. The cells are collected, lyophilized to dryness and kept in a desiccator. A fresh suspension of lyophilized cells is mixed with the enzyme sample and the time required for a decrease in absorbance of 0.10 at 450 nm (from about 0.65 to 0.55) is measured at room temperature.

The reciprocal of the reaction time is proportional to the concentration of the enzyme (Table 4). Activity in units, (u)/ml, is obtained using the formula (u) = $(T_s/T) \times 0.05$, where T is the time measured for the enzyme sample and T_s the time measured for a standard 0.05 mg/ml solution of egg-white lysozyme. This method gives an activity in terms of egg-white equivalent units. Because the specific activity of the enzyme is very high, dilution of the enzyme sample is usually required. Since the T4 phage enzyme is basic in character and consequently readily absorbed on glass surfaces, and is also unstable in very dilute solutions, the protein is diluted with a buffer containing 0.2% gelatin and 0.5 M NaCl.[108]

In a modified method the harvested cells are frozen for about a week and then lyophilized, with omission of the chloroform and heat treatment. When the cell powder is insufficiently sensitive to the enzyme, it may be transformed to a good substrate by storage at 37°C.

TABLE 4

Substrate Specificity of T4 Phage Lysozyme Compared with Egg-white Lysozyme

Substrate and enzyme	Amount of enzyme in reaction mixture (μg)	Activity†[1] (sec)	Ratio of T4 to egg-white activity†[2]
M. lysodeikticus			
Egg-white lysozyme	20	46.8	6.3
T4 phage lysozyme	23.8	6.1	
E. coli			
Egg-white lysozyme	10	26.9	
T4 phage lysozyme	0.119	9.0	250

†[1] In this case, the time required for an optical density decrease of 0.05 unit at 350 nm. The reciprocal of the values is a linear function of the activity.
†[2] Ratios are calculated on the basis of unit weight of the enzymes.

In an alternative method a lyophilized cell suspension is stored overnight and then incubated with the enzyme for 5 min at 37°C, and the absorbance at 350 nm followed. For each experiment, standard curves for phage lysozyme and egg-white lysozyme are constructed by plotting various concentrations of the lysozymes against absorbance. The relative activity of a given sample is determined by referring the absorbance measured to the standard curve.[81,108]

The apparent absorbance is due to light scattering. Several wave lengths have been used for the measurement, including 350 nm,[27,89] 450 nm,[27,106] 600 nm[20,100] and 650 nm.[90b] When the decrease of absorption at various wave lengths (at every 50 nm from 350 to 650 nm) was checked, no practical reason for preferring wave lengths between 450 nm and 650 nm was found. The decrease at 350 nm, however, was somewhat smaller due to interference by absorption of cell DNA. Since the purity of light at various wave lengths varied in the apparatus used, any wave length at which maximum decrease of absorbance after enzyme action occurs can be chosen for the assay.

C. Plate Spot Tests[50]

Although this method is both sensitive and simple, only rough estimations can be made. Top agar containing exponentially growing indicator bacteria was used for plating on plates containing bottom agar. If a more sensitive measurement is required, both agars are supplemented with sodium citrate.[8] After incubation the plates are exposed to chloroform vapor. Spot tests for measuring enzyme activity are carried out with reference to a standard solution of egg-white lysozyme. The spot containing enzyme activity shows noticeable clearing after further incubation. The dilution end point in comparison with the standard enzyme serves as an estimate of lysozyme activity.[96] By this method about 0.1 μg of egg-white equivalent enzyme activity can be determined.

D. Radiochemical Assay

Barrington and Kozloff[21] labeled *E. coli* cells with ^{15}N and prepared a cell-wall preparation according to Weidel and Katz[124] or Salton and Horn.[125] The lytic activity of T2 phage was measured in terms of the non-precipitable components formed during enzyme action.

Brown and Kozloff[69] labeled the cell with ^{14}C-glucose and ^{14}C-fructose or with a ^{14}C-totally labeled amino acid mixture. The cell-walls were prepared cautiously and subjected to measurement of lysozyme activity. The amount of liberated radioactivity is about 5% of the total counts in the cell-wall preparation. The ^{14}C cell-wall suspension was mixed with enzyme solution, incubated and centrifuged. Two aliquots of the supernatant are counted.

Recently, Schweiger and Gold[45] developed a sensitive method for the detection of lysozyme activity in an *in vitro* synthesis system. *E. coli* cells are grown to the stationary phase in the presence of ^{3}H-1,6-diaminopimelic acid. Aliquots of the cells are spotted on disks of filter paper. One half of the radioactive materials can be released by egg-white lysozyme and the other half by trypsin. The enzyme solution to be tested is incubated in a liquid scintillation counter. No quantitative differences are observed after comparison with conventional methods. The radioactive substrate is somewhat troublesome to prepare but seems to be more stable and is far more sensitive than chloroform-treated cells.

5.4.2. Substrate Specificity

The optimum pH for enzyme activity of the T4 phage lysozyme is 7.2–7.4, as shown in Fig. 11. The substrate specificity was examined by comparing the activities of egg-white lysozyme and phage lysozyme, using lyophilized cells of *E. coli* and *Micrococcus lysodeikticus* as substrates. The T4 phage lysozyme is about six times more active than egg-white lysozyme with *M. lysodeikticus* cells as substrate, and 250 times more active with *E. coli* cells as substrate.[108]

The enzymatic action of phage lysozyme was compared with egg-white lysozyme, using an *M. lysodeikticus* cell-wall preparation as substrate, prepared by the method of Hara and Matsushima.[126] After digestion of the cell-wall preparation by the enzymes, the digest was reduced with sodium borohydride and then hydrolyzed with strong acid. The amounts of free muramic acid, muramicitol, glucosamine, and glucosaminitol in the hydrolyzates were analyzed on an amino acid analyzer (Fig. 12, Table 5). From the results it can be seen that, as in the case of egg-white lysozyme, phage lysozyme breaks the cell-walls to give muramicitol, although the amount formed in the case of phage lysozyme (20%) was less than that in the case of egg-white lysozyme (64%). This suggests that the phage enzyme is a muramidase. Further, these values approximately paralleled the change in turbidity of the cell-wall suspension (measured as absorbancy). It is interesting to note that whereas both the phage enzyme and egg-white lysozyme lyse *Micrococcus* and *E. coli* preparations, the phage enzyme seems to manifest a specific affinity for *E. coli* cells, the natural host of the T4 phage.

Fig. 13 is a schematic representation of the bacterial cell-wall together with the action sites of lytic enzymes, including phage lysozyme. The mode of action of T2 phage lysozyme has been examined in some detail, several research groups being engaged in a study of its lytic mode as one representative example of general bacteriophage lysis. The enzyme activity is inhibited by anti-T4 lysozyme.[109] This work, together with other studies (including direct analysis with cell-wall preparations in the case of T4 phage lysozyme), has indicated that the enzyme specificities of T2 and T4 lysozyme are identical (that is, muramidase).[109]

The optimal pH for T2 lysozyme was found to occur between pH 6 and pH 7 in the preparation of Katz and Weidel,[34] and in our prepara-

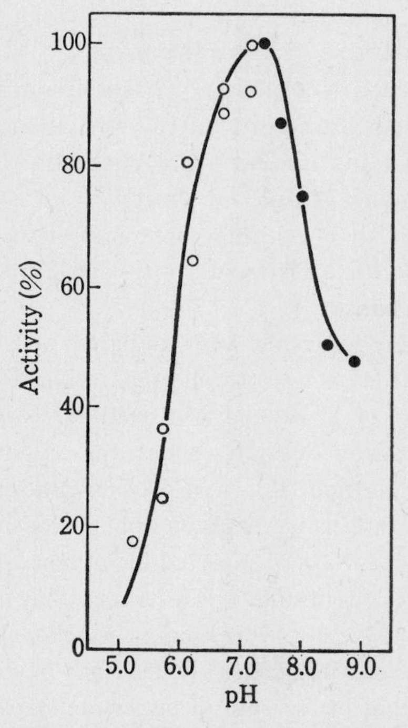

Fig. 11. pH-activity curve of T4 phage lysozyme. The activity assay was carried out as described in the text on *E. coli* substrate. Activities are expressed as a percent of the activity at pH 7.40. ○, Activities measured in 0.03 M phosphate buffer; ●, in 0.05 M Tris-HCl buffer.

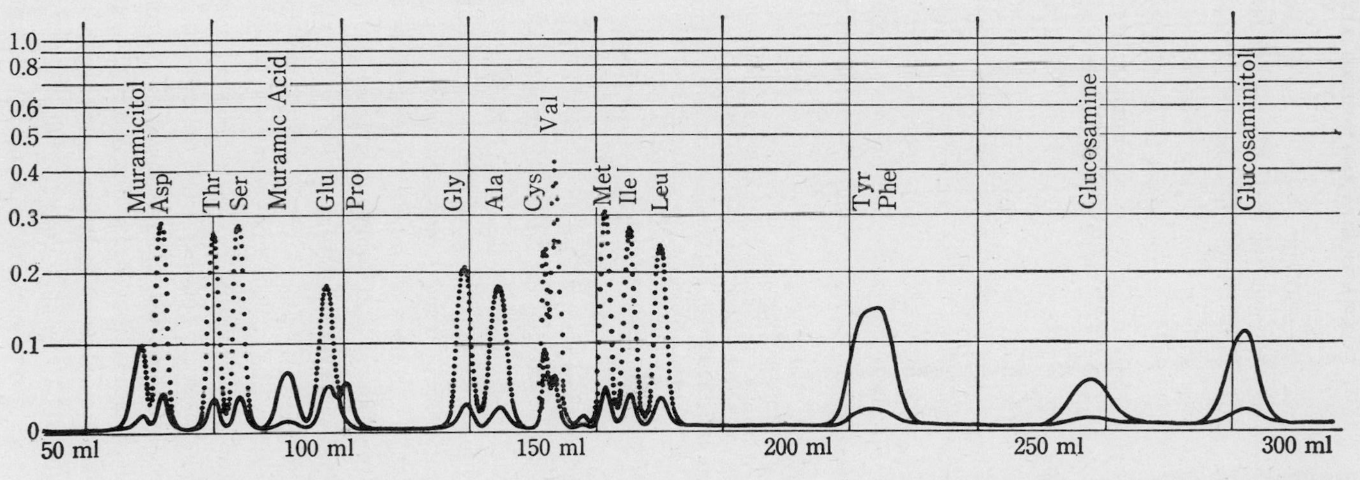

Fig. 12. Elution pattern of amino sugars. A mixture of muramic acid (0.121 μmole), muramicitol (2.32 μmole), glucosamine (0.186 μmole) and glucosaminitol (0.534 μmole) was applied to the 50 cm column of an amino acid analyzer (Beckman/Spinco) with a standard mixture of amino acids (0.25 μmole of each amino acid). The analysis was carried out under the usual conditions for amino acid analysis except that the buffer-change timer and temperature were set at 50 min and 45°C, respectively, in contrast to the usual 80 min and 55°C. Leucine equivalents were found to be 0.80, 0.049, 0.87, and 0.41 for muramic acid, muramicitol, glucosamine, and glucosaminitol, respectively.

TABLE 5
Hydrolysis of *M. lysodeikticus* Cell-walls by T4 Phage Lysozyme and Egg-white Lysozyme

Compound	Without enzyme	Egg-white lysozyme	Phage lysozyme
	(μmole/10 mg cell-wall)		
Muramic acid	7.65	2.55	5.50
Muramicitol	0	4.90	1.45
Glucosamine	10.0	9.05	9.15
Glucosaminitol	0	0	0
Glutamic acid	10.3	10.0	10.9
Glycine	10.0	10.2	10.2
Alanine	19.9	20.9	20.0
Lysine	10.3	10.3	10.4

Cell-walls of *M. lysodeikticus* (10 mg) were incubated with or without 0.2 mg of enzyme for 24 hr at 37 °C in 3 ml of 0.044 M phosphate buffer, pH 7.0. After incubation, the pH of the reaction mixture was adjusted to about 10 with sodium hydroxide, and 10 ml of sodium borohydride were added. The mixture was incubated for another 24 hr at room temperature. After reduction, the samples were dried and hydrolyzed with 6 N HCl for 24 hr at 105 °C. The hydrolysate was analyzed at 45 °C with use of an amino acid analyzer.

tion was about pH 7.2, which is almost equal to that of the T4 lysozyme. Koch and Weidel[25] also found that the T2 lysozyme splits off a component from the isolated bacterial cell-wall. This component was further investigated[28] and found to contain glucosamine, muramic acid, glutamic acid, diaminopimelic acid and alanine, and to have a molecular weight of about 10,000.[29,127] Koch and Dreyer[30] examined the effect of egg-white lysozyme on isolated and dinitrophenylated *E. coli* cell-walls[128] before and after incubation with the purified T2 enzyme. Egg-white lysozyme is only active on the cell-walls before and not after incubation with the enzyme. Similarly, the T2 enzyme was found to have no effect after incubation of the cell-walls with egg-white lysozyme. A comparison of the split products obtained from the cell-walls by the action of egg-white lysozyme and the T2 lysozyme was concomitantly carried out. The results indicated that identical compounds were liberated by the action of the enzymes. These data thus suggest that the action of the T2 phage enzyme is analogous to that of egg-white lysozyme.

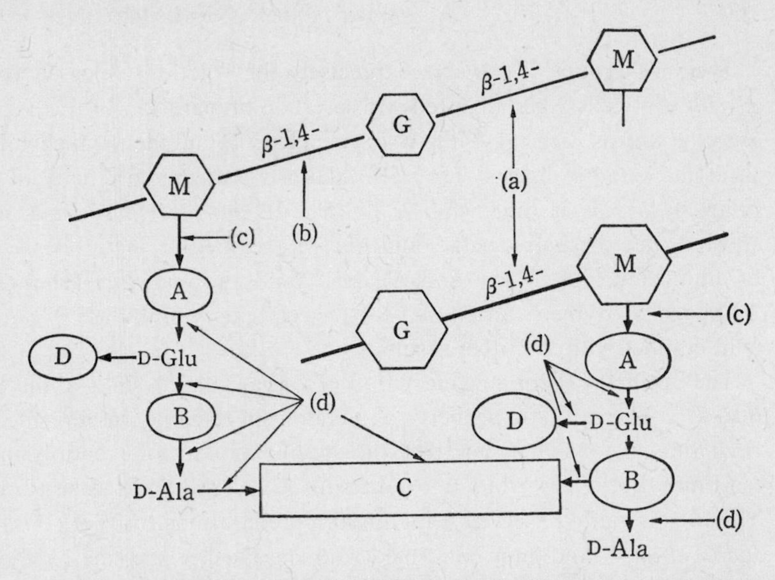

Fig. 13. Schematic presentation of the basic structure of cell-wall peptidoglycan, showing the action sites of lytic enzymes. (*Cf.* S. Kotani, *Protein, Nucleic Acid, Enzymes* (Japanese) **13,** 1136 (1968).)

Ⓐ L-Ala (mostly found)
L-Ser *Butyribacterium rettgeri*
Gly *Microbacterium lacticum*, etc.

Ⓑ L-Lys *Staphylococcus, Streptococcus, Pneumococcus, Micrococcus*, etc.
meso-
L-L- diaminopimelic acid (DAP); Gram-negative bacteria including *E. coli*; some Gram-positive bacteria such as *Bacillus*
D-D-
diaminobutyric acid; *Lactobacillus, Clostridium, Mycobacterium*
L-Orn: *Micrococcus radiodurans, Lactobacillus cellobiosus*
D-Orn: *Corynebacterium beta*
OH–L-Lys: *Streptococcus faecalis*, some strains of *S. aureu*

Ⓒ *Cross bridge*
D-Ala–Gly–Gly–Gly–Gly–Gly$\overset{\varepsilon}{-}$L-Lys; S. *aureus*
D-Ala–L-Ala–L-Ala–L-Ala$\overset{\varepsilon}{-}$L-Lys; S. *thermophilus, S. faecalis*
D-Ala–L-Ala–L-Ala$\overset{\varepsilon}{-}$L-Lys; Group A *Streptococcus*
D-Ala–D-Asp$\overset{\varepsilon}{-}$L-Lys; Group N *Streptococcus*
D-Ala–*meso*-DAP; *E. coli, B. megaterium* KM1

Ⓓ Amide (NH_2–); *S. aureus*, Group A *Streptococcus*, etc.
Gly; *M. lysodeikticus*
(D-Glu)–δ-Orn–α-(D-Ala); bridge as *B. rettgeri, C. poinsettiae*

(a) Site of action of endo-*N*-acetylglucosaminidase
(b) Site of action of endo-*N*-acetylmuramidase
{T2, T4} PHAGE LYSOZYME
(c) Site of action of *N*-acetylmuramyl–L-aelanin amidase
(d) Site of action of peptidoglycan peptidases

(M) : muramic acid   (G) : glucosamine

Weidel's group[37,129] analyzed precisely the fractions released from *E. coli* cell-walls when digested with a crude preparation of the T2 enzyme as well as with egg-white lysozyme. They found identical chemical products in both digests. They also claimed that the specificity of T2 phage lysozyme is quite similar to that of egg-white lysozyme, and showed that these products could not be obtained in the crude lysate of uninfected cells. Maass and Weidel[33] finally provided evidence that T2 phage lysozyme is an endo-β-1,4-*N*-acetylglucosaminidase,[21] which is in conflict with the other results.

The optimal pH for enzyme action of λ phage endolysin is about pH 6 to 7.[55] The catalytic property of λ endolysin remains obscure. It has been known for some time that the specific activity of λ endolysin is ten times that of egg-white lysozyme with *E. coli* cell-walls as substrate. This was initially observed using impure preparations by Work[130] and Jacob *et al.*,[46] and later confirmed with the purified enzyme by Black and Hogness.[55] From our observations purified λ endolysin was found to be 317 times more active than egg-white lysozyme against *E. coli* cell-walls,[57] and this is almost equal to the activity of T4 lysozyme.[108] Black and Hogness[55] tested tri- and penta-*N*-acetylglucosamine derived from chitin but no catalytic activity was found with the λ endolysin. In a previous study of an impure preparation of λ endolysin with cell-walls of various bacteria, Work and Lecadet[48,131] isolated products similar to those produced by egg-white lysozyme, but observed some dissimilar

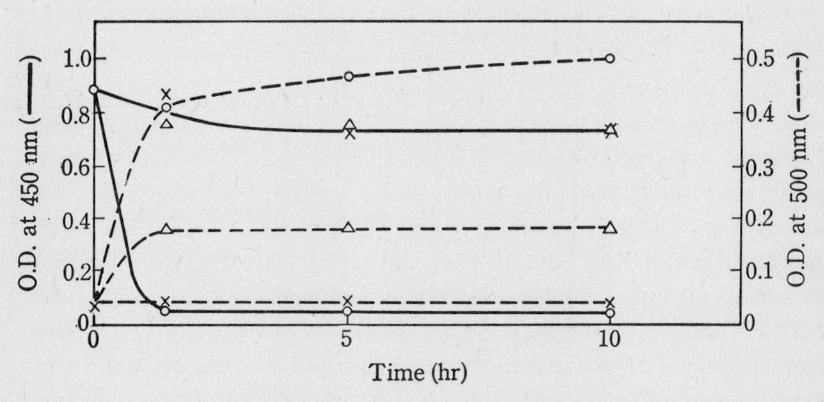

Fig. 14. Lytic activity of egg-white (○) and T4 phage (△) lysozyme and λ phage endolysin (×).

products when using cell-walls of *E. coli*. Work[130] also observed that a crude preparation of λ endolysin is active on cell-wall preparations of *E. coli* and many other kinds of bacteria. Black and Hogness found that the purified endolysin has no effect on *M. lysodeikticus* cell-walls; the catalytic rate in this enzyme is less than 0.001 of the rate observed for egg-white lysozyme. However, Imada in our laboratory[57] clearly observed a decrease in optical density of a suspension of an *M. lysodeikticus* cell-wall preparation,[126] incubated with purified enzyme as shown in Fig. 14. The rate of decrease in turbidity produced by λ endolysin is almost the same as that produced by the T4 lysozyme. No increase in the amount of reducing sugar, of muraminisitol after reduction, or of free NH_2 groups in the reaction mixture was found despite a careful search.

5.4.3. The Active Area of the Enzyme

The relationship between the enzyme structure and its activity has been vigorously examined by various methods including chemical modification, X-ray analysis, etc. Application of genetic methods is also useful in the study of this problem. Since most enzymes are known to be under the primary control of their corresponding genes, a modification of such genes, which may involve a base exchange and an addition or deletion of a base or bases, results in an alteration of the amino acid sequence controlled by the genes, and this in turn may subsequently affect their secondary and/or tertiary structures or even directly, the active area. The genetic approach to the active area of T4 lysozyme depends exclusively on the use of frameshift mutations and point mutations such as nonsense mutations.

A. Genetic Approach: Nonsense Mutations

It was reported by several groups[115,132,133] that Trp-62 in egg-white lysozyme plays an important role in its enzymatic action. In T4 phage lysozyme three tryptophan residues (126, 138 and 158) were found,[102] and it was of interest therefore to know if any of these residues are of special importance in the T4 phage lysozyme action mechanism. Experiments using a system of nonsense mutants and their revertants have been devised to answer this question. Streisinger isolated several amber (genetic code = UAG) mutants using a transitional mutagen, and from genetical analysis three were selected as being derived from the tryptophan residues (code = UGG). The locations of the amber mutations

TABLE 6

Mutants	Position	Suppressive bacteria su$^+$1 (Ser)	su$^+$2 (Gln)	su$^+$3 (Tyr)	su$^+$B (Gln?)	su$^+$C (Tyr)	Revertant*
Amber							
*e*M41	126	+	+	+(100)	+	+	+(100)(Tyr)
*e*M91	138	−	−	+ (60)	−	+	+ (60)(Tyr)
*e*M92	158	+	+	+ (60)	+	+	+(100)(Tyr)
*e*C1A	105	+	+	−	+	−	
*e*C2A	122	+	+	+	+	+	
*e*C3A	123	+	+	+	+	+	
*e*C4A	141	+	+	+	+	+	
Ochre							
*e*C1	105				+	−	+(100)(Trp)
*e*C2	122				+	+	+(100)(Trp)
*e*C3	123				+	+	
*e*C4	141				+	+	+ (20)(Arg)

The tests were carried out by plaque formation on plates using the respective suppressive bacteria at 37 °C overnight. + and − indicate positive and negative plaque formation, respectively, on the suppressive bacteria. In addition to these tests, amber mutants (*e*M41, *e*M91 and *e*M92), their revertants and the revertants of ochre mutants (*e*C2 and *e*C4) were directly tested using purified proteins. Numbers in parentheses indicate the specific activity of the lysozymes, which are corrected to a value of 100 for the e^+.

were confirmed from sequential studies of the suppressed proteins of these nonsense mutants[113] and the proteins of the spontaneous revertants of the amber mutants.[104,113]

Amber suppressive hosts, su$^+$1, su$^+$2 and su$^+$3 insert Ser, Gln and Tyr, respectively, for the amber codon. When harvested, the individual mutants in the above three kinds of host bacteria produced new lysozyme proteins, in which the original tryptophan residues were replaced by Ser, Gln or Tyr residues (Table 6). Thus, the tryptophan residues may be replaced by other (limited) kinds of amino acid residues one by one, and the activity of each newly formed enzyme can be measured. Experimental results showed that the Trp-126 and Trp-158 residues were replaceable by all the above three amino acids without a marked change in the enzyme activity, suggesting that these residues

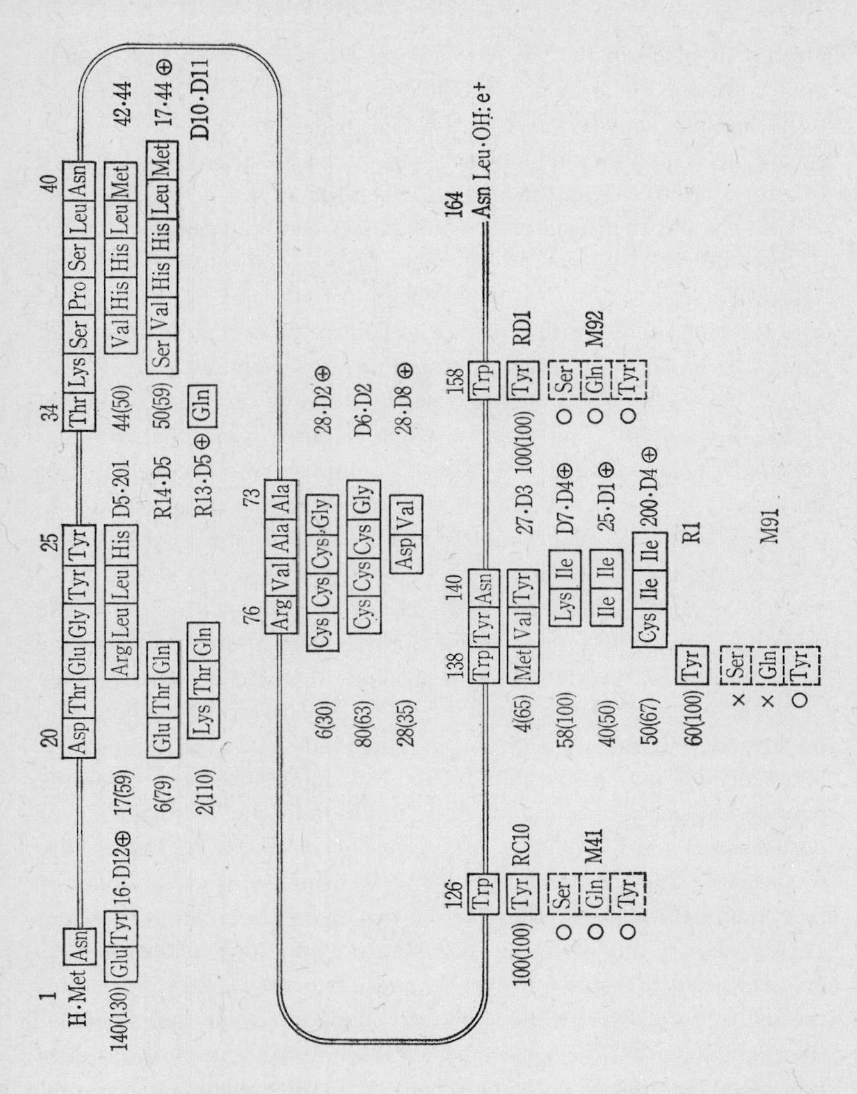

Fig. 15. Partial amino acid sequences of the e^+ lysozyme and the corresponding amino acid replacements in various mutants. Numbers indicate the enzyme activity and numbers in parentheses indicate the amount of lysozyme produced. M41, M91 and M92 are amber mutants, and RC10, R1 and RD1 are their corresponding revertants. The rest are frameshift mutants. Amino acid replacements by suppression of nonsense mutants are shown in broken-line rectangles, where ○ and × mean that active and inactive lysozyme, respectively, are produced by the particular amino acid inserted. ⊕ indicates that an additional amino acid was present.

are not involved in the catalytic action. However, when the Trp-138 amber mutant was grown in the suppressive hosts su$^+$1 (Ser) or su$^+$2 (Gln), no lytic activity was detected, and in su$^+$3 (Tyr), 40–50% of the specific activity of the wild type only was retained. The lysozyme protein of a revertant of this amber mutant, which has a tyrosine residue (code = UAU or UAC) instead of Trp-138, possessed about half the activity of the wild-type lysozyme.[105,113] These findings indicate that the Trp-138 residue is essential for the enzymatic activity and can be replaced by a tyrosine residue with retention of half of the original activity, but cannot be replaced by serine or glutamine with maintenance of activity.[104] The observations were confirmed by isolating a triple recombinant of the three amber revertants, in which all three tryptophan residues are replaced by tyrosine; the triple recombinant also retained 50% of the activity.[113] Further, Canfield[134] has reported that the essential Trp-62 of egg-white lysozyme is replaced by tyrosine in human lysozyme.

Five glutamine residues (code = CAA or CAG) are present at positions 69, 105, 122, 123 and 141 in the protein molecule. Four ochre mutants (UAA) have been isolated using transitional mutagens by Streisinger and Kneser and also in our laboratory.[106] From the mode of mutation, the mutants were thought to be derived only from glutamine residues (CAA). The genetic mapping indicated a correspondence between these four ochre mutants and four glutamine locations in the peptide sequence, with an ambiguity in the mutual correspondence of positions 121 and 122 (Kneser, *personal communication*). This ambiguity was recently clarified by sequential analysis using the revertant obtained from the opal mutant (UAG) derived from the ochre mutant (position 121).[107] The results of suppression studies and amino acid exchanges are also listed in Table 6. The 121- and 122-ochre mutants were suppressed by su$^+$ c (Tyr) without marked changes in their specific activities, indicating that the replacement (Gln/Tyr) at these positions does not affect the enzyme activity. When the 141-ochre mutant was suppressed by su$^+$ c, the same alteration (Gln/Tyr) caused about 50% decrease in specific activity. The 105-ochre mutant cannot be suppressed by the su$^+$ c host and no plaque is formed. This may indicate that this glutamine either plays an important role in the enzyme mechanism or is implicated in an active conformation.[106] The lysozyme proteins of revertants of opal mutant strains derived from the 122- and 141-ochre mutants were sequenced, and the results indicated the amino acid replacement, Gln_{122}/Trp, in both mutants. The specific activity of the revertant was almost identical with that of the wild type. Amino acid alteration in the revertant protein of an opal mutant derived from the 141-ochre mutant was found to be Gln/Arg, the specific activity being reduced by about 20%.[106] The code for the other glutamine residue, Gln-69, may be CAG. The results are further illustrated in the scheme, Fig. 15.

B. Genetic Approach: Frameshift Mutations

Pseudo-wild mutants carrying two (or three) frameshifts of bases are altered in more than one amino acid sequence between these mutations. Comparisons of the altered sequences in mutant lysozymes and their specific activities with those of the wild-type lysozyme should provide information about which amino acids are essential for enzyme activity.

The information derived from frameshift mutants together with that from nonsense mutants and their revertants is also summarized in Fig. 14. The specific activities of the individual frameshift mutants were obtained by direct measurements of the purified mutant proteins and/or by indirect methods from the enzyme activity and serological actitivy of the crude lysate.[9] In the figure, relative values of the specific activities of the mutant lysozymes compared with those of the wild-type enzymes are listed.

Among 15 frameshift mutants, eight (marked by ⊕) were found to contain an additional amino acid. In general, addition of an amino acid does not affect the catalytic activity, as can be seen in the mutant of *e*J16*e*J12, by comparison of *e*J42*e*J44 with *e*J17*e*J44,[2,6] in the mutants of *e*JD7*e*JD4, *e*J25*e*JD1 and *e*J200*e*JD4,[5,7] and in a set of mutants of *e*JR13*e*JD5 with *e*JR14*e*JD5.[4] The enzyme activity of *e*JR13*e*JD5, which is remarkably low, does not differ much from that of *e*JR14*e*JD5; it contains no additional amino acid but an altered amino acid common to *e*JR14*e*JD5 in the same (20 to 22) region. Thus, the addition of one amino acid in *e*JR13*e*JD5 seems not to affect the enzyme activity.

However, in a particular case, the enzyme activity was found to be markedly decreased with the addition of an amino acid (*e*J28*e*JD2), but no decrease was observed in the other relevant mutant (*e*J28*e*JD8).[10] It may appear that certain kinds of altered amino acids in *e*J28*e*JD2 cause the decrease in enzyme activity; however, the same alteration in the sequence without any addition of amino acid in the mutant *e*JD6-*e*JD2 does not cause any activity decrease. Thus, addition of one amino acid does not itself affect the enzyme activity, but addition together with a kind of replaced amino acid may affect the catalytic activity. A possible explanation for this particular example is that the configurational change caused by the addition of an amino acid may expose the cysteine residue on the surface of the enzyme protein, and this in turn may have a marked effect on its catalytic properties or stability.

The types of altered amino acids cause marked yet differing effects on the catalytic activity. When the e^+ sequence of Asp_{20}–Thr_{21}–Glu_{22} was replaced by Glu–Thr–Gln (*e*JR14*e*JD5) or by Asp–Lys–Thr–Gln (*e*JR13*e*JD5), the enzyme activity was found to be substantially decreased. From the replacements, one may predict that either Asp-20 or Glu-22, or a combination of them, is important for the enzyme activity.

Another alteration, Glu_{22}–Gly_{23}–Tyr_{24}–Tyr_{25} to Arg–Leu–Leu–His (*e*JD5*e*J201), involving replacement of Glu-22 but not Asp-20, again results in a low but somewhat higher activity than in the case of both *e*JR13*e*JD5 and *e*JR14*e*JD5. These differences indicate that modification of Asp-20 is probably at least in part responsible for the low enzyme activities of both *e*JR13*e*JD5 and *e*JR14*e*JD5. The considerably reduced activity of *e*JD5*e*J201, when compared with e^+, is caused by the replacement of four amino acids. The disappearance of Glu-22 and/or of other functional amino acid(s) such as tyrosine, as well as the appearance of the basic amino acids, arginine and histidine, is a plausible causal factor of the low activity. Although we cannot conclude which amino acid is the more important, Asp-20 together with other amino acids in this region (20–25) seems to play an important role in the enzyme activity.[9]

Observations of a set of alterations in the frameshift mutants around Trp_{138}–Tyr_{139}–Asn_{140} again suggest an important role for Trp-138. When the replacement is by Met–Val–Tyr (*e*J37*e*JD3) the enzyme activity is to a marked extent lost, whilst the following replacements (which do not involve tryptophan) do not greatly affect the enzyme activity: Trp–Tyr–Ile–Ile (*e*J25*e*JD1), Trp–Cys–Ile–Ile (*e*J200*e*JD4), and Trp–Tyr–Lys–Ile (*e*JD7*e*JD4).[5,7] In the last mutant protein, the basic amino acid lysine which is adjacent to tryptophan, has an influence on the function of the tryptophan residue. This is also comparable with the decrease of activity which results from a replacement (Gln/Arg) at position 141.[107] Relationships between the amino acid sequences from positions 138 to 141 and their enzymatic activity may be expressed as follows:

Trp_{138}–Tyr–Asn–Gln_{141} (100) = Trp–Tyr–*Ile*–*Ile*–Gln (100) = Trp–*Cys*–

Ile–Ile–Gln (100) > *Tyr*–Tyr–Asn–Gln (50) = Trp–Tyr–*Lys–Ile*–Gln (50)> Trp–Tyr–Asn–*Arg* (20) > *Met–Val–Tyr*–Gln (5) > *Ser*–Tyr–Asn–Gln (0) = *Gln*–Tyr–Asn–Gln (0), where values in parentheses indicate the changed specific activities.

C. Other Approaches

Lysozyme of T2 phage has the same specific enzyme activity as T4 lysozyme, but somewhat low activity against the antiserum of T4 lysozyme.[109] This would indicate that amino acid replacements exist which do not affect enzyme activity: these were found to be Ser_{40}/Asn, Val_{41}/Ala, and Ala_{151}/Thr (refer to Fig. 9).[109]

An additional preliminary experiment using exopeptidase was carried out to investigate any correlation between the terminal structure and the catalytic activity. Of the aminopeptidases, neither leucine aminopeptidase from a bovine kidney nor bacterial aminopeptidase M caused any liberation of amino acid without inactivation of the enzyme. On the other hand, carboxypeptidase liberated approximately one mole each of leucine and asparagine without noticeable loss of enzyme activity. The two *C*-terminal amino acids of the phage lysozyme were consequently suggested to be nonessential for enzyme activity (*unpublished result*).

The above results may be summarized as follows. Those amino acids or amino acid sequences which are not essential for catalytic activity or for polypeptide conformation are Asn-2, Gly_{23}–Tyr–Tyr_{25}, Thr_{34}–Lys–Ser–Pro–Ser–Leu–Asn–Ala_{41}, Ala_{73}–Ala–Val–Arg_{76}, Gln_{122}–Gln_{123}, Trp-126, Tyr-139, Thr-151, Trp-158 and Asn_{163}–Leu_{164}. It is suggested that the following amino acids are involved directly or indirectly in the essential area of catalytic activity: Asp-20, (Gln-22), Gln-105, Trp-138, (Asn-140) and (Gln-141), where the amino acids in parentheses are those considered to be more indirectly involved.

5.5. Immunochemical Properties

Antiserum of T4 phage lysozyme was prepared from an immunized rabbit using Freund's adjuvant.[109] The structural similarity between T4 lysozyme and lysozymes isolated from other sources has been investigated using the T4 lysozyme antiserum. The production of lysozyme proteins directed by various mutant genomes has also been studied by immunochemical methods.

5.5.1. Assay Method

Two different procedures for the measurement of immunochemical activity have been carried out.

(i) *Assay of the residual enzyme activity (neutralizing activity).* The reaction mixture, which consists of a definite amount of the lysozyme under test and various amounts of e^+ lysozyme antiserum, is incubated and the residual activity measured by conventional techniques.

(ii) *Quantitative inhibition test for cross-reacting material.*[135] Various amounts of mutant lysozyme are mixed with a sufficient quantity of the e^+ lysozyme antiserum to inhibit 0.07 enzyme unit of e^+ lysozyme, and after incubation for 10 min at 30°C, a slight excess of the e^+ lysozyme (0.08 unit) is added. After a further 10 min incubation period the remaining e^+ lysozyme activity is assayed.

5.5.2. The Structural Similarity between T4 Phage Lysozyme and Other Lytic Enzymes

Inouye in our laboratory has examined the effect of T4 lysozyme antiserum on the enzyme activities of T4 and T2 phages.[109] It was found that both enzyme activities decreased with increasing amounts of antibody, although the inhibitory effect of the antiserum (of T4 lysozyme) on the activity of T4 phage lysozyme was stronger than that on the activity of T2 phage lysozyme (see Fig. 16). This suggests that the structures of T2 and T4 phage lysozymes are closely related, and structural studies have shown that three amino acid replacements are involved.[109]

Black and Hogness[56] suggested the existence of structural similarities between λ phage endolysin and T4 and T2 lysozymes from a comparison of their amino acid compositions and the NH_2-termini. Double diffusion tests of λ endolysin and T4 lysozyme against anti-λ-endolysin serum and anti-T4 lysozyme–γ-globulin were carried out in agar. Homologous antigen-antibody precipitation lines were observed, but there was no evidence of heterologous interaction by this technique. The effects of λ-endolysin antiserum and anti-T4 lysozyme–γ-globulin on the enzyme activity were then examined. The T4 lysozyme activity was indeed inhibited by the λ-endolysin antiserum, although this cross-reaction was very weak and the amount of antiserum required was about 100-fold

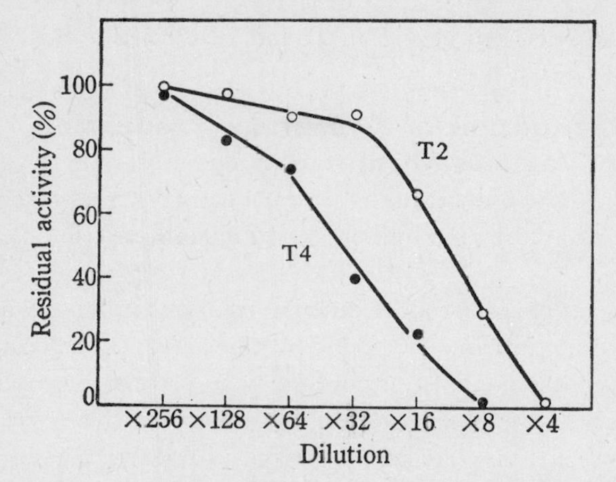

Fig. 16. Inhibitory effect of anti-T4 phage lysozyme on the activity of T2 phage lysozyme (○) and T4 phage lysozyme (●). Phage lysozymes were incubated with variously diluted antisera of T4 lysozyme (final dilution in the reaction mixture, 4 to 256 times) in 0.2 M phosphate buffer (pH 5.9) containing 0.5 M NaCl, 0.5 mg protamine/ml and 0.02% mercaptoethanol at room temperature for 10 min. Enzyme concentrations in the reaction mixture were 14 and 13 u/ml for T2 and T4 lysozymes, respectively. After incubation, the residual activities were measured with dried cells of *E. coli* as substrate.[102]

greater than that required for the heterologous reaction. No inhibition of the λ-lytic activity was observed by anti-T4 lysozyme–γ-globulin, even at levels about 20-fold greater than necessary for complete inhibition of an equivalent weight of T4 phage lysozyme. It was concluded that immunochemical cross-reactions occurred between the λ and T4 enzymes and that the weakness of the cross-reaction between configurationally similar chains might be due to different but functionally related side groups at various positions along the peptide chain.

Similar types of experiments have also been carried out in our laboratories (Inouye, *unpublished*) using λ endolysin and T4 lysozyme antiserum. The results clearly indicated that the antiserum did not inhibit enzyme activity, suggesting the absence of any structural similarity between λ endolysin and T4 lysozyme. In fact, when the amino acid sequences of λ endolysin and T4 or T2 lysozyme are compared, no apparent similarity can be observed except for a few short peptide sequences, such as Arg–Trp–Asp (70–72 in λ endolysin; 125–127 in T2 and T4 lysozyme), which are known to be nonessential areas for lysozyme activity in the lysozymes, Arg–Cys (116–117 in λ; 96–97 in T4) and Phe–Leu–Asp–Met–Leu (11–15 in λ) and Ile–Phe–Glu–Met–Leu (3–7 in T4).

5.5.3. Estimation of Lysozyme Production in Various Mutant Strains

Enzyme proteins which do not show enzyme activity (cross-reacting material) can be assayed by the immunological quantitative inhibition test.[135]

Various e^+ mutant phages produce different lysozyme proteins which differ from each other in terms of specific enzyme activity. The protein concentrations of these mutant lysozymes may be assayed by the immunological quantitative inhibition test, regardless of their specific activities. Evidence for this has been obtained as follows. The e^+ lysozyme antiserum was used, and each crude lysate and purified protein of *e*JR13*e*JD5 and *e*JR14*e*JD5 was assayed with those of e^+ as reference. The results are given in Table 7, in which the values are standardized by a plaque forming unit in the case of the measurements of the crude lysates, and by the protein concentrations measured by some other independent method such as ultraviolet absorbancy in the case of the purified protein. All these values are expressed as relative values compared with those of e^+ lysozyme.

Values both for the crude lysates and for the purified proteins are found to be in good agreement for the mutants listed here, as is also

TABLE 7
Enzyme Activity and Amount of Lysozyme Produced by *e*JR13*e*JD5, *e*JR14*e*JD5 and *e*JD5*e*J201

	Enzyme activity		Amount of enzyme produced	
	Lysate	Purified protein	Lysate	Purified protein
*e*JR13*e*JD5	2	1	110	70
*e*JR14*e*JD5	6	3	80	70
*e*JD5*e*J201	17	—	60	—
e^+	100	100	100	100

The amount of lysozyme was determined by the serological quantitative inhibition test. Assays for each mutant were carried out on both crude lysate and purified protein. The phage lysates were prepared as follows: bacteria (7 ml), grown to 8×10^8/ml in Fraser's medium at 37°C, were infected with mutant or e^+ phage at a multiplicity of 0.01. The infected cells were incubated for 5–7 hr and stored in ice with chloroform overnight to obtain the lysate. All results are normalized to a value of 100 for e^+.

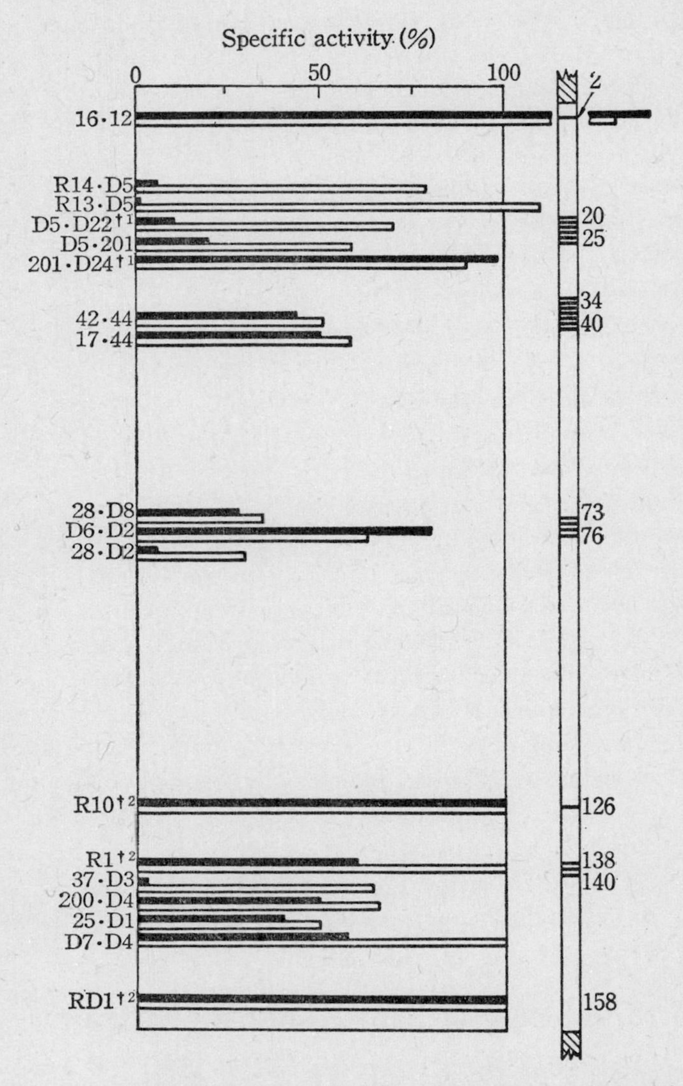

Fig. 17. Summary of amino acid replacements, enzyme activity and amount of lysozyme produced by frameshift and nonsense mutants. *e*J and *e* are omitted from the names of frameshift and nonsense mutants, respectively. Assays were carried out on crude lysates; the results are normalized to a value of 100 for e^+. Filled bars indicate enzyme activity and open bars indicate the amount of lysozyme, which was determined by the serological test. (†1 Amino acid replacements have not yet been determined. †2 Data reported previously by Inouye *et al.*)

found in many other cases. Hence, values derived from crude lysates can be considered to be equivalent to the relative activities observed with purified enzymes. On the other hand, we may assume that the immunological activities per plaque forming unit, i.e. the value for the crude lysate, correspond roughly with the amounts of enzyme protein produced in the cell, although in some exceptional cases the partial structural alterations in the various mutant proteins may affect the immunological activity. Thus we measured serological activities of the crude lysates of various available mutants, including three amber mutants[104] and their revertants, revertants derived from ochre mutants[107] and 13 frameshift mutants,[2-11] whose amino acid alterations cover 15% of the entire protein.[9] The results are summarized in Fig. 17. The production of mutant lysozymes was found to be almost the same as that from the e^+, although the enzyme activities varied in a wide range from 1% to 100% as discussed in section 5.4.3.

The results indicate that the alteration of certain amino acids in the lysozyme molecule has a great effect on the enzyme activity but a relatively small effect on enzyme production. However, in particular cases (*e*J28*e*JD2 and *e*J28*e*JD8) the immunological activities were found to be decreased by ⅓ of that of the e^+. In the relevant mutant (*e*JD6-*e*JD2) the immunological activity was found to be similar but slightly less in amount to that of e^+. In the former case, but not in the latter, an amino acid was inserted. This addition possibly causes a configurational change in the enzyme protein and has been discussed in section 5.4.3. with reference to the active area of the enzyme. One may confidently predict that addition of an amino acid, and the resulting configurational change, may affect the immunological behavior of the mutant lysozyme.

6. The Study of Some Other Biologically Interesting Problems with the T4 Phage Lysozyme System

6.1. The Genetic Code

Individual results obtained from various mutants carrying two or three frameshift mutations are summarized in Fig. 18, in which the altered amino acid sequences and corresponding base sequences for the

(a)

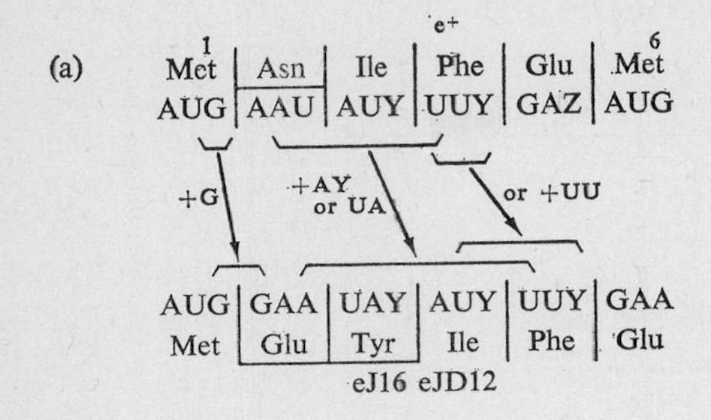

(b)

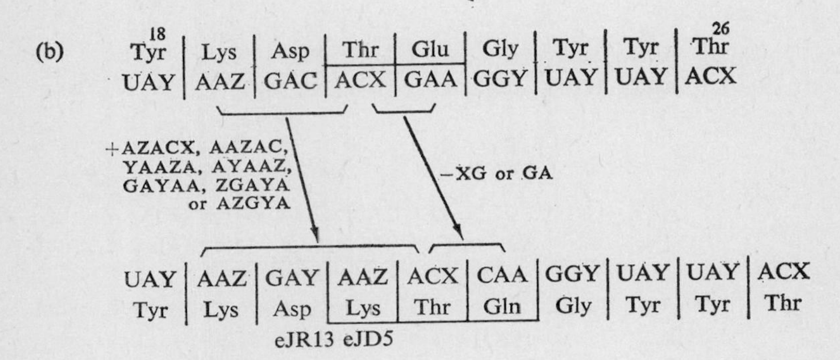

(c)

18 Tyr | Lys | Asp | Thr | Glu | Gly | Tyr | Tyr | 26 Thr

UAY | AAZ | GAC | ACX | GAA | GGY | UAY | UAY | ACX

+ZZ

−XG or GA

UAY | AAZ | GAZ | ACA | CAA | GGY | UAY | UAY | ACX

Tyr | Lys | Glu | Thr | Gln | Gly | Tyr | Tyr | Thr

eJR14 eJD5

(d)

18 Tyr | Lys | Asp | Thr | Glu | Gly | Tyr | Tyr | 26 Thr

UAY | AAZ | GAC | ACX | GAA | GGY | UAY | UAY | ACX

−XG or GA

+CX, XG, AU or UA

UAY | AAZ | GAC | ACX | AGG | YUA | CUX UUZ | CAY | ACX

Tyr | Lys | Asp | Thr | Arg | Leu | Leu | His | Thr

eJD5 eJ201

(e)

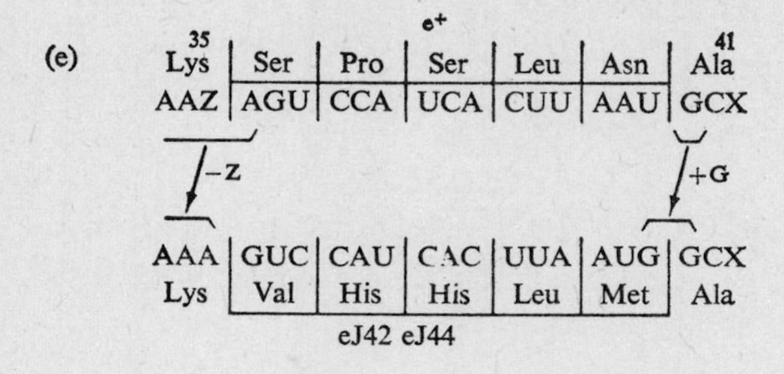

(f)

e+

35 Lys | Ser | Pro | Ser | Leu | Asn | 41 Ala

AAZ | AGU | CCA | UCA | CUU | AAU | GCX

+GU or UG   +G

AAZ | AGU | GUC | CAU | CAC | UUA | AUG | GCX

Lys | Ser | Val | His | His | Leu | Met | Ala

eJ17 eJ44

(g)

e+

32 Leu Leu | Thr | Lys | Ser | Pro | 38 Ser

. . . . . . | ACZ | AAZ | AGU | CCA | UCA

−1   −Z   +GU or UG

. . . . . . | CAA | AAG | UGU | CCA | UCA

(Leu Leu | Gln) | Lys | Cys | Pro | Ser

eJD10 eJ42 eJ17

(h)

e+

32 Leu Leu | Thr | Lys | Ser | 37 Pro

. . . . . . | ACZ | AAZ | AGU | CCA

−X   +Z

| CAZ | AAZ | AGU | CCA

(Leu Leu | Gln) | Lys | Ser | Pro

eJD10 eJD11

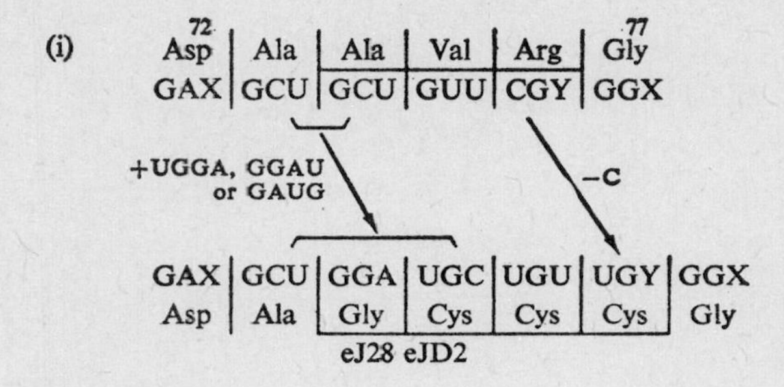

(j)

72 77

Asp | Ala | Ala | Val | Arg | Gly

GAX | GCU | GCU | GUU | CGY | GGX

+G −C

GAX | GGC | UGC | UGU | UGY | GGY

Asp | Gly | Cys | Cys | Cys | Gly

eJD6 eJD2

(k)

72 77

Asp | Ala | Ala | Val | Arg | Gly

GAX | GCU | GCU | GUU | CGY | GGX

−C +UGGA, GGAU or GAUG

GAX | GUG | GAU | GCU | GUU | CGY | GGX

Asp | Val | Asp | Ala | Val | Arg | Gly

eJD8 eJ28

(l)

137 141

Arg | Trp | Tyr | Asn | Gln

| UGG | UAU | AAY | CAX

+X −A

| AUG | GUA | UAU | CAX

Arg | Met | Val | Tyr | Gln

eJ37 eJD3

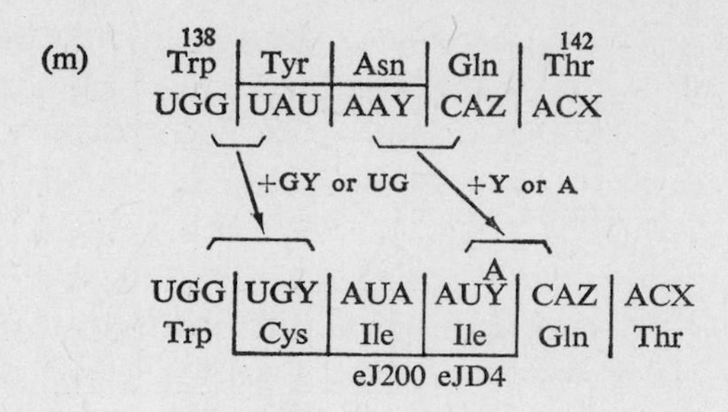

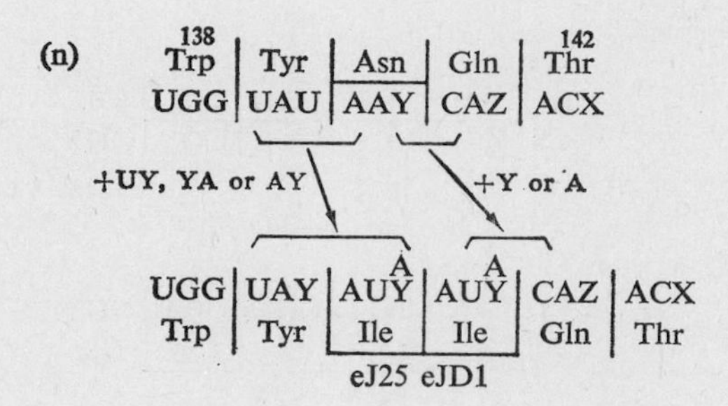

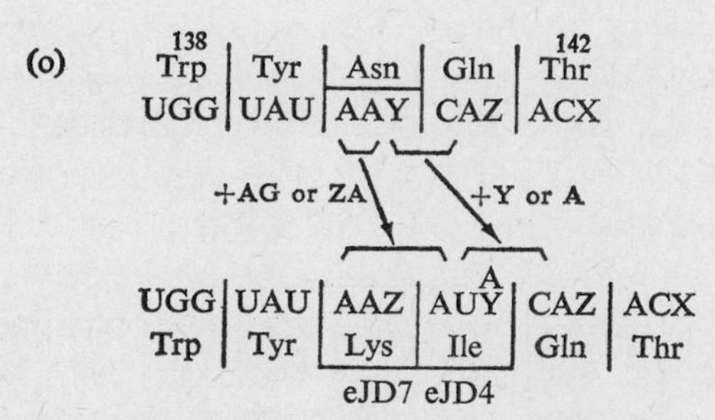

Fig. 18. Alteration of codon assignment with amino acid caused by frameshift mutation. X stands for U, C, A or G; Y for U or C; Z for A or G.

mutant strain lysozymes are presented in comparison with those of the e^+ strain. Eleven proflavin induced mutations and ten spontaneous mutations were present in the 15 pseudo-wild type revertants analyzed,[2–11] and in addition, seven nonsense mutants have also been analyzed.[105–107,112,113] In Fig. 19, the proposed codon assignments in the e^+ gene and in the mutants, respectively, are tabulated.

Twenty-nine codons out of 61 possible sense codons were identified and these covered 19 amino acids out of the 20 naturally occurring ones, the exception being phenylalanine.[9]

In the above assignments, ten codons identified in the e^+ strain mainly ended with either A or U, apart from UGG and AUG for tryptophan and methionine, respectively, each of which has only one codon ending with G. Fourteen codons found in the e^+ strain ended with either A or U, one ended with C, and three tryptophan codons and one methionine codon ended with G. However, codons in the mutant strains were found to end with random bases (see Table 8). Since the DNA of phage T4 is relatively rich in A-T base pairs (GC/AT = 0.53), the above observations would indicate that *the base composition of the DNA reflects the base composition at the third position of the codon.*

Several methods for sequencing the *m*RNA of a specific protein or viral RNA have been developed. Knowledge of the frameshift mutation system has also been of help in this field. *About 15% of the mRNA sequence corresponding to the T4 lysozyme gene was clarified* from a knowledge of the e^+ codon sequences.[9]

A plausible mechanism for the frameshift mutation was predicted from the detailed sequences of the mutated regions. *The mutation could occur either at the ends of DNA molecules, or internally in heterozygous regions.* At the end of a molecule an exonuclease is assumed to digest one of the two chains; the mutation occurs during resynthesis and is a consequence of mispairing. In a heterozygous region the mutation is also a consequence of mispairing and occurs during the synthesis that transforms a joint molecule into a hybrid molecule. The acridines can be mutagenic, not because they stretch the DNA molecule but rather because they stabilize it.[10,11]

The acridines seem predominantly to induce additions of bases and

TABLE 8

Frequencies of Third Bases in Codon Triplets Found in Wild-type and Mutant Strains

Base	Wild	Mutant
A	7	8
U	7	6
G	4†	5
C	1	4

†G was found only in methionine and tryptophan.

1st	2nd U	2nd C	2nd A	2nd G	3rd
U	Phe	Ser	Tyr ●	Cys ○○	U
	Phe	Ser	Tyr	Cys ○	C
	Leu ○	Ser ●	Ochre	Opal	A
	Leu	Ser	Amber	Trp ● ▲▲▲	G
C	Leu ●	Pro	His ○	Arg	U
	Leu	Pro	His ○	Arg	C
	Leu	Pro ○ ●	Gln ▲▲▲▲ ○○	Arg	A
	Leu	Pro	Gln	Arg	G
A	Ile	Thr	Asn ●	Ser ● ○	U
	Ile	Thr	Asn	Ser	C
	Ile ○	Thr ○	Lys ○	Arg	A
	Met ● ○○	Thr	Lys ○	Arg ○	G
G	Val ●	Ala ●● ○	Asp ○	Gly	U
	Val ○	Ala	Asp ●	Gly ○	C
	Val ○	Ala	Glu ●	Gly ○	A
	Val ○	Ala	Glu	Gly	G

Fig. 19. Codon dictionary: a summary of codon assignments and their corresponding amino acids which are utilized in T4–infected *E. coli*. The codons which have been found so far from our studies to be utilized *in vivo* are shown in frames. Codons with circles have been identified from studies on nonsense mutations; the frequency of observation is shown by the number of symbols. Filled symbols indicate the codons utilized in the e^+ strain and open ones those in mutant strains.

very rarely to induce deletions as observed not only in the first mutational events but also in secondary mutations. Further, proflavin induced mutations tend to induce the addition of more than two bases rather than a single base.[10]

These results would generally seem to support our predicted mechanism, namely that the acridines are mutagenic not solely because they may induce stretching but because they have a stabilizing effect on hydrogen bonding in the mispaired region. In addition, the frameshift study finalized the direction of translation of the lysozyme gene of phage T4 relative to the linkage map.[11]

6.2. The Ending Code in Protein Synthesis

The variation of suppression efficiencies was analyzed among two sets of nonsense mutations located at different positions on the lysozyme gene of phage T4.[105–107] In one case the three amber mutants derived from tryptophan residues were suppressed by bacterial suppressors which insert tyrosine as the suppressed amino acid, and in addition the individual revertants derived from these nonsense mutants were all found to be replaced with tyrosine residues. Comparative estimates of the immunological activities and enzymatic activities of these three sets of suppressed nonsense mutants and their revertants were made, when a clear variation in the individual suppression efficiencies was observed due to the locations of the nonsense mutations.[105]

In another set, suppression efficiencies of four nonsense mutants derived from the glutamine residues, which were able to be suppressed by glutamine suppressor bacteria (Su II and Su B), were analyzed with reference to the wild-type strain. Noticeable variations of the suppression efficiencies were observed along the locations of the nonsense mutants.[106] The variation could be considered as being due to the effect of the base sequences adjacent to the nonsense mutations. This "adjacent" effect represents a new and important characteristic of the ending mechanism.[105,106]

Further studies are clearly required if molecular details of the adjacent codon effect on the linear genetic information are to be defined, and perhaps support the concept that not only triplet base sequences but also sextuplet sequences in *m*RNA may interact with other cell constituents such as the ribosome and its active environment and so express genetic control of the ending mechanism for protein translation. The codon context such as the adjacent cluster may play other important roles in the controlling mechanism in living cells.

The above kinds of study can be carried out only with a biological system that is furnished with well established genetic information and a known protein sequence. Since the work on the T4 phage lysozyme was initiated in the field of genetics, other possible and more conventional approaches used in enzyme investigation have been left undiscussed in this review. Also, several new, unusual chemical modifications are now in progress with the phage lysozymes, in which alterations of activity, the location of modified amino acids, accompanying physical conformational changes and variations in serological activity are to be investigated. A new genetic approach to the characterization of the catalytic region of enzyme proteins, as exemplified by T4 phage lysozyme, has been developed.

Acknowledgments

The original investigations cited in this review were supported in part by research grants from the National Institute of Health, U.S.A. and The Jane Coffin Childs Memorial Fund for Medical Research, U.S.A. Parts of this review have been covered by an article appearing in *The Enzymes, III* (ed. P. Boyer), Academic Press. The authors wish to thank Dr. M. Imada for making available his results prior to publication. Miss M. Kobayashi is acknowledged for her excellent secretarial help in the preparation of this paper.

References

1. F. H. C. Crick, L. Barnett, S. Brenner and R. J. Watts-Tobin, *Nature*, **192**, 1227 (1961).
2. E. Terzaghi, Y. Ocada, G. Streisinger, J. Emrich, M. Inouye and A. Tsugita, *Proc. Natl. Acad. Sci. U.S.*, **56**, 500 (1966).
3. Y. Ocada, E. Terzaghi, G. Streisinger, J. Emrich, M. Inouye and A. Tsugita, *ibid.*, **56**, 1692 (1966).
4. M. Inouye, E. Akaboshi, A. Tsugita, Y. Ocada and G. Streisinger, *J. Mol. Biol.*, **30**, 39 (1967).
5. M. J. Lorena, M. Inouye and A. Tsugita, *Mol. Gen. Gen.*, **102**, 69 (1968).
6. Y. Ocada, G. Streisinger, J. Emrich, J. Newton, A. Tsugita and M. Inouye, *Science*, **162**, 807 (1968).
7. A. Tsugita, M. Inouye, T. Imagawa, T. Nakanishi, Y. Ocada, J. Emrich and G. Streisinger, *J. Mol. Biol.*, **41**, 349 (1969).
8. Y. Ocada, G. Streisinger, J. Emrich, A. Tsugita and M. Inouye, *ibid.*, **40**, 299 (1969).
9. Y. Ocada, S. Amagase and A. Tsugita, *ibid.*, **54**, 219 (1970).
10. M. Imada, M. Eda, M. Inouye and A. Tsugita, *ibid.*, **54**, 199 (1970).
11. G. Streisinger, Y. Ocada, J. Emrich, J. Newton, A. Tsugita, E. Terzaghi and M. Inouye, *Cold Spring Harb. Symp. Quant. Biol.*, **31**, 77 (1966).

12. F. W. Twort, *Lancet*, **2**, 1241 (1915).
13. F. d'Herelle, *Compt. Rend. Soc. Biol.*, **165**, 373 (1917).
14. F. d'Herelle, *The Bacteriophage and its Behavior*, Williams and Wilkins Co., Baltimore, 1926.
15. J. Bronfenbrenner and R. Muckenfuss, *J. Exptl. Med.*, **45**, 887 (1927).
16. V. Steric, *Zentr. Bakteriol. Parasitenk. Abt. I. Orig.*, **110**, 125 (1929); *Compt. Rend. Soc. Biol.*, **126**, 1074 (1936).
17. C. J. Schuurman, *Zentr. Bakteriol. Parasitenk. Abt. I. Orig.*, **137**, 438 (1936).
18. E. Wollman and E. Wollman, *Compt. Rend. Soc. Biol.*, **112**, 164 (1933).
19. E. Wollman, *ibid.*, **115**, 1616 (1934).
20. A. Pirie, *Brit. J. Exptl. Pathol.* **29**, 1939 (1939).
21. L. F. Barrington and L. M. Kozloff, *J. Biol. Chem.*, **223**, 615 (1956).
22. L. F. Barrington and L. M. Kozloff, *Science*, **120**, 110 (1954).
23. L. M. Kozloff and M. Lute, *ibid.*, **234**, 539 (1959).
24. A. Brown, *J. Bacteriol.*, **71**, 482 (1956).
25. G. Koch and W. Weidel, *Z. Naturforsch.*, **11b**, 345 (1956).
26. G. Koch and E. M. Jordan, *Biochim. Biophys. Acta*, **25**, 437 (1957).
27. D. D. Brown and L. M. Kozloff, *J. Biol. Chem.*, **225**, 1 (1957).
28. G. Koch, *Fortschr. Botan.*, **19**, 412 (1957).
29. W. Weidel and J. Primosigh, *Z. Naturforsch.* **12b**, 421 (1957).
30. G. Koch and W. J. Dreyer, *Virology*, **6**, 291 (1958).
31. W. Weidel, H. Frank and H. H. Martin, *J. Gen. Microbiol*, **22**, 158 (1960).
32. J. Primosigh, H. Pelzer, D. Maass and W. Weidel, *Biochim. Biophys. Acta*, **46**, 68 (1960).
33. D. Maass and W. Weidel, *ibid.*, **78**, 369 (1963).
34. W. Katz and W. Weidel, *Z. Naturforsch.*, **16b**, 363 (1961).
35. W. Katz, *ibid.*, **19b**, 129 (1964).
36. R. R. Burgess, A. A. Travers, J. J. Dunn, and E. K. Bautz, *Nature*, **221**, 43 (1969).
37. E. K. F. Bautz and J. J. Dunn, *Biochem. Biophys. Res. Commun.*, **34**, 230 (1969).
38. A. A. Travers, *Nature*, **223**, 1107 (1969).
39. E. K. F. Bautz, F. A. Bautz and J. J. Dunn, *ibid.*, **223**, 1022 (1969).
40. A. A. Travers, *ibid.*, **225**, 1009 (1970).
41. T. Kasai and E. K. F. Bautz, *J. Mol. Biol.*, **41**, 401 (1969).
42. T. Kasai, E. K. F. Bautz and W. Szybalski, *ibid.*, **34**, 709 (1968).
43. W. Salser, R. F. Gesteland and A. Bolle, *Nature*, **215**, 588 (1967).
44. W. Salser, R. Gesteland and B. Ricard, *Cold Spring Harb. Symp. Quant. Biol.*, **34**, 771 (1969).
45. M. Schweiger and L. M. Gold, *Proc. Natl. Acad. Sci. U.S.*, **63**, 1351 (1969).
46. F. Jacob, C. Fuerst and E. Wollman, *Ann. Inst. Pasteur*, **93**, 724 (1957).
47. F. Jacob and C. R. Fuerst, *J. Gen. Microbiol.*, **18**, 518 (1958).

48. E. Work, *ibid.*, **25**, 167 (1961).
49. A. Campbell, *Virology*, **14**, 22 (1961).
50. A. Del Campillo-Campbell and A. Campbell, *Biochem. Z.*, **342**, 485 (1965).
51. D. S. Hogness, W. Doerfler, J. B. Egan and L. W. Black, *Cold Spring Harb. Symp. Quant. Biol.*, **31**, 129 (1966).
52. P. M. Naha, *ibid.*, **29**, 676 (1966).
53. R. Thomas, C. Leurs, C. Dambly, D. Parmenter, L. Lambert, P. Brachet, N. Lefebvre, S. Mousset, J. Porcheret, J. Szpirer and D. Wauters, *Mutation Res.*, **4**, 735 (1967).
54. L. W. Black and D. S. Hogness, *J. Biol. Chem.*, **244**, 1968 (1969).
55. L. W. Black and D. S. Hogness, *ibid.*, **244**, 1972 (1969).
56. L. W. Black and D. S. Hogness, *ibid.*, **244**, 1982 (1969).
57. M. Imada and A. Tsugita, *Nature*, in press.
58. J. Emrich and G. Streisinger, *Virology*, **36**, 387 (1968).
59. R. Josslin, *ibid.*, **40**, 719 (1970).
60. L. D. Simon and T. F. Anderson, *Virology*, **32**, 279 (1967).
61. M. A. Jesaitis and W. F. Geobel, *J. Exptl. Med.*, **96**, 409 (1952).
62. W. Weidel, *Ann. Rev. Microbiol.*, **12**, 27 (1958).
63. S. Brenner, G. Streisinger, R. W. Horne, S. P. Champe, L. Barnett, S. Benzer and M. W. Roes, *J. Mol. Biol.*, **1**, 281 (1958).
64. R. Takata and A. Tsugita, *ibid.*, **54**, 45 (1970).
65. S. Imada and A. Tsugita, *Mol. Gen. Gen.*, **109**, 338 (1970).
66. R. S. Edgar and I. Lielausis, *Genetics*, **52**, 1187 (1965).
67. J. King and W. B. Wood, *J. Mol. Biol.*, **39**, 583 (1969).
68. J. H. Wilson, R. B. Luftig and W. B. Wood, *ibid.*, in press.
69. R. S. Edgar and W. B. Wood, *Proc. Natl. Acad. Sci. U.S.*, **55**, 498 (1966).
70. C. To, *Ph.D. Thesis*, Kansas State Univ., 1968.
71. C. Veldhuisen and E. B. Goldberg, *Methods in Enzymol.*, **XIIB**, 858 (1968).
72. W. Weidel and E. Kellenberger, *Biochim. Biophys. Acta*, **17**, 1 (1955).
73. P. P. Dukes and L. M. Kozloff, *J. Biol. Chem.*, **234**, 534 (1959).
74. J. King, *J. Mol. Biol.*, **32**, 231 (1968).
75. M. Delbruck, *J. Gen. Physiol.*; **23**, 643 (1940).
76. K. K. Mark, *New Asia College Acad. Ann.*, **XI**, 31 (1969).
77. T. T. Puck and H. H. Lee, *J. Exptl. Med.*, **99**, 481 (1954).
78. T. T. Puck and H. H. Lee, *ibid.*, **101**, 151 (1955).
79. F. Mukai, G. Streisinger and B. Biller, *Virology*, **33**, 398 (1967).
80. A. H. Doermann, *J. Bacteriol.*, **55**, 257 (1948).
81. J. Emrich, *ibid.*, **35**, 158 (1968).
82. G. S. Stent and O. Maale, *Biochim. Biophys. Acta*, **10**, 55 (1953).
83. A. H. Doermann, *J. Gen. Physiol.*, **35**, 645 (1952).
84. J. Sechaud, E. Kellenberger and G. Streisinger, *Virology*, **33**, 402 (1967).

85. S. Silver, *J. Mol. Biol.*, **29**, 191 (1967).
86. R. C. French and L. Siminovitch, *J. Microbiol.*, **1**, 757 (1955).
87. R. M. Herriott and J. L. Barlow, *J. Gen. Physiol.*, **40**, 809 (1957); **41**, 307 (1957).
88. I. R. Lehman and R. M. Herriott, *ibid.*, **41**, 1067 (1958).
89. V. Bonifas and E. Kellenberger, *Biochim. Biophys. Acta*, **16**, 330 (1955).
90. *a.* D. H. Duckworth, *J. Virol.*, 3, 92 (1969); *b. Virology*, **40**, 673 (1970).
91. R. W. Reader and L. Siminovitch, *Virology*, **43**, 607, 623 (1971).
92. C. A. Hutchison III and R. L. Sinsheimer, *ibid.*, **25**, 88 (1966).
93. R. Baker and I. Tessman, *Proc. Natl. Acad. Sci. U.S.*, **58**, 1438 (1967).
94. N. D. Zinder and L. B. Lyons, *Science*, **159**, 84 (1967).
95. R. Fujimura and P. Kaesberg, *Biophys. J.*, **2**, 433 (1962).
96. G. Streisinger, F. Mukai, W. J. Dreyer, B. Miller and S. Horiuchi, *Cold Spring Harb. Symp. Quant. Biol.*, **26**, 25 (1961).
97. R. H. Epstein, A. Bolle, C. M. Steinberg, E. Kellenberger, E. Boy de la Tour, R. Chvalley, R. S. Edgar, M. Susman, G. H. Denhardt and A. Leilausis, *ibid.*, **28**, 375 (1963).
98. S. Brenner, L. Barnett, F. H. C. Crick and A. Orge, *J. Mol. Biol.*, **3**, 121 (1961).
99. H. Dinitzis, *Proc. Natl. Acad. Sci. U.S.*, **47**, 247 (1961).
100. J. Bishop, J. Leahy and R. Schweet, *Proc. Natl. Acad. Sci. U.S.*, **46**, 1030 (1960).
101. A. Goldstein and B. J. Brown, *Biochim. Biophys. Acta*, **53**, 438 (1961).
102. A. Tsugita and M. Inouye, *J. Mol. Biol.*, **37**, 201 (1968).
103. M. Inouye and A. Tsugita, *ibid.*, **22**, 193 (1966).
104. M. Inouye, E. Akaboshi, M. Kuroda and A. Tsugita, *ibid.*, **50**, 71 (1970).
105. A. Tsugita, *Proc. 12th Intl. Congr. Genetics*, **II**, 35 (1968).
106. H. Yahata, Y. Ocada and A. Tsugita, *Mol. Gen. Gen.*, **106**, 208 (1970).
107. N. Ohta and A. Tsugita, in preparation.
108. A. Tsugita, M. Inouye, E. Terzaghi and G. Streisinger, *J. Biol. Chem.*, **243**, 391 (1968).
109. M. Inouye and A. Tsugita, *J. Mol. Biol.*, **37**, 213 (1968).
110. W. Moo-Penn and H. Wiesmeyer, *Biochim. Biophys. Acta*, **178**, 330 (1969).
111. R. Kretsinger, *J. Mol. Biol.*, in press.
112. M. Inouye, E. Akaboshi, M. Kuroda and A. Tsugita, *ibid.*, **50**, 71 (1970).
113. M. Inouye, H. Yahata, Y. Ocada and A. Tsugita, *ibid.*, **33**, 957 (1968).
114. C. C. F. Blake, D. F. Koenig, G. A. Mair, A. C. T. North, D. C. Phillips and V. R. Sarma, *Nature*, **206**, 757 (1965).
115. C. C. F. Blake, G. A. Mair, A. C. T. North, D. C. Phillips and V. R. Sarma, *Proc. Roy. Soc. (London)*, **B167**, 365 (1967).
116. K. Hayashi, T. Inoto, G. Funatsu and M. Funatsu, *J. Biochem. (Tokyo)*, **58**, 227 (1965).

117. R. E. Canfield and A. K. Liu, *ibid.*, **240**, 1997 (1965).
118. P. Jolles, *Proc. Roy. Soc.* (*London*), **B167**, 350 (1967).
119. M. Inouye, Y. Ocada and A. Tsugita, *J. Biol. Chem.*, **245**, 3439 (1970).
120. M. Inouye, M. Imada, E. Akaboshi and A. Tsugita, *ibid.*, **245**, 3455 (1970).
121. M. Inouye, M. J. Lorena and A. Tsugita, *ibid.*, **245**, 3467 (1970).
122. M. Inouye, M. Imada and A. Tsugita, *ibid.*, **245**, 3479 (1970).
123. M. Sekiguchi and S. S. Cohen, *J. Mol. Biol.*, **8**, 638 (1964).
124. W. Weidel and W. Katz, *Z. Naturforsch*, **166**, 156 (1961).
125. M. R. J. Salton and R. W. Horn, *Biochim. Biophys. Acta*, **7**, 177 (1951).
126. S. Hara and Y. Matsushima, *Bull. Chem. Soc. Japan*, **39**, 1826 (1966).
127. W. Weidel and J. Primosigh, *J. Gen. Microbiol.*, **18**, 513 (1958).
128. M. R. J. Salton, *Bacteriol. Rev.*, **21**, 82 (1957).
129. W. Weidel, *Z. Naturforsch.*, **60**, 251 (1951).
130. E. Work, *Biochem. J.*, **76**, 38 (1960).
131. E. Work and M. Lecadet, *ibid.*, **76**, 39 (1960).
132. L. N. Johnson and D. C. Phillips, *Nature*, **206**, 761 (1965).
133. J. A. Rupey, *Proc. Roy. Soc.* (*London*), **B167**, 41 (1967).
134. R. E. Canfield, *Brookhaven Symp. Biol.*, No. **21**, 137 (1968).
135. P. Lerner and C. Yanofsky, *J. Bacteriol.*, **74**, 494 (1957).

Ribonuclease T_1

Kenji Takahashi
Department of Biophysics and Biochemistry, Faculty of Science,
University of Tokyo
Bunkyo-ku, Tokyo, Japan

1. Introduction

Ribonuclease T_1 (RNase T_1, EC 2.7.7.26, ribonucleate guaninenucleotido-2′-transferase (cyclizing)) was discovered by Sato and Egami[1] in 1957 in Takadiastase, a commercial enzyme mixture from *Aspergillus oryzae*. They demonstrated the unique specificity of the enzyme; it only cleaves the phosphodiester linkages involving 3′-guanylic acid, in contrast with pancreatic ribonuclease A (RNase A, EC 2.7.7.16) which is specific for pyrimidine nucleotides. Because of this specificity, the enzyme has found wide use as a reagent in the study of the chemical structure of ribonucleic acids.

Despite the differences in specificity, however, the enzymes probably have some mechanisms in common. In both cases hydrolysis proceeds in two steps. The first is a transphosphorylation reaction which leads to the formation of a cyclic intermediate. In both cases it is the 2′-ester bond that is hydrolyzed in the second stage to yield the 3′-nucleotide (Fig. 1).

RNase T_1 is a fairly stable protein with a molecular weight of about 11,000.[2,3] Since it is also a very acidic protein with an isoelectric point at about pH 3, it was effectively purified by chromatography on a column of diethylaminoethyl cellulose.[2,4] The complete amino acid

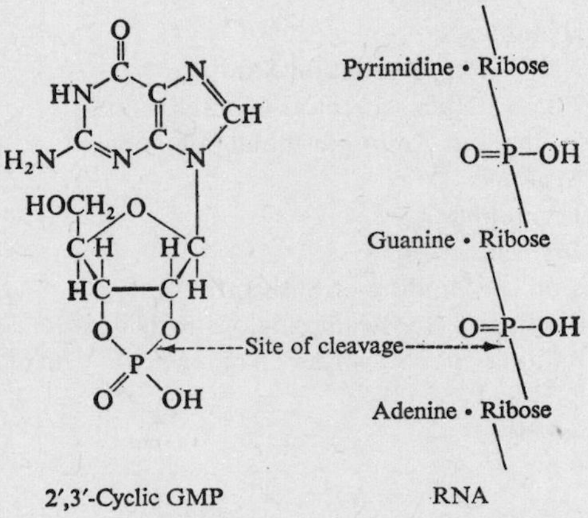

Fig. 1. Substrate specificity of RNase T_1.

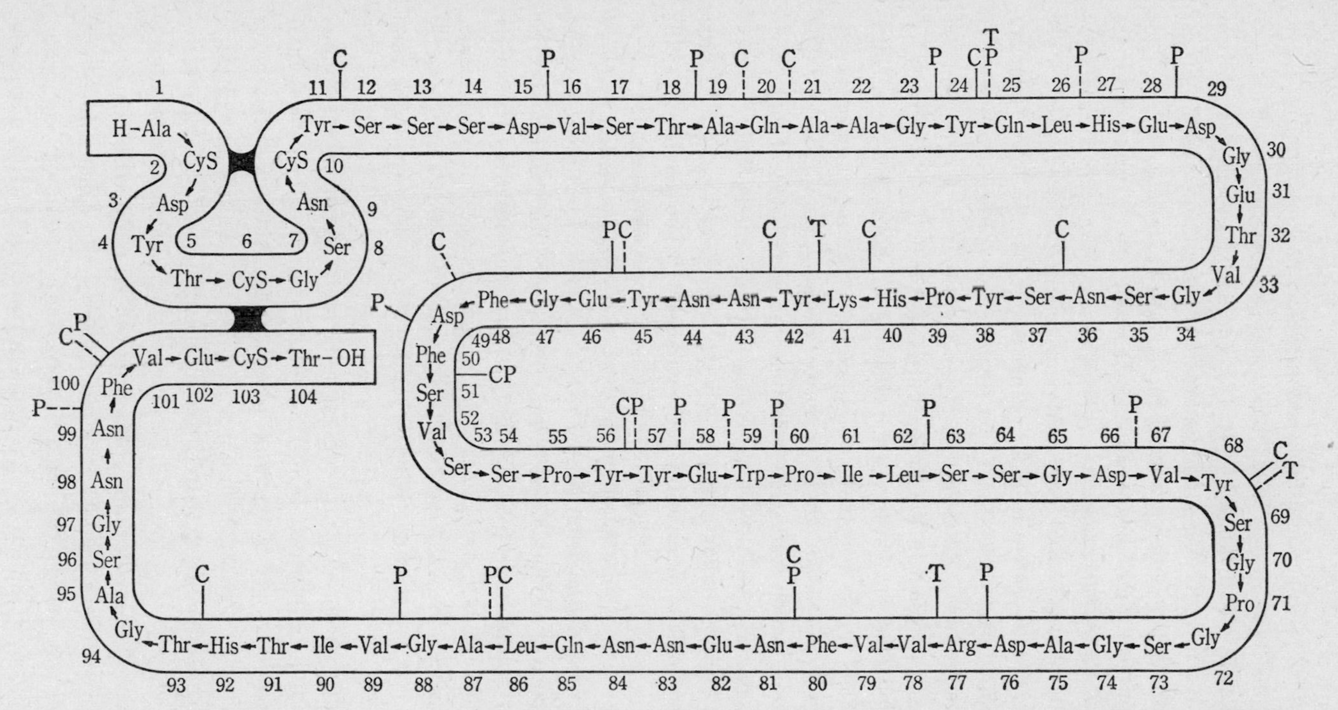

Fig. 2. The complete amino acid sequence of RNase T_1. The points of hydrolysis by trypsin and chymotrypsin in the performic acid-oxidized protein and by pepsin in the heat-denatured protein are marked by T, C, and P, respectively. The *solid* lines represent extensive or rapid hydrolyses; the *broken* lines, incomplete or slower hydrolyses.

sequence of RNase T_1 was established in 1965[5,6] as shown in Fig. 2. The enzyme consists of a single polypeptide chain of 104 amino acid residues cross-linked by two disulfide bonds. This sequence is quite different from that of pancreatic RNase A and, apart from some minor resemblances, no close similarity exists between them. RNase T_1 contains only a small number of basic amino acid residues (one lysine, one arginine and three histidine residues per molecule), while RNase A (Fig. 3) is a basic protein containing ten lysine, four arginine and four histidine residues per molecule.[7,8] The content of half-cystine residues is also low. There are two disulfide bonds in RNase T_1, whereas RNase A has four. RNase T_1 has no methionine; RNase A has four methionines. One residue of tryptophan is present in RNase T_1, but absent in RNase A.

Among the unique features of the chemical structure of RNase T_1 are the locations of the two disulfide bonds in the amino- and carboxyl-terminal regions, where they form two loops of peptide chain. The smaller loop, which somewhat resembles the small loop of eight residues present in RNase A, is composed of nine amino acid residues. Cleavage of the disulfide bonds in RNase T_1 by reduction in 8 M urea results in complete inactivation of the enzyme.[9,10] The inactive protein, however, is capable of regaining full activity upon reoxidation[9,10] as in the case of RNase A. The sequence of residues 12–18 is rich in hydroxy amino acids and somewhat similar to the sequence 15–23 in RNase A. The sequence of residues 55–62 merits special attention, because it is the one most abundant in hydrophobic residues and involves the single tryptophan residue and two of the four proline residues in the molecule. This sequence somewhat resembles the sequence 114–120 in RNase A that contains the essential His-119 residue and again two of the four proline residues.

The only two amino groups present in the molecule, one at the amino-terminus and the other at Lys-41, have been found to be nonessential for enzymatic activity by chemical modifications, such as deamination by nitrous acid,[11] polyalanylation with *N*-carboxyalanine anhydride[12] and trinitrophenylation with trinitrobenzenesulfonate.[13] The chemically modified enzyme is still active despite the fact that it has no amino groups. The carboxyl-terminal threonine can be removed by carboxypeptidase A without loss of enzymatic activity.[14] In other studies[15–17] the enzyme was found to be inactivated by bromoacetate or iodoacetate but the nature of amino acid residues involved remained unknown.

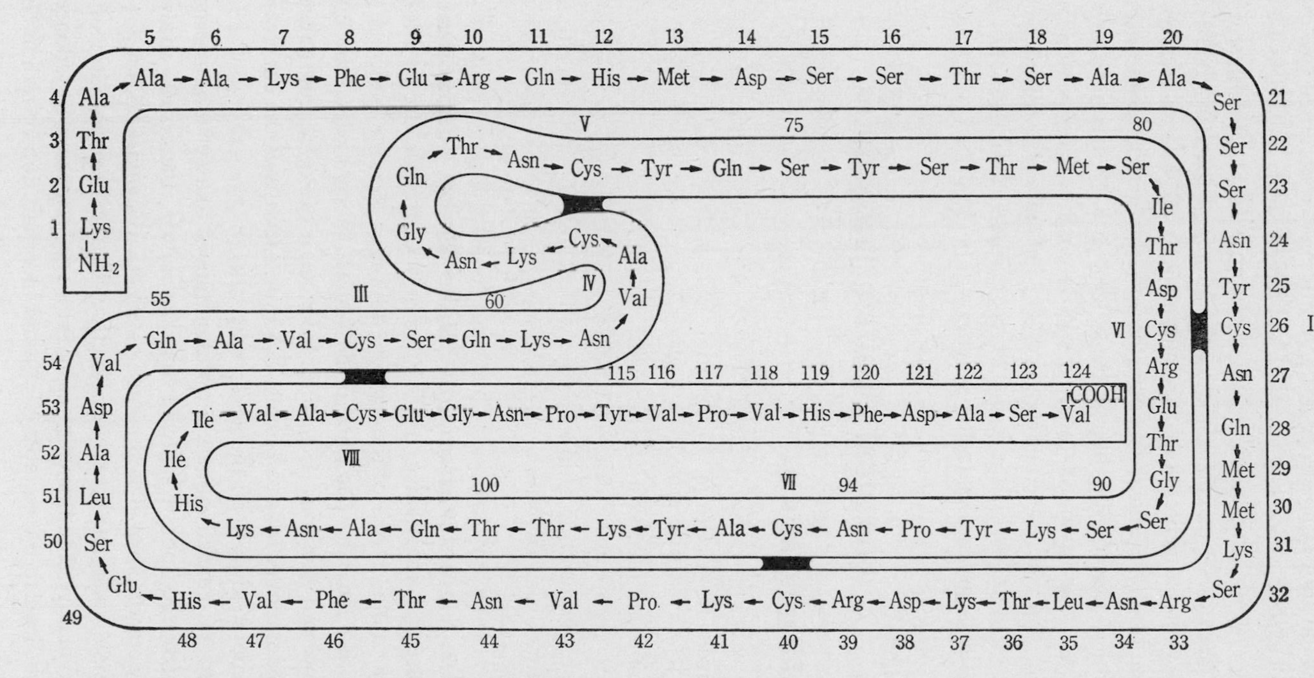

Fig. 3. The complete amino acid sequence of RNase A.

With the above knowledge at hand, attention was turned toward the fuller characterization of the residues in the molecule most directly concerned with its function as a catalyst. From 1965 to 1968, part of this series of studies was performed by the author at the Rockefeller University with the generous collaboration of Dr. S. Moore and Dr. W. H. Stein on the chemical aspect of the structure-activity relationship of RNase T_1. The results obtained since 1965 will be described in some detail with particular reference to specific amino acid residues in the enzyme. Part of the following review has already appeared elsewhere.[18,19] In addition, the results of recent studies on the mode of binding of substrate analogs to the enzyme will be described.

2. Chemical Modification Studies

2.1. Glutamic Acid

It is well known that when RNase A is treated with iodoacetate or bromoacetate, a specific inactivation results from the alkylation of either His-12 or His-119 at the active center of the enzyme.[20] A third alkylation can involve a crucial lysine residue.

The experiments[21] described below were designed to explore the course of the reaction of iodoacetate with RNase T_1 and hence to learn the nature of the amino acid residue or residues involved in the inactivation reaction.

2.1.1. Inactivation of RNase T_1 by Iodoacetate

RNase T_1 is inactivated at pH 5.5 and 37°C by exposure to a 180- to 300-fold molar excess of iodoacetate, although the rate is slower than that with RNase A. The time for 50% inactivation under these conditions is about one hour (Fig. 4). Bromoacetate is slightly more reactive than iodoacetate. Iodoacetamide, however, inactivates the enzyme very much more slowly than the acidic reagents.

When ^{14}C-labeled iodoacetate was employed to ascertain the amount of reagent incorporated during the inactivation, the results shown in Fig. 5 were obtained. The percent loss of activity against both RNA and 2′,3′-cyclic guanylate correlates exactly with the incorporation of one carboxymethyl group per molecule. Prolonged exposure to iodoacetate does not lead to appreciable additional incorporation of radioactivity.

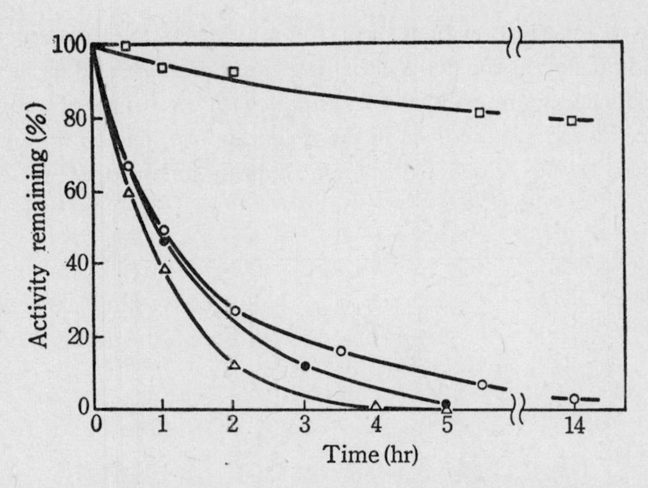

Fig. 4. Rates of inactivation of RNase T_1 by iodoacetate, bromoacetate and iodoacetamide at pH 5.5 and 37°C. Activity was measured toward RNA as substrate. ○, Iodoacetic acid, 0.6%; protein, 0.1%; μ, 0.11. ●, Iodoacetic acid, 1.5% protein, 0.5%; μ, 0.19. △, Bromoacetic acid, 0.7%; protein, 0.1%; μ, 0.15. □, Iodoacetamide, 0.5%; protein, 0.1%; μ, 0.10.

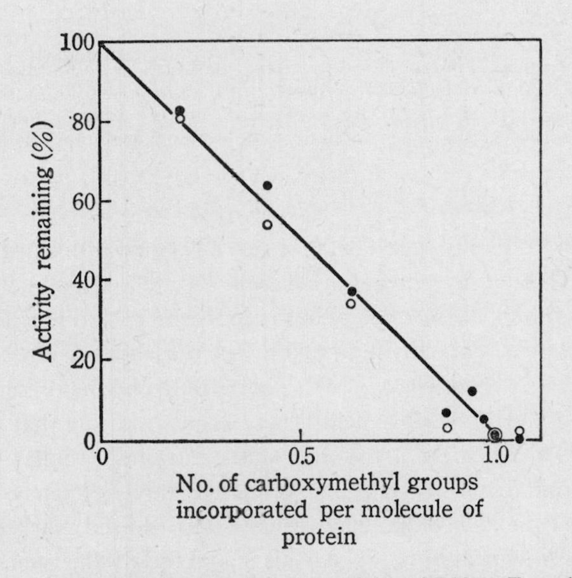

Fig. 5. Relationship between activity of RNase T_1 and extent of carboxymethylation. ●, Activity toward RNA; ○, activity toward 2′,3′-cyclic GMP.

The rate of inactivation of RNase T_1 by iodoacetate is maximal at pH 5.5, and falls off as the pH is either increased or decreased as shown in Fig. 6. The curve implies that two groups with pK values of about 4 and 7, respectively, are involved in the reaction. The inactivation reaction is slowed by the presence of metallic cations such as Cu^{2+} or Zn^{2+}

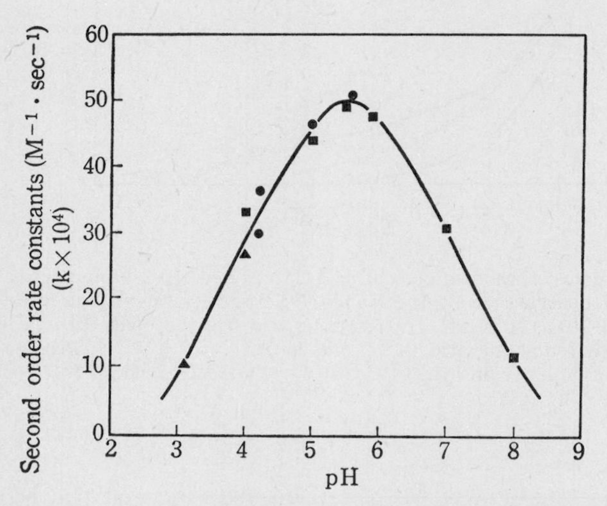

Fig. 6. pH dependence of rate of inactivation of RNase T_1 by iodoacetate at 37°C. Iodoacetic acid, 0.5%; protein, 0.1%; μ, 0.12–0.13. Activity was determined toward RNA as substrate. ▲, Sodium formate buffers; ●, sodium acetate buffers; ■, sodium acetate—Tris-HCl buffers.

which are known[17,22] to inhibit RNase T_1 (Fig. 7). A comparable degree of protection is exhibited by polyanions such as citrate and phosphate under similar conditions. It has already been reported[15,16] that inactivation by bromoacetate is markedly inhibited by the presence of substrate analogs such as 2′- or 3′-guanylic acid, but not inhibited by the presence of 5′-guanylic acid and 2′-cytidylic acid. These results indicate that the reaction probably involves the active site of the enzyme. Finally, the inactivation reaction seems to require the native three dimensional structure of the enzyme, since the inactivation is also slowed markedly in the presence of 8 M urea (Fig. 7). At this stage, the results seem to parallel very closely the reaction which leads to the alkylation of His-12 or His-119 in pancreatic ribonuclease.

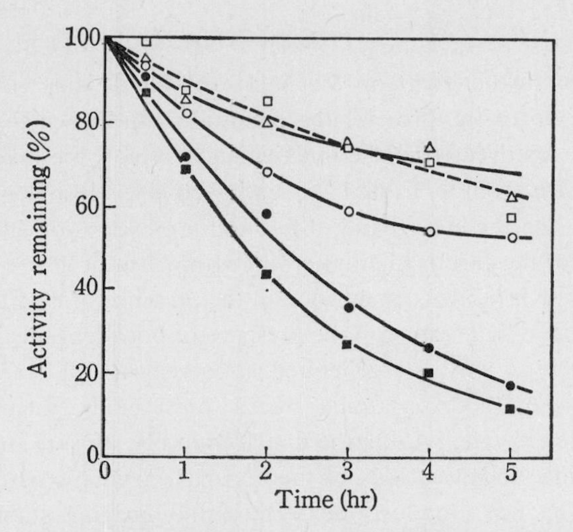

Fig. 7. Effect of Zn^{2+}, Cu^{2+}, and 8 M urea upon inactivation of RNase T_1 by iodoacetate at pH 5.5 and 37°C. Iodoacetic acid, 0.5%; protein, 0.1%. Activity was determined toward RNA as substrate. △, 0.043 M Zn^{+2}; μ, 0.25. ○, 0.043 M Cu^{2+}; μ, 0.25. □, 8 M urea; μ, 0.12. ●, No addition; μ, 0.27. ■, No addition; μ, 0.13.

2.1.2. Properties of Monocarboxymethylated, Inactive RNase T_1

Amino acid analysis proved the dissimilarity between pancreatic RNase and RNase T_1. An acid hydrolyzate of RNase T_1 inactivated by iodoacetate was analyzed chromatographically, and the amino acid composition was indistinguishable from that of the parent enzyme. There was no evidence for the presence of any carboxymethyl histidine or carboxymethyl lysine. Spectroscopic and colorimetric tests provided no evidence for the involvement of the single tryptophan residue in the molecule. Whatever the product of the reaction, the altered residue or residues seem to be acid-labile. The acid hydrolyzate of the enzyme inactivated by ^{14}C-labeled iodoacetate was therefore examined for the possible presence of ^{14}C-labeled glycolic acid and one mole per mole of protein was found. It is thus concluded that an acid-labile carboxymethyl group is present which liberates glycolic acid upon hydrolysis. It was thought that the alkylation of the carboxyl group of one of the aspartic or glutamic acid residues in the protein, with the formation of an ester, was involved as shown in Eq. (1).

$$\text{Protein–COO}^- + \text{ICH}_2\text{COO}^- \longrightarrow \text{Protein–COOCH}_2\text{COO}^- + \text{I}^- \quad (1)$$

In order to shed further light on the nature of the linkage of the carboxymethyl group to the protein, the conditions under which the ^{14}C-label could be removed from the inactivated enzyme were investigated. The results are given in Table 1. No radioactivity is lost at pH 9 and 37°C in 48 hr. On the other hand, 0.1 N sodium hydroxide effects complete removal of the carboxymethyl group within 7 hr at 25°C. The carboxymethyl group is labilized somewhat in the presence of 8 M urea or hydroxylamine, and still further in the presence of both reagents. No detectable enzymatic activity is regenerated upon removal of the carboxymethyl group by hydroxylamine which presumably forms a hydroxamate of the enzyme. During the first 10 min of exposure to 0.1 N sodium hydroxide, however, 10% of the original enzymatic activity was regenerated. This was soon lost due to the lack of enzymic stability to alkali (see below, Fig. 16). The radioactive label was removed even

TABLE 1

Behavior of ^{14}C-Labeled Carboxymethyl Group in Inactivated RNase T_1 under Various Conditions†[1]

Treatment†[2]	Time (hr)	No. of carboxymethyl groups per molecule remaining
pH 9, 37°C	48	1.0
0.1 N NaOH, 25°C	7	0.0
pH 9, 8 M urea, 25°C	19.5	0.8
pH 7.6, 0.5 M NH_2OH, 25°C	19	1.0
pH 9, 2 M NH_2OH, 25°C	18.5	0.8
pH 9, 8 M urea, 1 M NH_2OH, 25°C	19.5	0.3
pH 8.5, 8 M urea, 1% mercaptoethanol plus carboxymethylation (1.5 hr), 25°C	20.5	0.7

†[1] The protein sample used (0.003–0.09%) originally contained 1.05 carboxymethyl groups per molecule.

†[2] After each treatment, the mixture was passed through a column (0.9 × 50 cm) of Sephadex G-25 equilibrated with 0.01 M NH_4HCO_3, pH 7.4, at room temperature to separate protein from reagents. This procedure took about one hour. The carboxymethyl content of the protein was calculated from the ratio of radioactivity to absorbance at 280 nm.

more readily when the disulfide bonds were ruptured. The reduced, carboxymethylated inactivated protein was kept at pH 9 and 37°C for 49 hr, and 90% of the label was lost. In the presence of 2 M hydroxylamine, the label completely disappeared after 21 hr at pH 9 and 25°C. Under these conditions label would not have been lost had alkylation occurred at histidine or lysine residues. The formation of an ester by iodoacetate alkylation of a carboxyl group is strongly indicated by the above evidence.

2.1.3. Identification of Residue Alkylated by Iodoacetate

In order to isolate a purified peptide containing a labeled aspartic or glutamic acid residue, the inactivated protein was hydrolyzed enzymatically. The label is, however, lost much more easily from peptides than from either the intact or the reduced, carboxymethylated protein. An extensive degradation frequently occurs during hydrolysis by enzymes such as chymotrypsin and during subsequent attempts to isolate ^{14}C-labeled peptide by chromatography on Dowex 50 or Sephadex G-25.

In the method which ultimately led to a pure peptide, the inactivated protein was hydrolyzed with Nagarse at pH 8 and 37°C for 3 hr. The

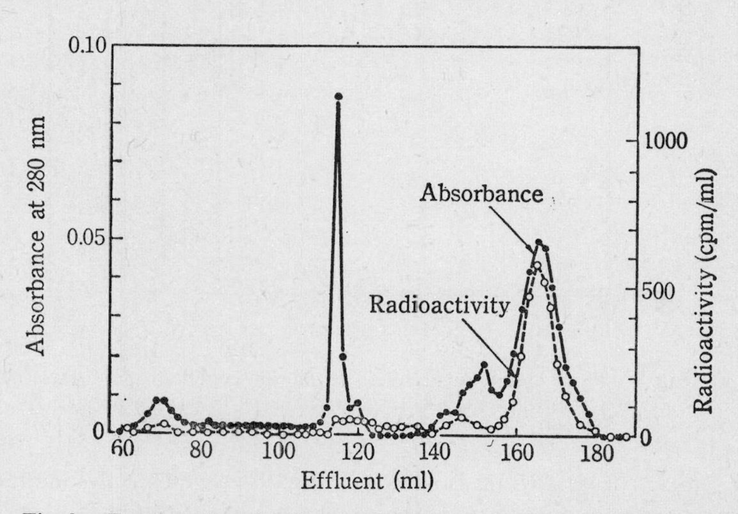

Fig. 8. Fractionation of a Nagarse hydrolyzate of ^{14}C-labeled iodoacetate-inactivated RNase T_1 on a column of Sephadex G-25 (0.9 × 220 cm) equilibrated with 0.05 M acetic acid. Flow rate, about 6 ml/hr; 1.90 ml effluent fractions collected; 4°C. The protein was hydrolyzed with Nagarse at pH 8 and 37°C for 3 hr.

hydrolyzate was immediately passed through a long column (0.9 × 220 cm) of Sephadex G-25 at 4°C with 0.05 M acetic acid as the eluent (Fig. 8). There is a single radioactive peak (radioactivity recovery, 79%) which emerges fairly late, at a position near to that of inorganic salts. This peptide peak has by far the strongest absorption at 280 nm. Amino acid analysis of the material in the single radioactive peak shows that a mixture of peptides is present. Further purification was carried out by paper chromatography in butanol–pyridine–acetic acid at pH 5.5 and 4°C, which yielded one radioactive zone with an R_f of 0.7 to 0.9. In some cases, an additional minor zone was obtained (R_f 0.5 to 0.6) which corresponded in position to and gave a color reaction for glycolic acid.

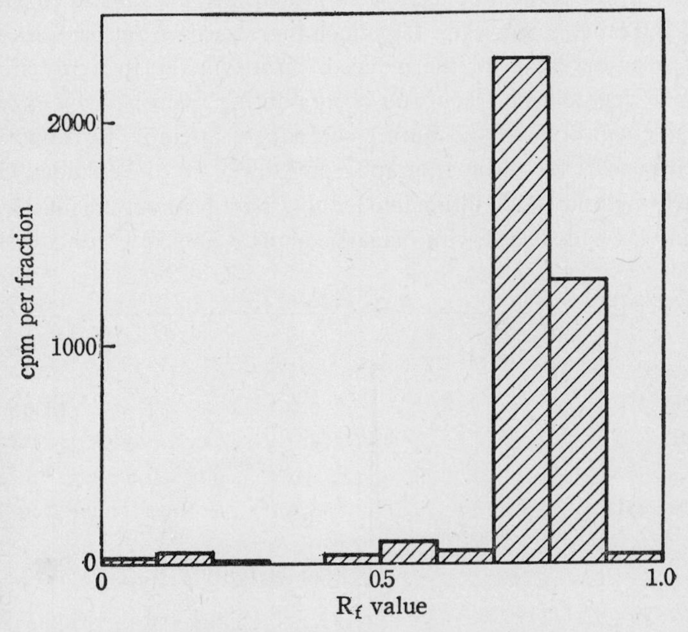

Fig. 9. Paper chromatography of ^{14}C-labeled peptide from Nagarse hydrolyzate. Solvent, *n*-butanol–pyridine–acetic acid–water (15:10: 3:12, by vol); 4°C.

A typical chromatogram is shown in Fig. 9, in which only one radioactive zone can be seen near the solvent front, separated from several ninhydrin-positive components not shown here. The radioactive zone was eluted from the paper and a portion of the eluate submitted to acid hydrolysis and amino acid analysis. The analysis indicates that a homogeneous peptide has been obtained (overall yield, 73%) and that five amino acids (tyrosine, glutamic acid, proline, isoleucine, and leucine) are present in equimolar amounts. A single radioactive peak is observed at the position of glycolic acid and the amount of radioactivity corresponds approximately to one carboxymethyl group per molecule of peptide.

The only portion of the RNase T_1 molecule from which this group of amino acids could come is the sequence, Tyr_{57}–Glu–Trp–Pro–Ile–Leu_{62}. This sequence contains tryptophan, which, of course, would not survive acid hydrolysis. Since the carboxymethylated amino acid also decomposes and loses its label on acid hydrolysis, other methods of degradation were required in order to localize the site of attachment of the carboxymethyl group.

For this purpose, aminopeptidase M was found useful. The action of this enzyme at pH 7 and 37°C for 24 hr yields most of the amino acids, but there was essentially no glutamic acid. Instead a new radioactive, ninhydrin-positive component was eluted ahead of aspartic acid on the chromatogram. The liberation of this component was almost complete in 3 hr, but it was unstable to longer periods of digestion, decomposing to yield increasing amounts of glycolic acid. To characterize the radioactive component further, it was isolated from an aminopeptidase M digest by chromatography on a 60-cm column of the amino acid analyzer (yield over 70% from a 3-hr aminopeptidase digest). Upon hydrolysis in 6 N HCl at 25°C for 8 hr, it gave glutamic acid and radioactive glycolic acid at yields 82% and 104% of theory, respectively, based on a monoester of glutamic acid. Alkaline hydrolysis at pH 12 and 25°C for 4 hr yielded 0.8 residue of glycolic acid but no ninhydrin-positive substance. Presumably pyroglutamic acid was formed.

The component released by aminopeptidase M appears to be the γ-carboxymethyl ester of glutamic acid. For final proof of the structure, this compound was prepared by the reaction of sodium glycolate and glutamic acid in dioxane in the presence of thionylchloride. The synthetic γ-carboxymethyl ester of glutamic acid and the ^{14}C-labeled compound obtained from inactivated RNase T_1 were identical in all properties tested. They are chromatographically indistinguishable on the amino acid analyzer and on paper. Both yield equimolar amounts of glutamic and glycolic acids on hydrolysis with acid, and both exhibit the same behavior on saponification with alkali.

2.1.4. Conclusion: the Role of the Carboxyl Group of Glu-58

The evidence presented so far seems to permit the conclusion that when RNase T_1 is inactivated by iodoacetate at pH 5.5, a single carboxymethyl residue is introduced into the enzyme by a reaction at the γ-carboxyl group of the Glu-58 residue, according to Eq. (2).

$$
\begin{array}{c} CH_2COO^- \\ | \\ CH_2 \\ | \\ \text{---NHCHCO---} \\ \text{Glu-58} \end{array}
+ ICH_2COO^- \longrightarrow
\begin{array}{c} CH_2COOCH_2COO^- \\ | \\ CH_2 \\ | \\ \text{---NHCHCO---} \\ \text{Glu-58} \\ \gamma\text{-carboxymethyl ester} \end{array}
+ I^- \qquad (2)
$$

This is the first time that iodoacetic acid has been found to yield an ester in a protein. In order for this unexpected alkylation to take place, the carboxyl group in the enzyme must be exceptionally reactive, in the same sense that the His-12 and 119 residues in RNase A are unusually reactive. These unusual properties must be conferred upon it by the three dimensional structure of the native protein. The decrease in reactivity of this group upon exposure of the enzyme to 8 M urea (Fig. 7) is further evidence in support of this contention. The CD spectrum of the carboxymethylated enzyme is almost indistinguishable from that of the native protein,[23] indicating that little significant change occurs in the protein conformation by this modification. The enzyme is thus inactivated by moving the critical carboxyl group at position 58 about 4 Å, that is by changing the residue in question from glutamic acid to γ-carboxymethylglutamic acid.

The role of the carboxyl group of Glu-58 in the catalytic activity of RNase T_1 is not yet fully known. It seems possible that, in the ionized form, it may function as a proton acceptor, a role similar to that postulated[24,25] for the unprotonated form of an imidazole group in the activity of RNase A. In this enzyme, two histidine residues (12 and 119) are thought to work in concert in the catalysis, one as a general base and the other as a general acid. It is tempting to postulate that the unusual carboxyl group in RNase T_1 also acts in synchrony with a strategically placed basic group in the catalyst.

In that case, the basic group should be a histidine residue since one or two histidine residues have been shown to be implicated in the active site, as will be described later. A proposed mechanism for the action of

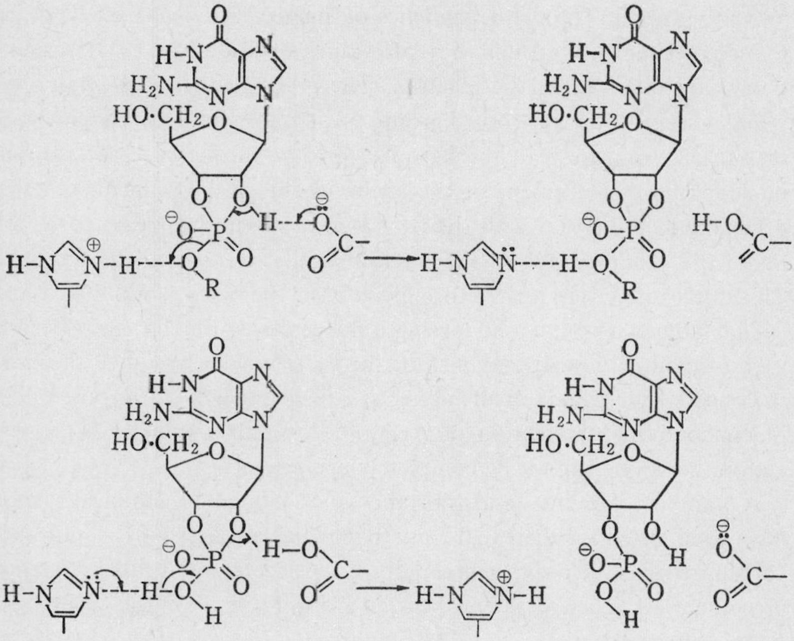

Fig. 10. A proposed mechanism for the action of RNase T_1.

RNase T_1 is shown in Fig. 10. In this mechanism, Glu-58 acts as a general base in the transphosphorylation step and as a general acid in the hydrolysis step. The mechanism in which Glu-58 acts in a reverse manner in each step is also conceivable. The possibility cannot be excluded as yet that two histidine residues constitute the catalytic site, as in the case of RNase A, in which case Glu-58 must be part of the binding site.

When RNase T_1 is treated with iodoacetate at pH 7.5 or 8.5, the alkylation occurs primarily with Glu-58 and to a minor extent with histidine.[26)] No other amino acids are altered. Thus the alkylation of the γ-carboxyl group of Glu-58 continues to be the predominant reaction and hence the major cause of the inactivation of the enzyme at pH 7.5 and 8.5. Iodoacetate ion therefore appears to be strongly oriented toward reaction with Glu-58 over a wide pH range.

Since iodoacetamide does not react in the same way, it is tempting to assume the presence of a favorably placed basic group or groups in the vicinity of the active center which may attract and orient the acidic alkylating agent. The pH dependence of inactivation of RNase T_1 (Fig. 6) indicates that two groups with pK values of about 4 and 7, respectively, are involved in the reaction. The group with the pK value of about 4 seems to be a carboxyl group, presumably the γ-carboxyl group of Glu-58. Ionization of the carboxyl group will increase the rate of nucleophilic displacement of the iodine atom of the reagent. On the other hand, the group with the pK value of about 7 appears to be an imidazole group of one of the three histidine residues in the enzyme. This imidazole group is therefore likely to be the basic group which has been assumed to bind and orient iodoacetate anion for the specific carboxymethylation. It may also be the same residue assumed to work in concert with Glu-58 at the active site of the enzyme. The possibility of concomitant implication of Arg-77 in the carboxymethylation reaction, however, cannot be completely excluded.

A number of mono- and dibromo acids related to iodoacetic acid have been tested for their reactivity toward RNase T_1.[23] β-Bromopyruvate inactivates the enzyme as rapidly as iodoacetate, but other reagents are much less reactive as can be seen from Table 2. In the case of α-

TABLE 2
Second Order Rate Constants for the Reactions of Halo-acids with RNase T_1 at pH 5.5

Reagent	$k \times 10^4 (M^{-1}.\ sec^{-1})$
Bromoacetate	57
Iodoacetate	49
β-Bromopyruvate	34
DL-α-Bromopropionate	8.0
DL-α-Bromo-n-butyrate	6.5
DL-α-Bromovalerate	5.8
DL-α-Bromophenylacetate	3.2
β-Bromopropionate	2.1
α,α'-Dibromoadipate	10
α,α -Dibromoacetate	1.5
α,α'-Dibromosuccinate	0.7
α,β -Dibromopropionate	0.6

The reactions were performed in the dark at pH 5.5 in sodium acetate buffer (μ= 0.07 to 0.1). The ionic strength of the reaction mixtures varied from 0.08 to 0.15 depending on the reagents used.

bromo acids, substitution of one of the hydrogens on the α-carbon atom with an alkyl or an aryl group appears to decrease greatly its reactivity toward RNase T_1. This effect may be due partly to steric hindrance, reflecting the unique conformation of the active site, although it may be partly inherent in the reagents themselves. Dibromo acids are also not highly reactive except α,α'-dibromoadipic acid. Similar but slower inactivation by these reagents is also observed at pH 8.0. The amino acid composition of acid hydrolyzates of RNase T_1 inactivated by reaction with β-bromopyruvate (at pH 5.5), α,α'-dibromoacetate (at pH 5.5) or α,α'-dibromoadipate (at pH 5.5 and 8.0) is indistinguishable from that of the native protein as in the case of the carboxymethylated enzyme. This indicates that these mono- and dibromo acids react only with a carboxyl group in the enzyme, presumably with the γ-carboxyl group of Glu-58.

In addition to the alkylating reagents described above, use has been made[26] of some other reagents which are potentially capable of modifying carboxyl groups. These include N^{α}-diazoacetylnorleucine methyl ester, diazouracil, *p*-bromophenacyl bromide and bromophenacyl bromide. When ribonuclease T_1 is treated with N^{α}-diazoacetylnorleucine methyl ester at pH 4.5 and 14°C for 24 hr, no significant inactivation takes place. Amino acid analysis of an acid hydrolyzate of the reaction product shows, however, that about one residue of norleucine is incorporated into the protein. Except for this, the amino acid composition of the protein derivative is indistinguishable from that of the native protein. The indication is that some carboxyl group or groups in the protein are nonessential for the enzyme activity. When the reaction is performed in the presence of Cu^{2+} at 30°C, more extensive reaction as well as inactivation takes place with the concomitant formation of protein precipitate. The other three reagents show no significant effect on the enzymatic activity of RNase T_1.

2.2. Histidine

Attention was turned toward the characterization of other amino acid residues which may be involved in the active center of the enzyme. The amino acid residues of special interest in this connection were the basic ones. Those present in RNase T_1 are three histidine, one lysine, and one arginine residue. Some may be part of the binding site of the enzyme, since the substrate is negatively charged. As discussed in the preceding section, one of them, possibly a histidine residue, appears to be the basic residue which binds and orients iodoacetate anion to react with Glu-58. Some of the histidine residues also appear to be involved in the active site of the enzyme as in the case of RNase A.

2.2.1. Photooxidation Studies

In earlier experiments,[27] RNase T_1 was photooxidized in the presence of methylene blue, a cationic dye. One histidine residue was destroyed and about 90% inactivation occurred. Activity is lost almost completely when two histidine residues have disappeared. The tryptophan residue is oxidized at the same time, but the rate of oxidation of this residue is slower than that of the decrease in enzymatic activity. These results demonstrate the primary importance of one or two of the histidine residues as involved in the active site of the enzyme. The results of photooxidation of the enzyme in the presence of riboflavin are in support of this conclusion. Moreover, it has been found[28] that the inactivation by photooxidation at pH 6 in the presence of methylene blue is markedly inhibited by the presence of substrate analogs, such as 2′- or 3′-guanylic acid.

Recently photooxidation of RNase T_1 has been carried out [29] in the presence of rose bengal, an anionic dye, in the hope that the oxidation will take place more specifically with histidine residues than in the case of methylene blue-catalyzed photooxidation, since the negatively charged dye is expected to be bound to imidazole groups more strongly and specifically than neutral or positively charged dyes. When RNase T_1 (0.035% solution) is irradiated at pH 8.5 and 37°C in the presence of 0.002% rose bengal with a 200 W incandescent lamp from a distance of 20 cm, inactivation takes place rapidly with the concomitant loss of histidine and tryptophan residues. The time for 50% inactivation is about 15 min and complete inactivation occurs within 2 hr. The relationship between the loss of enzymatic activity and the extent of modification of histidine and tryptophan residues is shown in Fig. 11. Over 70% inactivation takes place when one histidine residue is oxidized. Oxidation of two histidine residues leads to nearly complete inactivation. On the other hand, the rate of loss of the tryptophan residue is much slower than that of activity loss. When about 80% inactivation had occurred, only about 0.2 residue of the tryptophan was lost.

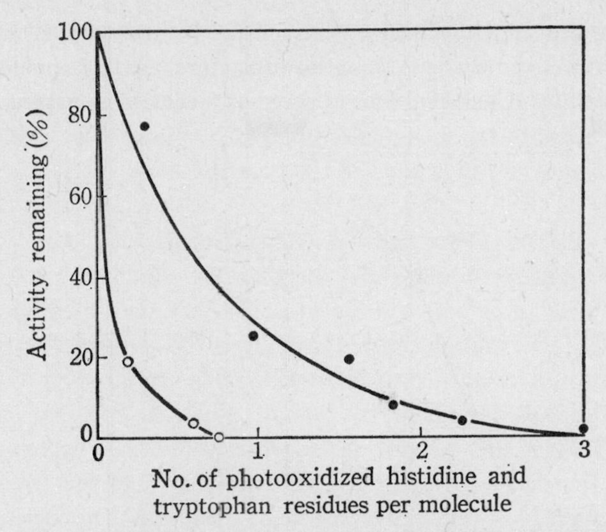

Fig. 11. Relationship between activity of RNase T_1 and extent of photooxidation of histidine and tryptophan residues. ●, Histidine; ○, tryptophan. Activity was determined toward RNA as substrate.

The rate of inactivation is pH-dependent; the extent of inactivation in one hour is 97% at pH 8.5 (Tris–HCl buffer), 25% at pH 7.0 (phosphate buffer) and 8% at pH 5.5 (acetate buffer). The pH dependence of the inactivation of RNase T_1 investigated more recently[30] is shown in Fig. 12. The indication is that there is at least one critical histidine residue with an apparent pK value of about 7.5. The rate of inactivation by photooxidation is markedly slowed by the presence of substrate analogs. As can be seen in Fig. 13, the protective effect was in the order: 2′(3′)-GMP> 5′-GMP> 2′(3′)-AMP> 2′(3′)-CMP. 2′(3′)-GMP is most effective, and almost no effect is observed with 2′(3′)-CMP. This is consistent with the known base specificity of the enzyme. Binding of these nucleotides to the active center of the enzyme will protect the critical amino acid residue or residues from photooxidation.

The unusual reactivity of the γ-carboxyl group of Glu-58 toward iodoacetate is lost in parallel with the loss of activity as shown in Table 3.[29]

These lines of evidence strongly indicate the implication of some histidine residues in the active center of the enzyme.

In order to identify the photooxidized histidine residues, photooxidized RNase T_1 was oxidized with performic acid and hydrolyzed with chymotrypsin.[30] The digest was submitted to high voltage paper electrophoresis at pH 3.6 (Fig. 14). In this procedure, the three histidine-containing peptides from the oxidized protein[5,6] can be separated into

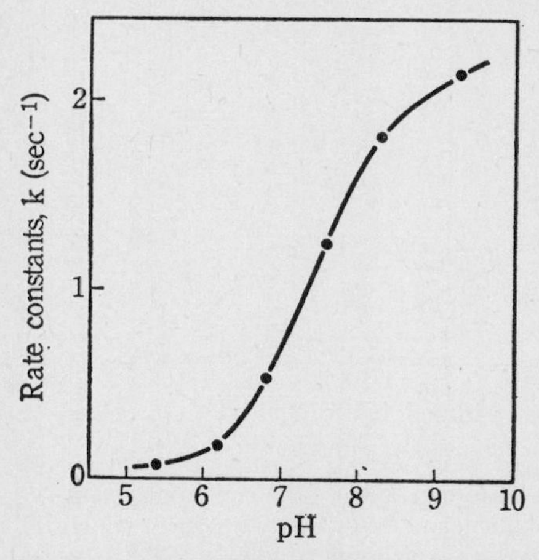

Fig. 12. pH dependence of rate of inactivation of RNase T_1 by rose bengal-catalyzed photooxidation. Photooxidation was performed at 37°C at a protein concentration of 0.029% and a rose bengal concentration of 0.0013% in Tris chloride—sodium acetate buffers; μ, 0.1. Activity was determined toward RNA as substrate.

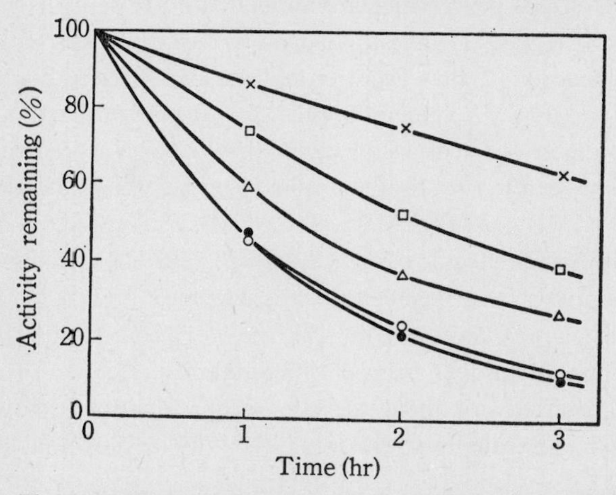

Fig. 13. Effect of substrate analogs on rate of inactivation of RNase T_1 by rose bengal-catalyzed photooxidation. Photooxidation was performed under similar conditions to Fig. 12 at pH 6.2 and a nucleotide concentration of 0.05%. Activity was determined toward RNA as substrate. ×, 2′(3′)-GMP; □, 5′-GMP; △, 2′(3′)-AMP; ○, 2′(3′)-CMP; ●, no addition.

TABLE 3

Reaction of ^{14}C-Labeled Iodoacetate†1 with Chemically Modified RNase T_1

Treatments	Activity toward RNA (%)	Extent of modification at given residues (no. of residues per molecule of protein)†2						Number of carboxymethyl groups introduced per molecule of protein
		Ala-1	Lys-41	His	Arg-77	Trp-59	Tyr	
Untreated	100							1.02
Maleic anhydride (pH 9, 0°C, 1 hr)	*ca.* 100	*ca.* 1.0	*ca.* 1.0					0.90
Photooxidation in the presence of rose bengal (pH 8.5, 37°C, 0.5 hr)	23			1.9		0.3		0.41
(pH 8.5, 37°C, 2 hr)	2			3.0		0.87		0.01
Phenylglyoxal (pH 8.0, 25°C, 4 hr)	52	1.0			0.42			0.79
(pH 8.0, 25°C, 23.5 hr)	5	1.0			0.86			0.19
2-Hydroxy-5-nitrobenzyl bromide in 8 M urea (pH 3.5, 25°C, 0.5 hr)	3					1.0		0.19
Tetranitromethane (pH 8.0, 25°C, 2 hr)	44						3.1	0.96
(pH 8.0, 25°C, 7.5 hr)	21						5.6	0.82

†1 The reaction of ^{14}C-labeled iodoacetate was performed at pH 5.5 and 37°C for 5 hr.
†2 Calculated assuming the presence of 1.00 residue each of Ala-1, Lys-41, Arg-77 and Trp-59, 3.00 residues of histidine and 9.00 residues of tyrosine per molecule of native protein.

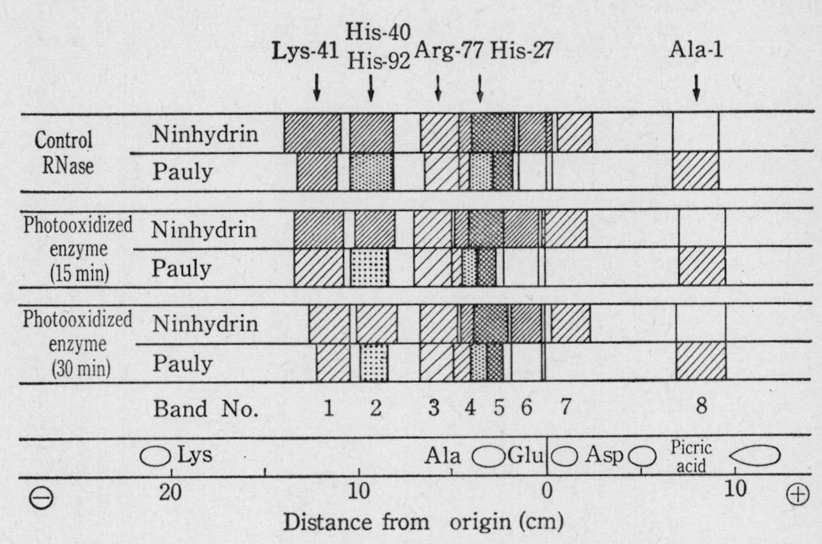

Fig. 14. Separation of peptides obtained by chymotryptic hydrolysis of photooxidized, performic acid-oxidized RNase T_1 by high voltage paper electrophoresis at pH 3.6. The protein samples were prepared by photooxidation of RNase T_1 (0.29%) in the presence of rose bengal (0.013%) at pH 8.5 and 37°C. Oxidized proteins were hydrolyzed at pH 8.0 and 25°C with 1–2% (by wt) of chymotrypsin for 24 hr. Electrophoresis was carried out at pH 3.6 on strips of Whatman No. 3 MM paper at a potential gradient of 45 V/cm for 60 min. Band 1, Lys_{41}–Tyr_{42}; band 2, Ser_{37}–Tyr–Pro–His_{40} + Ala_{87}–Gly–Val–Ile–Thr–His_{92}; band 3, Ser_{69}–Gly–Pro–Gly–Ser–Gly–Ala–Asp–Arg–Val–Val–Phe_{80}; band 4, Gln_{25}–Leu–His–Glu–Asp–Gly–Glu–Thr–Val–Gly–Ser–Asn_{36} + some other peptides; bands 5 and 6, mixtures of peptides; band 7, Thr_{93}–Gly–Ala–Ser–Gly–Asn–Asn–Phe–Val–Glu–$CySO_3H$–Thr_{104} + some minor peptides; band 8, Ala_1–$CySO_3H$–Asp–Tyr–Thr–$CySO_3H$–Gly–Ser–Asn–$CySO_3H$–Tyr_{11}. In the Pauly test on paper, the dotted areas represent the reddish coloration characteristic of histidine and the shaded areas, the brownish coloration characteristic of tyrosine.

two fractions: one containing Ala_{87}–Gly–Val–Ile–Thr–His_{92} and Ser_{37}–Tyr–Pro–His_{40}, and the other containing Gln_{25}–Leu–His–Glu–Asp–Gly–Glu–Thr–Val–Gly–Ser–Asn_{36} together with some other peptides. Simultaneously, native RNase T_1 was treated in the same way as the control. From the amino acid analyses of the peptide fractions thus separated, the relative yields of each histidine containing peptide were calculated (and hence, the extent of photooxidation at a given histidine residue). As shown in Table 4, the analyses reveal that at pH 8.5 His-92 is lost most rapidly and almost in parallel with the loss of enzymatic

TABLE 4

Yields of Histidine Containing Peptides Obtained by Chymotryptic Hydrolysis of Performic Acid-oxidized Photooxidized RNase T_1

Peptides		Relative yields†[1]				Extent of photooxidation at a given histidine residue	
		Exp. 1		Exp. 2			
Band†[2]	Structure	Control	Sample 1†[3]	Control	Sample 2†[3]	Sample 1 (Activity lost: 76%†[4])	Sample 2 (Activity lost: 84%†[4])
		(%)	(%)	(%)	(%)	(%)	(%)
2	α: Ala–Gly–Val–Ile–Thr–His_{92}	99	36	111	20	64	82
	β: Ser–Tyr–Pro–His_{40}	96	69	79	39	28	51
4	Gln–Leu–His_{27}–Glu–Asp–Gly–Glu–Thr–Val–Gly–Ser–Asn	107	79	112	58	26	48

†[1] Calculated by assuming the yield of the arginine containing peptide, Ser_{69}–Gly–Pro–Gly–Ser–Gly–Ala–Asp–Arg–Val–Val–Phe_{80} to be 100% in each case. This peptide was present in band 3 (see Fig. 14) and its actual overall yield from oxidized RNase T_1 was about 50% without correction for losses during the procedure. †[2] See Fig. 14.

†[3] Photooxidation of RNase T_1 (0.26% solution) was performed at pH 8.5 and 37°C in the presence of rose bengal (0.013%) for 15 min (Sample 1) and 30 min (Sample 2). Before performic acid oxidation, Sample 1 contained 1.40 residues of histidine, and Sample 2, 1.24 residues of histidine per molecule of protein. The histidine content did not change on performic acid oxidation.

†[4] Determined toward RNA before performic acid oxidation.

activity. The other two histidine residues are lost much more slowly. His-92 is therefore probably one of the essential histidine residues in the active center of the enzyme and appears to correspond to the residue with a p*K* value of about 7.5 found in the pH-rate profile study. Interestingly, the amino acid sequence around His-92, i.e. Leu_{86}–Ala–Gly–Val–Ile–Thr–His–Thr–Gly–Ala_{95}, is unique in that it is the portion second richest in hydrophobic amino acid residues,[5,6] the first being Pro_{55}–Tyr–Tyr–Glu–Trp–Pro–Ile–Leu_{62} which involves the active site Glu-58. Part of the former sequence (residues 86–95) appears to be involved in the architecture of the active center which may be located in a fairly hydrophobic region of the enzyme.

There is not much evidence regarding which of the remaining two histidine residues might be involved in the active center. The enzymatic activity is known, however, not to be affected by several chemical modifications of Lys-41,[11–13,18] which is adjacent to His-40. This suggests that His-40 may be neither essential nor involved in the active center of the enzyme. If this is true, then His-27 would be, by elimination, the only other histidine residue involved in the active center.

Thus, His-92 and probably one additional histidine residue are involved in the active center of the enzyme and may constitute parts of the catalytic and binding sites, respectively. The implication of a histidine residue in the substrate binding has been suggested from studies on the binding of substrate analogs to the enzyme as will be described later.

Iida and Ooi[31] determined by potentiometric titration the p*K* values of the three histidine residues in RNase T_1 to be 6.6, 6.6, and 7.5, respectively. The histidine residue with a p*K* value of 7.5 probably corresponds to His-92. Irie[32] also reported the involvement of a histidine residue with a p*K* value of 7.5 in the inactivation of the enzyme by methylene blue-catalyzed photooxidation. Rüterjans *et al.*[33] investigated the p*K* values of the histidine residues by proton magnetic resonance spectroscopy. One histidine residue was found with a p*K* value of around 8, which appears to interact with a carboxylate anion of an acidic amino acid residue of the protein as well as with the phosphate moiety of 3′-GMP. Another histidine, the p*K* value of which has not been determined, was suggested to reside near the binding site of the enzyme. The third histidine with a p*K* value of about 7.1 was suggested to interact with a carboxylate anion even more strongly than the first, but the p*K* value was only slightly affected by the presence of 3′-GMP. The first and the second histidine residues probably correspond to His-92 and the additional histidine suggested to be present in the active center of the enzyme, respectively.

Waku and Nakazawa[34] and Irie[32] report that two of the three histidine residues are protected from methylene blue-catalyzed photooxidation by the presence of 2′(3′)-GMP and that the photooxidation of the third did not lead to any significant inactivation. These results are also compatible with those obtained in the rose bengal-catalyzed photooxidation.

2.2.2. Other Chemical Modifications

Chemical modifications of histidine residues have also been attempted[26] through the use of a number of alkylating reagents which are potentially capable of modifying this amino acid residue. When RNase T_1 is treated with N^{α}-bromoacetyl-L-arginine methyl ester, ω-bromoacetamidoethylnicotinic acid amide or iodoacetamide at pH 8.0 or 8.5 at 25°C for 23–46 hr, about 80% or more inactivation takes place with the concomitant alkylation of approximately two histidine residues in the molecule, the alkylation occurring mainly at the N-3 position of the imidazole ring. Under these conditions, 4-(iodoacetamido)salicylic acid is similarly effective, and iodoacetic acid also reacts with histidine residues, although much more slowly. Fig. 15 shows the time course of

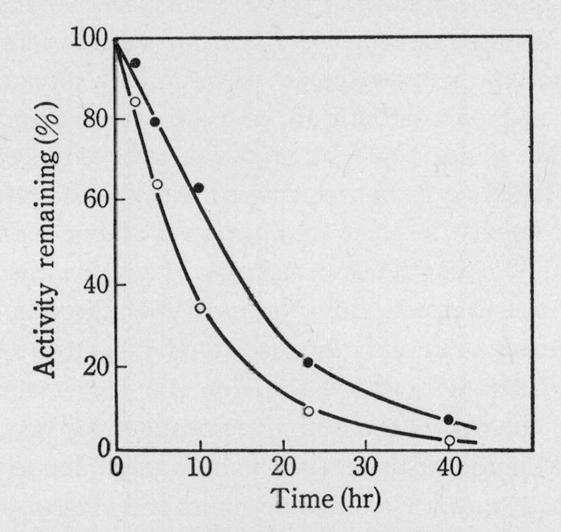

Fig. 15. Rate of inactivation of RNase T_1 by reaction with N^{α}-bromoacetyl-L-arginine methyl ester. Protein, 0.17%; reagent, 1%; pH 8.0; 25°C. ●, Activity toward RNA; ○, activity toward 2′,3′-cyclic GMP.

inactivation of RNase T_1 with a 170-fold molar excess of N^{α}-bromoacetyl-L-arginine methyl ester at pH 8.0 and 25°C. In this case, the enzymatic activity toward 2′,3′-cyclic GMP is lost somewhat faster than the activity toward RNA. The rate of inactivation is slowed somewhat in the presence of 3′-GMP, a competitive inhibitor of the enzyme. It is also slower at lower pH values. When the enzyme (0.005% solution) is treated with the reagent (1% solution) at 25°C, the loss of activity toward RNA was 86% at pH 8.0, but only 25% at both pH 7.0 and 5.5. Generally, the rates of inactivation by the other reagents tested are also much slower at pH 5.5 than at higher pH values except for iodoacetic acid.

At pH 8.0 or 8.5, the reactions of these reagents are not so specific to histidine residues and in most cases the amino-terminal alanine residue is alkylated concomitantly. In addition, the possible reaction with Glu-58 cannot be excluded in most cases. However, chemical modifications of the amino-terminal alanine residue with various reagents are known to have little effect on the enzymatic activity of RNase T_1.[11-13,18] Moreover, when the enzyme inactivated by reaction with iodoacetamide or N^{α}-bromoacetyl-L-arginine methyl ester is kept in 0.01 N NaOH at 25°C for 3 hr, no increase in activity is observed, whereas, under similar conditions, the enzyme inactivated by iodoacetate at pH 5.5 regains significant activity owing to the hydrolysis of the γ-carboxymethyl ester of Glu-58 (Fig. 16). These lines of evidence, therefore, seem to indicate that there are two histidine residues which are preferentially alkylatable by these reagents and that one or both of them are essential for the enzymatic activity. The third histidine residue appears to be buried in the interior of the protein molecule or bound to some other residues, thus becoming less reactive. Otherwise, there may be two histidine residues which are alkylatable only in a mutually exclusive way as is the case for carboxymethylation of His-12 and 119 in RNase A.[20] The second order rate constants for the inactivation of the enzyme at pH 8.0 and 25°C are 4.7×10^{-4}, 4.4×10^{-4}, 3.2×10^{-4} and 0.5×10^{-4} $M^{-1} \cdot sec^{-1}$ for N^{α}-bromoacetyl-L-arginine methyl ester, ω-bromoacetamidoethylnicotinic acid amide, 4-(iodoacetamido)salicylic acid and iodoacetamide, respectively. Although these values are only approximate owing to polyalkylation which takes place at pH 8.0, they show that the three charged reagents inactivate the enzyme more rapidly than iodoacetamide despite bulky structures which might cause steric hindrance. This may indicate

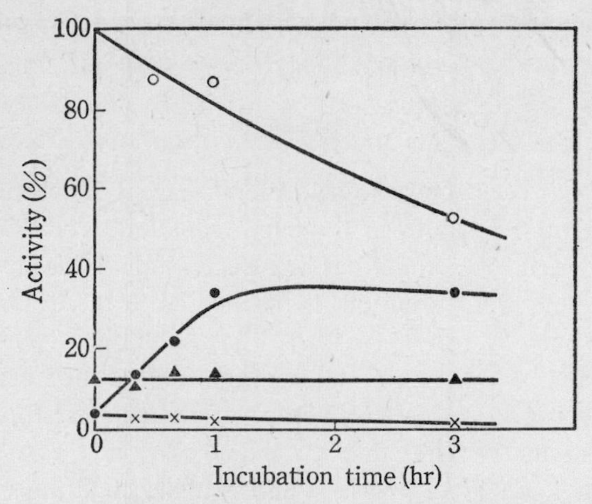

Fig. 16. Reactivation of inactive alkylated derivatives of RNase T_1 by alkali treatment. Proteins were kept in freshly prepared 0.01 N NaOH at 25 °C. Activity was determined toward RNA as substrate. ○, Native RNase T_1; protein, 0.003%. ●, Enzyme inactivated by reaction with iodoacetate at pH 5.5 and 37 °C for 5 hr; protein, 0.007%. ▲, Enzyme inactivated by reaction with iodoacetamide at pH 8.0 and 25 °C for 46 hr; protein, 0.001%. ×, Enzyme inactivated by reaction with N^{α}-bromoacetyl-L-arginine methyl ester at pH 8.0 and 25 °C for 40 hr; protein, 0.001%.

that there are several charged groups, including Glu-58 and some histidine residue(s) at or near the active center, which may bind and orient these charged reagents.

Bromoethylamine, ethyleneimine and 1-chloro-3-tosylamido-7-amino-2-heptanone (TLCK) show no effect on the enzymatic activity at pH 5.5 and 8.0 at 37°C.

2.3. Lysine and Amino-terminal Alanine

As described previously, the two amino groups of the Lys-41 and amino-terminal alanine residues in RNase T_1 are not involved in the active site of the enzyme.[11-13]

Recently maleic anhydride has been used[18,23] to modify these amino groups. The protein derivative so formed possesses full activity. There is no significant change in the reactivity of the γ-carboxyl group of Glu-58 toward iodoacetate (see Table 3). Similar results have been obtained[35] with trinitrophenylated RNase T_1. The results are thus in full support of the conclusion that these groups do not participate at the active center of the enzyme.

2.4. Arginine

Phenylglyoxal[36] has been shown to be a useful reagent for chemical modification of arginine residues in proteins under mild conditions, such as at pH 7 to 8 and 25°C, and it has been successfully utilized with RNase A. The enzyme (0.5% solution) is inactivated to the extent of 80–90% within 30 min by reaction with a 1.5% solution of phenylglyoxal at pH 8.0 and 25°C. Analyses indicate that the reaction occurs primarily at Arg-39 and 85, which have been shown to be closest to the active site of the enzyme by X-ray diffraction studies.

When the reagent is used[37] on RNase T_1 under similar conditions, the inactivation also takes place, although much more slowly than in the case of RNase A. Activity toward both RNA and 2′,3′-cyclic GMP is lost almost in parallel. When the enzyme is treated with a 760-fold molar excess of reagent, the time for 50% inactivation is about 4 hr and nearly complete inactivation takes place in 24 hr as can be seen in Fig. 17.

Amino acid analysis of the acid hydrolyzate of the reaction product

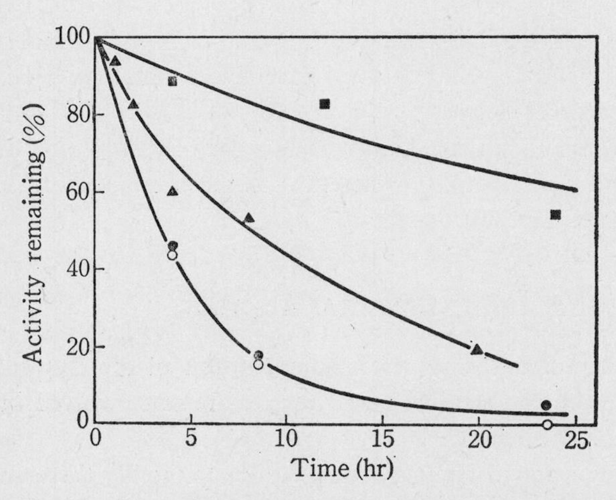

Fig. 17. Rates of inactivation of RNase T_1 by reaction with phenylglyoxal at pH 8.0 and 25°C. ●, ○, Protein, 0.16%; reagent, 1.5%. ▲, Protein, 0.03%; reagent, 0.6%. ■, Protein, 0.09%; reagent, 0.15%. Activity was determined toward RNA (●, ▲, ■) and 2′, 3′-cyclic GMP (○).

shows that the only changes involved are the loss of the single arginine residue and the amino-terminal alanine residue. The loss of activity against both RNA and 2′,3′-cyclic GMP parallels the loss of the arginine residue as shown in Fig. 18. The loss of the amino-terminal residue was found to be independent of the inactivation. The reactivity of the γ-carboxyl group of Glu-58 is lost almost in parallel with the loss of the arginine residue (see Table 3).

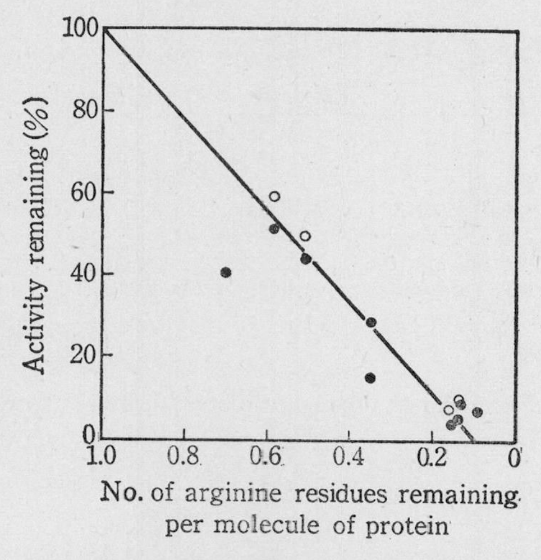

Fig. 18. Relation between loss of the arginine residue and extent of inactivationof RNase T_1 by reaction with phenylglyoxal at pH 8.0 and 25 °C. Activity was determined toward RNA (●) and 2′,3′-cyclic GMP (○).

When glyoxal is used in place of phenylglyoxal, a similar inactivation is observed (Fig. 19).[37] In this case, however, the activity toward RNA is lost somewhat faster than that toward 2′,3′-cyclic GMP. This may be due in part to the difference in steric effects of the groups introduced into protein. The rate of inactivation is retarded in the presence of 3′-GMP.

These lines of evidence point to the modification of the arginine residue as the major cause of inactivation of RNase T_1. The role of the arginine residue in the activity is not yet clear but its location appears to be at or near the active center of the enzyme. The modification of the arginine may change the pK values of adjacent charged groups. Arg-77

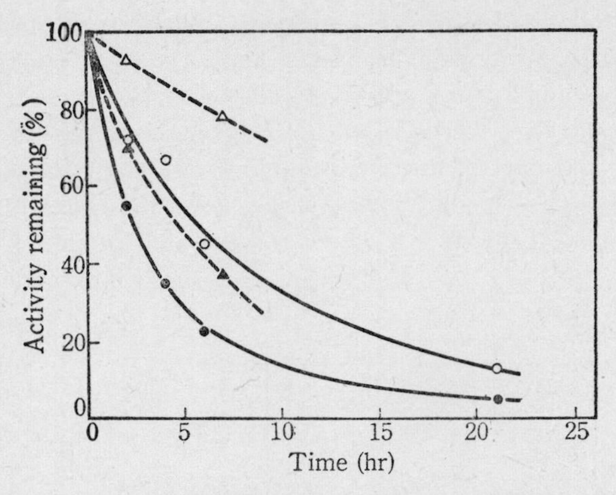

Fig. 19. Rates of inactivation of RNase T_1 by reaction with glyoxal at pH about 8.5 and 25 °C in the presence and absence of 3′-guanylic acid. Protein, 0.03%; reagent, 0.6%; 3′-GMP, 0% (○, ●) or 0.25% (△, ▲). Activity was determined toward RNA (●, ▲) and 2′,3′-cyclic GMP (○, △).

may constitute part of the binding site for negatively charged substrates or may be important in the structure of the active center of the enzyme.

2.5. Tryptophan

The location of the single tryptophan residue (Trp-59) is special because it is situated adjacent to the reactive Glu-58. In an earlier study, RNase T_1 was treated with *N*-bromosuccinimide at pH 4.0.[4] The tryptophan residue, however, was found to be inaccessible to this reagent, and was therefore concluded to be either buried in the interior of the protein molecule or bound to some other residue(s), thus becoming unreactive. The tryptophan residue is also modified by photooxidation.[27,29] In these cases, however, the inactivation occurs in parallel with the modification of histidine residues and thus no conclusion can be drawn from the results on the role of the tryptophan residue in the enzyme.

A more specific chemical modification of this residue was desired. The reagent used[38] was 2-hydroxy-5-nitrobenzyl bromide, which is known to be specific for tryptophan residues. The reaction was carried out at pH 3.5 or 5.0 in the presence and absence of 8 M urea. It was soon found that the reaction occurs only in the presence of 8 M urea, when the protein molecule is unfolded. (The native protein is known to be reversibly denatured in 8 M urea.[17]) When exposed to 8 M urea, the protein molecule unfolds, although it still retains some activity toward RNA.[23]) Upon removal of urea, the protein molecule refolds to the native state with full restoration of activity.) When the reaction was carried out in 8 M urea, the enzyme was rapidly and irreversibly inactivated as shown in Fig. 20. The protein modified in 8 M urea prob-

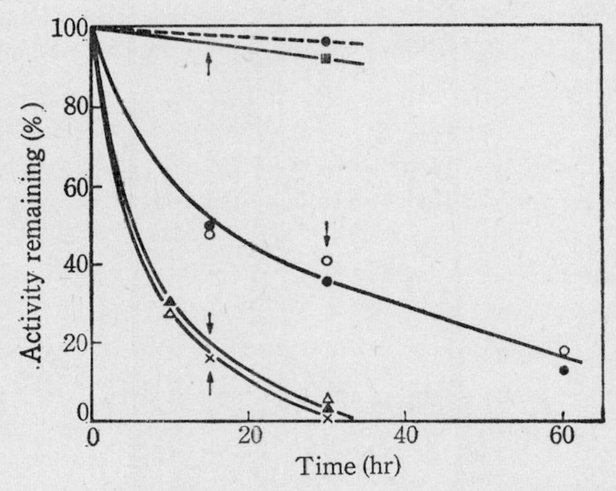

Fig. 20. Rates of inactivation of RNase T_1 by 2-hydroxy-5-nitrobenzyl bromide at 25°C.---●---, Control; protein, 0.18%; reagent, none; urea, 8 M; pH 3.5. ■, Protein, 0.08%; reagent, 0.8% × 2; urea, none; pH 5.0. ●, ○, Protein, 0.08%; reagent, 0.8% × 2; urea, 8 M; pH 5.0. ▲, △, Protein, 0.29%; reagent, 0.8% × 2; urea, 8 M; pH 3.5. ×, Protein, 0.18%; reagent, 0.9% × 2; urea, 8 M; pH 3.5. Activity was determined toward RNA (●, ■, ▲, ×) and/or 2′, 3′-cyclic GMP (○, △). Each arrow indicates a point where an additional amount of reagent was supplied to the reaction mixture.

ably cannot refold to form the native conformation and hence is inactive. The rate of reaction varies somewhat depending on the reaction conditions used. The enzyme treated at pH 5.0 for 60 min retains 10–20% activity and contains about one residue of 2-hydroxy-5-nitrobenzyl group per molecule. On the other hand, the enzyme treated at pH 3.5 for 30 min retains almost no activity and contains nearly two residues of the 2-hydroxy-5-nitrobenzyl group per molecule. The loss of activity appears to parallel the incorporation of the first residue of the 2-hydroxy-5-nitrobenzyl group into the indole ring, although the indole ring appears to accommodate maximally two residues of 2-hydroxy-5-nitrobenzyl group per molecule. The results of amino acid analyses of the inactivated enzyme and its Nagarse peptide fraction containing the 2-hydroxy-5-nitrobenzyl group are in support of the conclusion that the reaction occurs specifically with the single tryptophan residue in the enzyme.

The reaction of this modified protein with iodoacetate occurs slowly when compared with the native protein (see Table 3). The reactivity of Glu-58 with iodoacetate therefore seems to be diminished markedly by the modification of Trp-59. This may be due to steric hindrance and/or to a conformational change induced by the bulky group(s) introduced into the indole ring.

Since the tryptophan residue does not react in the absence of urea, the indole ring is probably not on the surface of the enzyme, or may be bound to some other residue, thus becoming unreactive. Hence, it is not a group actually involved in the catalytic process. However, it still seems likely that this residue is one of the essential building blocks of the active center of the enzyme.

The reaction of 2-hydroxy-5-nitrobenzyl bromide with RNase T_1 was also investigated by Terao and Ukita[39] and by Irie.[40] The results obtained by the latter author are consistent with those described above. Under the conditions employed by the former authors, however, the enzyme was not inactivated completely. This is probably due to incomplete reaction owing to rapid hydrolysis of the reagent in aqueous solution. Kawashima and Ando[41] carried out the reaction of *N*-bromosuccinimide with RNase T_1 in 8 M urea. The enzyme was inactivated in parallel with the modification of the tryptophan residue as observed in the case of 2-hydroxy-5-nitrobenzylbromide in 8 M urea.[38]

On the other hand, Waku and Nakazawa[34] reported that the tryptophan residue is not protected significantly from methylene blue-catalyzed photooxidation by the presence of 2′(or 3′)-GMP and that 90% of the activity remains after oxidation of most of the tryptophan residue. Such a mild modification may not lead to destruction of the active site of the enzyme. From fluorescence studies, Irie,[40] Pongs[42] and Yamamoto and Tanaka[43] obtained evidence to suggest that the tryptophan residue is buried in the interior of the protein molecule. In addition, Irie[40] showed that the state of the tryptophan residue in carboxymethylated RNase T_1 is similar to that in the native enzyme.

2.6. Tyrosine

RNase T_1 has nine tyrosine residues per molecule. When the enzyme is treated [18,23] with tetranitromethane at pH 8.0 and 25°C, nitration of tyrosine residues takes place slowly with the progressive loss of activity. After 2 hr of reaction, the enzyme has lost 56% of the original activity, when 3.1 tyrosine residues have been modified. After 7.5 hr, 79% of the original activity is lost with the modification of 5.6 tyrosine residues. The reactivity of Glu-58 toward iodoacetate, however, was not lost in parallel with the loss of activity (see Table 3). These results indicate that there are probably no tyrosine residues which are directly involved in the active site of the enzyme. Some tyrosine residues, however, may be important primarily for the maintenance of the enzymatically active conformation of the enzyme.

Use has also been made[18,23] of acetylimidazole, another reagent specific for tyrosine residues, at pH 7 to 7.5 and 25°C. The enzyme is inactivated slowly by this reagent. Further experiments[44] reveal that 3–4 tyrosine residues can be acetylated relatively easily. At this point, the enzyme retains about 65% of the original activity. When the reaction is carried out in 8 M urea, nearly all the tyrosine residues are modified with almost complete loss of activity. The inactivated protein is fully reactivated by treatment with hydroxylamine.

Diazo-1*H*-tetrazole and dinitrofluorobenzene have also been used[45] in a similar manner. Diazo-1*H*-tetrazole has been found to react preferentially with 1–2 tyrosine residues, mainly Tyr-4 and 45, as well as the two amino groups in the molecule. Interestingly, the enzyme which had lost most of its activity toward RNA by this reaction retained about 50% of the activity toward 2′,3′-cyclic GMP. These tyrosine residues are thought to be in the vicinity of the active site of the enzyme. Dinitrofluorobenzene reacts similarly.

The reactivities of the tyrosine residues toward several reagents thus indicate that a few tyrosine residues are exposed on the surface of the enzyme, but that most of the other tyrosine residues are in the interior of the protein molecule. A similar conclusion has been drawn recently by Iida and Ooi[31] from spectrophotometric titration and by Yamamoto and Tanaka[43] from ultraviolet absorption spectroscopy.

3. Studies on the Binding of Substrate Analogs

In the preceding pages, the results of recent chemical modification studies have been described. These were performed in order to gain information on the roles of some specific amino acid residues in the enzymatic activity of RNase T_1. In order to understand the molecular mechanism of action of the enzyme, it is equally important to elucidate the mode of binding of substrates to the enzyme and hence the nature of the substrate binding site. Because RNase T_1 is unique in its substrate specificity, it is an interesting object for such studies. From work on the competitive inhibition of RNase T_1 by various nucleotides, it has been shown by Irie[46,47] that guanylates have a higher affinity toward the enzyme than other nucleotides, consistent with the known specificity of the enzyme, though binding affinities of other nucleotides are not always in parallel with the base specificity of RNase T_1. The dissociation constants of the enzyme-inhibitor complex were shown to increase in the order of 2′-GMP < 3′-GMP < 5′-GMP.

This has been confirmed by the spectroscopic studies on the binding of RNase T_1 with 2′-GMP and related compounds by Sato and Egami.[16] From investigation of the difference spectrum related to the interaction of RNase T_1 and 2′-GMP at 290 nm, it was concluded that one molecule of RNase T_1 binds one molecule of 2′-GMP, indicating that the RNase T_1 molecule has one active site responsible for the binding. This was confirmed by a gel-filtration experiment. The absorbancy difference shows a maximum at about pH 5 and decreases at both higher and lower pH values, indicating the participation of certain groups with pK values of about 3 and 6 in the interaction of RNase T_1 and the nucleotide. 2′-GMP protects RNase T_1 from inactivation by carboxymethylation with iodoacetate, and RNase T_1 carboxymethylated in the absence of 2′-GMP has no more absorbancy difference at 290 nm when mixed with 2′-GMP although the binding without an absorbance difference cannot be excluded. Since both 2′-guanosine monosulfate and the benzyl ester of 2′-GMP, which have no undissociated hydroxyl groups, do not bind to RNase T_1, it was suggested that an undissociated hydroxyl group of phosphomonoester is essential for the strong binding with RNase T_1

and for the appearance of the difference spectrum. From the analysis of kinetic parameters of RNase T_1 measured at various pH values by using guanosine cyclic phoshate or guanylyl nucleosides as substrates and 2′-GMP as a competitive inhibitor, Irie[47,48] also derived the similar conclusion that the preferable binding form of 2′-GMP with the enzyme is presumably a monoionic species and that at least three dissociable groups of the enzyme with pK values of 3.7, 5.7 and about 7.5 participate in the active center of RNase T_1. The results of these experiments have already been reviewed in more detail.[29]

In the following pages, the results of some more recent studies[23,49,50] on the binding of substrate analogs to RNase T_1 will be described. These were obtained by using a gel-filtration method essentially the same as that developed by Hummel and Dreyer.[51] In this method, RNase T_1 or its derivative dissolved in a buffer solution containing a substrate analog is passed through a column (0.9 × 20 cm) of Sephadex G-25, equilibrated and eluted at 25°C with the same buffer solution containing the substrate analog. The total protein eluted from the column is determined by amino acid analysis of an acid hydrolyzate of a portion of the combined protein fraction. The amount of the substrate analog bound to the protein is estimated from the trough area in the elution profile by using a suitable molar extinction coefficient. From these values, the number of molecules of the analog bound per molecule of protein can be calculated, and the dissociation constant (K_i) of the enzyme-analog complex obtained.

3.1. Binding of Substrate Analogs to Native RNase T_1

A typical elution profile is shown in Fig. 21. Under the specified conditions, RNase T_1 binds 0.90 molecule of 3′-GMP per molecule with a K_i value of 0.0077 mM. The extent of binding of 3′-GMP to RNase T_1 was also measured at varying concentrations of 3′-GMP at pH 5.5 and 25°C. The results (shown in Fig. 22) indicate that one molecule of 3′-GMP binds to one molecule of RNase T_1, and hence, the presence of one binding site per molecule of protein. At pH 5.5 and 25°C, the order of binding strengths of some mononucleotides is 3′-GMP$>$ 5′-GMP$>$ 3′-AMP$>$ 3′-CMP and 3′-UMP. The approximate K_i values are given in Table 5. These results are consistent with the known base specificity of the enzyme and also accord with the order of the protective effect on

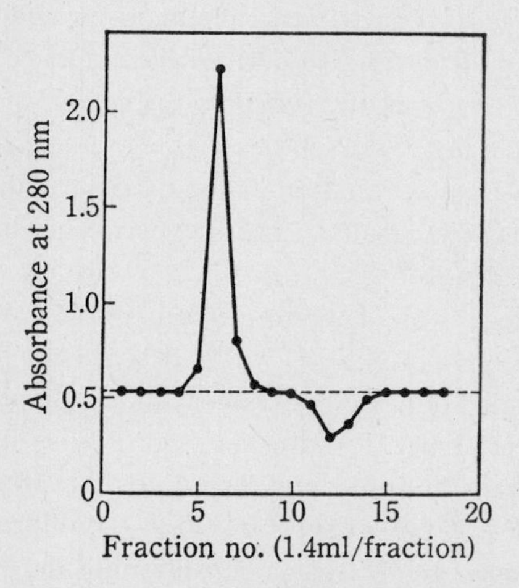

Fig. 21. Elution profile accompanying passage of RNase T_1 through a Sephadex G-25 column. Column, 0.9 × 20 cm; protein, 0.1 μmole; 3′-GMP, 0.068 mM; pH 5.5; 25°C.

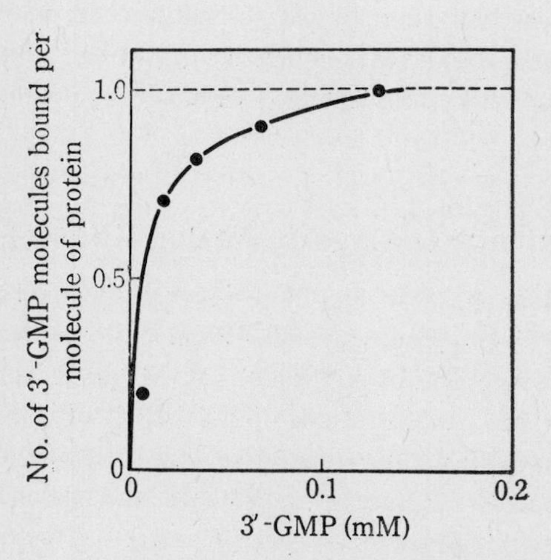

Fig. 22. Number of 3′-GMP molecules bound per molecule of RNase T_1. Gel-filtration on Sephadex G-25; column, 0.9 × 20 cm; protein, 0.1 μmole; pH 5.5; 25°C.

TABLE 5
Binding of Some Nucleotides to RNase T_1 (pH 5.5, 25°C)

	3′-GMP	5′-GMP	3′-AMP	3′-CMP (UMP)
K_i (mM)	0.0077	0.068	0.30	≳3∼4

inactivation of the enzyme by reaction with iodoacetate[16,35] or by photooxidation in the presence of rose bengal[30] (see Fig. 12). They are also compatible with the data obtained by competitive inhibition experiments[46,47] and with those obtained recently by Rüterjans *et al.* from fluorescence spectroscopy.[33] Similar results have also been obtained by Campbell and Ts'o from a gel-filtration study.[52]

The binding has also been measured with guanosine, deoxyguanosine and guanine, and the order of the binding strengths is 3′-GMP> guanosine> deoxyguanosine> guanine, as can be seen in Table 6. More than 10-fold decreases in K_i values were observed with changes from guanine to guanosine and from guanosine to 3′-GMP. The difference in binding affinity between guanosine and deoxyguanosine is also notable, indicating the importance of the 2′-OH group for effective binding. The binding of D-ribose alone, however, appears to be very weak, since the binding of 3′-GMP to the enzyme is only slightly weakened by the presence of a 100-fold molar excess of D-ribose over the nucleotide. These results indicate that each of the guanine, ribose, and phosphate portions is involved in the binding of 3′-GMP to the enzyme and that the integrity of these components is important for strong binding. As described above, 3′-CMP (or UMP) does not bind to the enzyme to any significant extent, although it has ribose and phosphate portions in common with 3′-GMP. The guanosine portion, therefore, must be of primary importance for the recognition by, and hence the specific binding to the enzyme; the phosphate portion so helps to strengthen the binding further.

TABLE 6
Binding of Guanine Containing Compounds to RNase T_1 (pH 5.5, 25°C)

	3′-GMP	Guanosine	Deoxyguanosine	Guanine
K_i (mM)	0.0077	0.12	0.7	>2

In contrast to the fairly strong binding of guanosine to the enzyme, the other nucleosides examined (Table 7) bind to the enzyme only very weakly or not at all under the conditions employed at pH 5.5. The K_i values of these nucleosides except guanosine are all above 1 mM. The structural requirements for the base portion must be quite rigorous and at least the N-1 and N-7 positions and the 2-amino and 6-oxo (or hydroxy) groups on the purine ring are important for strong binding of the base portion to the enzyme. In Table 7, the known susceptibility of 3′-phosphodiester bonds of these nucleosides toward the enzyme is also shown (see the recent review[19]). This susceptibility is generally in accord with the binding strength of each nucleoside to the enzyme. 3′-Phosphodiester bonds of inosine and xanthosine are known to be hydrolyzed by RNase T_1, whereas the nucleosides bind only very weakly to the enzyme. The 2-amino group on guanine appears to contribute greatly to the binding, although it is not necessarily required for the cleavage of 3′-phosphodiester bonds. The N-8 position of guanine does not appear to be important at least for hydrolysis, since the 2′,3′-cyclic phosphate of 8-bromoguanosine[53] is known to be hydrolyzed by RNase T_1.

The binding also needs the native conformation of the enzyme. Thus the binding of 3′-GMP to the enzyme is weakened gradually to nil with increasing urea concentration to 8 M as can be seen from Fig. 23, where the enzyme is known to be reversibly denatured.[17] Rupture of the two disulfide bonds also abolishes the binding ability as will be described later.

Table 7

Binding of Some Nucleosides (pH 5.5, 25°C) and Susceptibility of Their 3′-Phosphodiesters to RNase T_1

	guanosine	N^7-methylguanosine	N^1-methylguanosine
(a) †1	++	±	±
(b) †2	+++	−	±

	N^2-methylguanosine	N^2-dimethylguanosine	thioguanosine
(a) †1	±	±	±
(b) †2	not determined	±	−

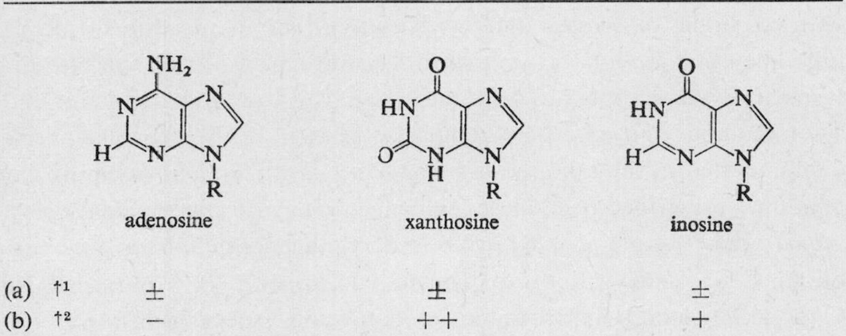

	adenosine	xanthosine	inosine
(a) †1	±	±	±
(b) †2	−	++	+

†1 Binding of nucleosides

†2 Susceptibility of nucleoside 3′-phosphodiesters[19)]

(a) Binding of nucleosides: R, ribose; ++, well bound; ±, not well bound (K_i >1 mM).

(b) Susceptibility of nucleoside 3′-phosphodiesters: R, ribose-3-phosphodiester; +++, attacked very easily; ++, attacked easily; +, attacked; ±, attacked difficultly; −, not attacked.

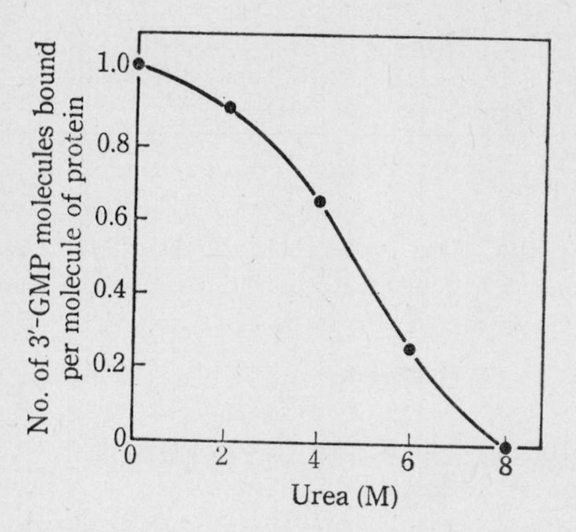

Fig. 23. Effect of urea on binding of 3′-GMP to RNase T_1. Gel-filtration on Sephadex G-25; column, 0.9 × 20 cm; protein, 0.1 μmole; 3′-GMP, 0.066 mM; pH 5.5; 25°C.

The binding of 3′-GMP and guanosine to RNase T_1 was measured over a wide pH range. As can be seen from Fig. 24, the optimum pH for binding of these compounds lies around pH 5. In this pH region, the enzyme binds maximally one molecule of 3′-GMP per molecule of protein as described previously. These results are in accord with the spectroscopic data obtained for the binding of 2′-GMP to the enzyme.[16] The pH dependence of guanosine binding indicates that two groups with p*K* values of about 4 and 7, respectively, are involved in the binding. They are presumably a carboxyl group and an imidazole group, respectively, and appear to be implicated in the protein–base (or nucleoside) binding. Indeed, some imidazole group involvement has been demonstrated in the binding of 3′-GMP to the enzyme as noted in the previously described results of rose bengal-catalyzed photooxidation of the enzyme.[30] A pH profile fairly similar to that for guanosine was also obtained for deoxyguanosine. The pH dependence of binding of 3′-GMP differs somewhat from that of guanosine, especially in the acidic pH region. This is probably due mainly to the presence of the phosphate group in the former. The first dissociation (p*K*, *ca.* 0.7) of the phosphate group is thought to strengthen the binding, whereas the second dissociation appears rather to weaken the binding as pointed out earlier.[16] The latter may be due to electrostatic repulsion between the phosphate dianion and a carboxylate anion in the active center of the enzyme.

Participation of a carboxyl group in protein-base binding has also been suggested recently by Uenishi *et al.*[54] from spectroscopic studies on the interaction of RNase T_1 and its substrate analogs.

Zinc ion is known to be a strong inhibitor of the enzyme.[17] The binding of 3′-GMP to the enzyme, however, was not affected significantly by the presence of a 10-fold molar excess of zinc over the nucleotide. This ion, therefore, may interact directly with some amino acid(s), possibly a carboxyl or an imidazole group, in the catalytic site rather than in the binding site.

3.2. Binding of Substrate Analogs to Chemically Modified RNase T_1

As can be seen from Fig. 24, a pH dependence similar to that with native enzyme has been observed for the binding of guanosine to the enzyme inactivated by carboxymethylation of the γ-carboxyl group of Glu-58 in the active center. The pH profile for binding of 3′-GMP to the carboxymethylated enzyme is also similar to that for the native enzyme, except that the binding ability is somewhat lowered (Table 8) especially in the acidic pH region. Thus, carboxymethylation of Glu-58 appears to affect mainly the binding of the phosphate portion, and only slightly the

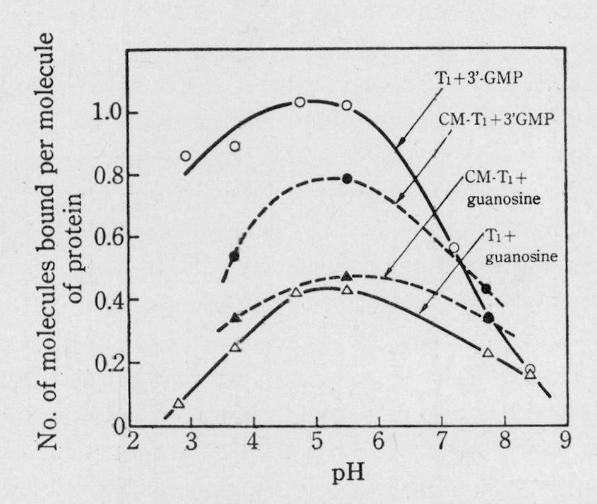

Fig. 24. pH dependence of binding of 3′-GMP and guanosine to native and carboxymethylated (CM-) RNase T_1. Gel-filtration on Sephadex G-25; column, 0.9 × 20 cm; protein, 0.1 μmole; 3′-GMP, 0.13 mM; guanosine, 0.09 mM; deoxyguanosine, 0.1 mM; 25°C.

TABLE 8

Binding of 3′-GMP†[1] to Chemically Modified RNase T_1

Treatment	Activity toward RNA (%)	Extent of modification at given residues (no. of residues per molecule of protein†[2])							Number of 3′-GMP molecules bound per molecule of protein	K_I (mM)
		Glu-58	Ala-1	Lys-41	His	Arg-77	Trp-59	Cys/2		
Untreated	100								0.90	0.0076
Iodoacetate										
(pH 5.5, 37°C, 5 hr) ...	0.2	*ca.* 1.0							0.79	0.018
Photooxidation in the presence of rose bengal										
(pH 8.5, 37°C, 0.5 hr) ..	28				1.32		0.23		0.22	0.23
(pH 8.5, 37°C, 2 hr) ...	2				3.00		0.78		~0	>10
Phenylglyoxal										
(pH 8.0, 25°C, 24 hr) ..	8		0.70	(?)		>0.7			~0	>10
Reduction with mercapto-ethanol and carboxy-methylation	1							3.3	0.16	0.69

†[1] The extent of binding was determined by the gel-filtration method using about 1 mg of protein and 0.068 mM (or 0.13 mM) 3′-GMP at 20–25°C.

†[2] Calculated assuming the presence of 1.00 residue each of Glu-58, Ala-1, Lys-41, Arg-77 and Trp-59, 3.00 residues of histidine and 4.00 residues of half-cystine per molecule of native protein.

binding of the nucleoside portion to the enzyme. Similar pH profiles have also been obtained for the binding of 2′,3′-cyclic GMP and guanylyl-3′,5′-adenosine to the carboxymethylated enzyme. These results therefore imply that Glu-58 is part of the catalytic site rather than part of the binding site of the enzyme, although the latter possibility cannot be excluded. Indeed, the participation of Glu-58 in the binding site has recently been suggested from NMR[55] and difference spectral studies.[56]

The binding ability of the enzyme toward 3′-GMP was markedly decreased by photooxidation of some histidine residues,[30] and also by modification of the single arginine residue (Arg-77) with phenylglyoxal[37] as shown in Table 8. Thus some histidine residue(s) appear to be implicated in the active center as part of the binding site, as suggested from the pH dependence of binding of substrate analogs to the enzyme. Arg-77 may also constitute part of the binding site or may be situated in the vicinity of the active center. These results are consistent with those obtained from chemical modification studies. As already described, the native three dimensional structure is also important for the binding. Cleavage of the disulfide bonds in RNase T_1 with mercaptoethanol in 8 M urea results in a considerable loss of the binding ability of the enzyme (Table 8). Further studies on the interaction of substrate analogs with native and chemically modified RNase T_1 are now in progress.

4. Conclusion

In the decade or so since the discovery of RNase T_1 in 1957, its covalent structure has been determined and information has been accumulated about its conformational structure, the nature of the active site, and the mode of action of the enzyme. As described, modification studies (mainly by chemical methods) have indicated the participation of Glu-58 and probably two histidine residues at the active site of the enzyme. The single arginine and tryptophan residues appear to be important in some manner for enzymatic activity, although they may not be directly involved in catalysis. Some of these residues may also be important in building up the architecture of the active site. Information on the mode of interaction between the enzyme and its substrates has also accumulated. The pH dependence of binding of substrate analogs to the enzyme suggests the involvement of a carboxyl and an imidazole group in the protein-base binding. The structural requirements for the substrate molecule have been found to be fairly rigorous. Thus at least the N-1 and N-7 positions and the 2-amino and 6-oxo (or hydroxy) groups on the purine ring are important for the strong binding of the base portion to the enzyme. In addition, D-ribose and phosphate portions are also involved in the binding, and the integrity of these components is important to the strength of the bond.

Fig. 25 shows a model for the active center of RNase T_1 based on the knowledge so far obtained. The binding site for the 3′-guanylyl portion of the substrate can be assumed to consist of three subsites, which are responsible for binding of the base, the sugar, and the phosphate portions. The guanine residue is thought to be hydrogen-bonded to the protein at some or all of the four points—the N-1 and N-7 positions and 2-amino and 6-oxo groups. The N-7 position has been suggested to be protonated by a certain protondonating group in the enzyme from the difference spectral study of Oshima and Imahori.[56] The 6-oxo group be hydrogenbonded to an imidazole group (protonated form) and N-1 to a carboxyl group (unprotonated form). The nature of the ribose binding site is not clear. The 2′-OH may be hydrogenbonded to some

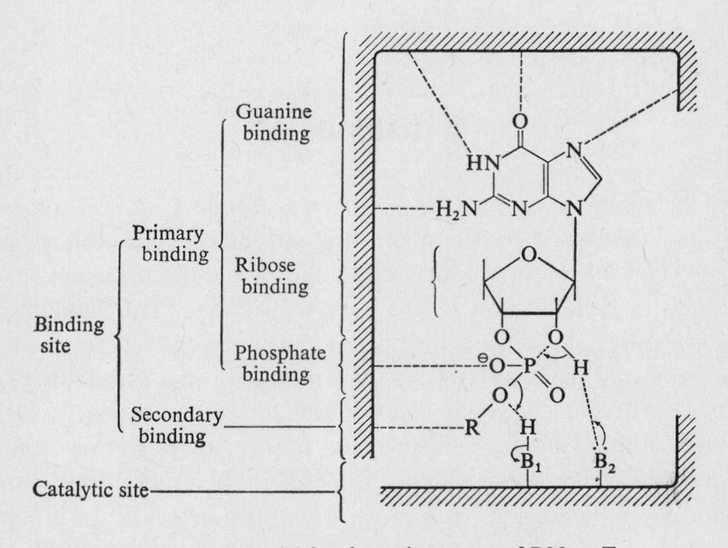

Fig. 25. A proposed model for the active center of RNase T_1.

group in the protein. The phosphate anion may be bound in the phosphate binding site to the guanidinium group of Arg-77, or to the imidazole group assumed to be present at the catalytic site. In addition, the site responsible for binding of the nucleotide or nucleoside portion of the balance of the nucleic acid (see –R in Fig. 25) is also thought to be present in this region (i.e. secondary binding site). The catalytic site is assumed to involve most simply two groups—an acidic and a basic group (B_1H and $\ddot{B}_2$ in Fig. 25) working cooperatively as a proton donor and a proton acceptor, respectively, and vice versa. The most likely candidates for these groups are His-92 and Glu-58 (or His-27 or 40).

Such a model is still a tentative one and further elaborate studies are required for the full understanding of the nature of the active site and the molecular mechanism of catalysis. For this purpose, the elucidation of the three dimensional structure of the enzyme by X-ray diffraction analysis will be necessary. Chemical synthesis of the enzyme will be another important approach toward this goal. In addition, comparative studies on the structure-function relationships of other RNases, especially those with similar substrate specificity (such as RNases U_1 and N_1) appear promising, and are already in progress.

References

1. K. Sato and F. Egami, *J. Biochem.* (*Tokyo*), **44**, 753 (1957).
2. K. Takahashi, *ibid.*, **49**, 1 (1961).
3. N. Ui and O. Tarutani, *ibid.*, **49**, 9 (1961).
4. K. Takahashi, *ibid.*, **51**, 95 (1962).
5. K. Takahashi, *J. Biol. Chem.*, **240**, 4117 (1965).
6. K. Takahashi, *Protein, NucleicAcid* and *Enzyme* (Japanese), **10**, 1277 (1965).
7. C. H. W. Hirs, S. Moore and W. H. Stein, *J. Biol. Chem.*, **235**, 633 (1960).
8. D. C. Smyth, W. H. Stein and S. Moore, *ibid.*, **238**, 227 (1963).
9. S. Yamagata, K. Takahashi and F. Egami, *J. Biochem.* (*Tokyo*), **52**, 272 (1962).
10. K. Kasai, *ibid.*, **57**, 372 (1965).
11. Y. Shiobara, K. Takahashi and F. Egami, *ibid.*, **52**, 267 (1962).
12. Y. Kuriyama, *ibid.*, **59**, 596 (1966).
13. H. Kasai, K. Takahashi and T. Ando, *ibid.*, **66**, 591 (1969).
14. K. Takahashi, *ibid.*, **52**, 72 (1962).

15. S. Saigusa, K. Takahashi and F. Egami, *Symposium on Enzyme Chemistry*, vol. XV, p. 48, Nankodo, 1961.
16. S. Sato and F. Egami, *Biochem. Z.*, **342**, 437 (1965).
17. K. Takahashi, *J. Biochem.* (*Tokyo*), **60**, 239 (1966).
18. K. Takahashi, *Protein, Nucleic Acid* and *Enzyme* (Japanese), **14**, 1127 (1969).
19. K. Takahashi, T. Uchida and F. Egami, *Advan. Biophys.* (*Tokyo*), **1**, 53 (1970).
20. A. M. Crestfield, W. H. Stein and S. Moore, *J. Biol. Chem.*, **238**, 2413 (1963).
21. K. Takahashi, W. H. Stein and S. Moore, *ibid.*, **242**, 4682 (1967).
22. F. Egami, K. Takahashi and T. Uchida, *Nucleic Acid Research and Molecular Biology*, vol. 3, p. 59, Academic Press, 1964.
23. K. Takahashi, *J. Biochem.* (*Tokyo*), to be published.
24. D. Findley, D. G. Herries, A. P. Mathias, B. R. Rabin and C. A. Ross, *Biochem. J.*, **85**, 152 (1962).
25. F. H. Westheimer, *Advan. Enzymol.*, **24**, 441 (1962).
26. K. Takahashi, *J. Biochem.* (*Tokyo*), **68**, 517 (1970).
27. S. Yamagata, K. Takahashi and F. Egami, *ibid.*, **52**, 261 (1962).
28. S. Sato, *Unpublished.*
29. K. Takahashi, *J. Biochem.* (*Tokyo*), **67**, 833 (1970).
30. K. Takahashi, *ibid.*, **69**, 331 (1971).
31. S. Iida and T. Ooi, *Biochemistry*, **8**, 3897 (1970).
32. M. Irie, *J. Biochem.* (*Tokyo*), **68**, 69 (1970).
33. H. Rüterjans, H. Witzel and O. Pongs, *Biochem. Biophys. Res. Commun.*, **37**, 247 (1969).
34. K. Waku and Y. Nakazawa, *J. Biochem.* (*Tokyo*), **68**, 63 (1970).
35. T. Terao and T. Ukita, *Biochim. Biophys. Acta*, **149**, 613 (1967).
36. K. Takahashi, *J. Biol. Chem.*, **243**, 6171 (1968).
37. K. Takahashi, *J. Biochem.* (*Tokyo*), **68**, 659 (1970).
38. K. Takahashi, *ibid.*, **67**, 541 (1970).
39. T. Terao and T. Ukita, *Biochim. Biophys. Acta*, **181**, 347 (1969).
40. M. Irie, *J. Biochem.* (*Tokyo*), **68**, 31 (1970).
41. S. Kawashima and T. Ando, *Intern. J. Protein Res.*, **1**, 185 (1969).
42. O. Pongs, *Biochem. Biophys. Res. Commun.*, **38**, 431 (1970).
43. Y. Yamamoto and J. Tanaka, *Biochim. Biophys. Acta*, **207**, 522 (1970).
44. H. Kasai, K. Takahashi and T. Ando, *Seikagaku* (Japanese), **42**, 503 (1970).
45. H. Kasai, K. Takahashi and T. Ando, *J. Biochem.* (*Tokyo*), to be published.
46. M. Irie, *ibid.*, **56**, 496 (1964).
47. M. Irie, *ibid.*, **61**, 550 (1967).
48. M. Irie, *ibid.*, **63**, 649 (1968).
49. K. Takahashi, *Seikagaku* (Japanese), **42**, 526 (1970).
50. K. Takahashi, *J. Biochem.* (*Tokyo*), **68**, 941 (1970).

51. J. P. Hummel and W. J. Dreyer, *Biochim. Biophys. Acta*, **63**, 530 (1962).
52. M. K. Campbell and P. O. P. Ts'o, *ibid.*, **213**, 222 (1971).
53. R. Yuki and H. Yoshida, *Seikagaku* (Japanese), **42**, 461 (1970).
54. N. Uenishi, Y. Yabuuchi, T. Oshima and K. Imahori, *ibid.*, **42**, 526 (1970).
55. H. Rüterjans and O. Pongs, *European J. Biochem.*, **18**, 313 (1971).
56. T. Oshima and K. Imahori, *J. Biochem.* (*Tokyo*), **69**, 987 (1971).

Ribonuclease T_1-Substrate Analog Complexes

Tairo OSHIMA and Kazutomo IMAHORI
Department of Agricultural Chemistry, Faculty of Agriculture, University of Tokyo
Bunkyo-ku, Tokyo, Japan

1. Introduction

Since ribonuclease T_1 (RNase T_1) was isolated from Takadiastase powder, a commercial product of *Aspergillus oryzae*,[1] many experiments on various aspects of the enzyme have been carried out. RNase T_1 can be regarded as a guanylate specific endoribonuclease, but the cleavage of the internucleotide bonds of RNA by the ribonuclease is essentially a transphosphorylation followed by hydrolysis. The enzyme catalyzes the transfer of the phosphodiester bonds of 3′-guanylic acid groups in RNA to yield oligo- and mononucleotides with terminal 2′,3′-cyclic phosphodiester groups, which are further hydrolyzed by the enzyme to the corresponding 3′-phosphate compounds.[2] RNase T_1 is stable to heat and acid. It is a small protein without a prosthetic group and consists of a single polypeptide chain of 104 amino acid residues, whose sequence was determined by Takahashi.[3] This enzyme should be one of the most suitable enzymes for the study of structure-function relationships, as well as being useful in the analysis of nucleotide sequences.

Many attempts have already been made to elucidate the mechanism of enzyme action, and important information has been accumulated, although it is not yet possible to explain the molecular basis of this action. Modifications of various amino acid residues of the RNase T_1 molecule have been studied in order to determine their roles in the catalytic action and the results are well summarized in review papers[4-7] along with the detailed physical and chemical natures of the enzyme protein and the catalytic reaction. The base specificity of the enzyme has been studied using various modified RNA's, synthetic dinucleoside monophosphates and nucleoside-2′,3′-cyclic phosphates, in an attempt to discover the essential structural requirements of the base for susceptibility to attack. Egami and his co-workers[4-6] and Whitfeld *et al.*[8] suggested the requirement of a lactam structure at the 1,6 positions or an oxo group at the 6 position of the purine base. From the results of recent investigations, Egami *et al.*[6] proposed in addition to the above hypothesis the participation of N-7 of the purine base in the susceptibility to the enzyme.

The examination of the enzyme-substrate complex is of great importance to work on the mechanism of enzyme action. Due to the instability of enzyme-substrate complexes, substrates have frequently been replaced with substrate analogs which fit into the enzyme but are not cleaved by it. Sato and Egami[9] have reported the formation of RNase T_1-guanosine-2′ or 3′-phosphate complexes. They measured the difference spectrum of mixtures of RNase T_1 and guanosine-2′-phosphate in solution (the latter is known to be a competitive inhibitor for the enzyme), and concluded that one molecule of the nucleotide combines with one molecule of the enzyme. The binding was also confirmed by gel-filtration experiments.[9] The nucleotide is presumed to fit into the substrate binding site of the enzyme to form the enzyme-substrate analog complex.

As has been emphasized by Egami and his co-workers,[4,5,9] the ribonuclease-substrate analog complexes can be a useful tool for the elucidation of the mechanism of RNase T_1 action. These complexes will be important in the deduction of a mechanism for the unique substrate specificity of the enzyme. At the same time, research on both the chemical and physical basis of enzyme-analog complex formation would be of value for a comprehensive understanding of the modes of interaction and recognition between protein and nucleotide. In this context, an investigation has recently been made on the optical nature of the RNase T_1-substrate analog complexes. The present article presents the results of the study and gives a discussion of the mechanism of complex formation.

2. Studies on Difference Spectra Observed upon Binding of Various Guanine Derivatives to RNase T_1

2.1. Difference Spectrum Observed upon Binding

An ultraviolet absorption spectrum of a mixture of RNase T_1 and guanosine-2′ or 3′-phosphate is different from the graphical summation of the respective spectra of these components observed independently.[9]

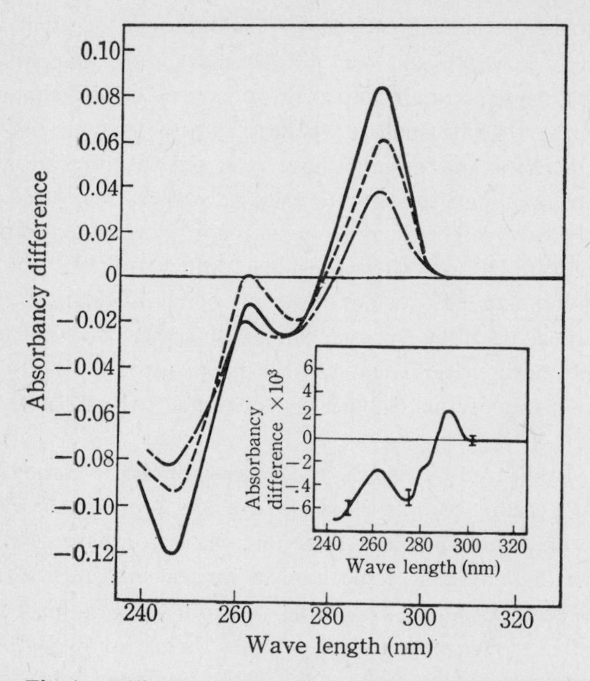

Fig. 1. Difference spectra of interaction of RNase T_1 and guanine derivatives, recorded at 25°C using cells of 3 mm path length. ——, RNase T_1, 9.5×10^{-5} M; guanosine-3′-phosphate, 2.65×10^{-4} M in 0.014 M sodium acetate buffer; pH 5.6. ------, RNase T_1, 7.0×10^{-5} M; guanosine-2′-phosphate, 2.2×10^{-4} M in 0.014 M sodium acetate; pH 5.6. —-—, RNase T_1, 9.2×10^{-5} M; guanosine, 3.26×10^{-4} M in 0.014 M sodium acetate; pH 5.6. *Inset:* RNase T_1, 8.45×10^{-5} M; 9-methylguanine, 1.03×10^{-4} M in 0.08 M sodium acetate; pH 5.6.

As shown in Fig. 1, a strong peak at 291 nm and a deep trough at 247 nm were characteristic of the difference spectra observed upon binding of the indicated nucleotides to the enzyme. When guanosine or 9-methylguanine was added to the enzyme solution instead of guanosine-2′ or 3′-phosphate, similar difference spectra were obtained[10] (Fig. 1), although the peak height at 291 nm was reduced in the case of guanosine or 9-methylguanine, indicating that the affinities of these two analogs to the enzyme are lower than those of guanosine-2′ or 3′-phosphate. Similarly, the interaction of guanosine-5′-phosphate or guanine deoxyriboside (deoxyguanosine) and RNase T_1 yielded the characteristic changes in absorbancy, as shown in Fig. 2.[11] The wave lengths of the peaks and troughs of the spectra in this figure are identical with those observed

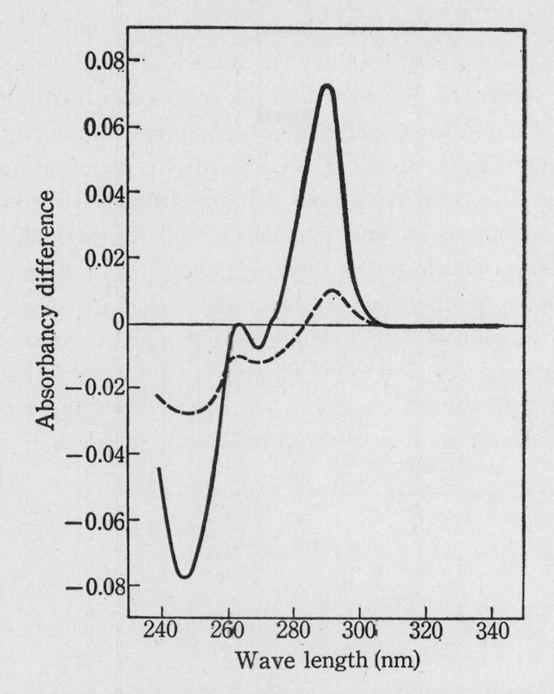

Fig. 2. Difference spectra of interaction of RNase T_1 and guanosine-5′-phosphate or deoxyguanosine, recorded at 25 °C using cells of 3 mm path length. —— 9.5×10^{-5} M RNase T_1 was mixed with 2.2 $\times 10^{-4}$ guanosine-5′-phosphate in 0.013 M sodium acetate, pH 5.6. ------ 9.5×10^{-5} M RNase T_1 was mixed with 2.7×10^{-4} M deoxyguanosine in 0.013 M sodium acetate, pH 5.6.

upon binding of guanosine-2′ or 3′-phosphate to RNase T_1. These results indicate that the enzyme interacts with the guanine base even in the absence of phosphate and ribose groups, and that the difference spectra shown in Fig. 1 and 2 represent the nature of this interaction. It is interesting to note that the enzyme is a ribonuclease and does not split the internucleotide bonds of DNA,[5] whereas the binding of deoxyguanosine to RNase T_1 was spectrophotometrically observed.

2.2. Assignment of Difference Spectrum

To assign these peaks and troughs of the difference spectra observed by mixing guanine derivatives with RNase T_1, absorbancy changes of guanine derivatives were measured by lowering the pH of the solution.

The difference spectra obtained are shown in Fig. 3 and 4, which strikingly resemble those in Fig. 1 and 2. The wave lengths of the peaks and troughs of the spectra in Fig. 1 and 2 are in good accord with those in Fig. 3 and 4, i.e. the difference spectra observed upon complex formation are qualitatively similar to those observed upon protonation of guanine derivatives. It is worthy of note that the difference spectrum observed on the interaction of 9-methylguanine and the enzyme (the inset in Fig. 1) differs somewhat from those of guanylates or guanosine but is qualitatively in good agreement with the difference spectrum obtained upon protonation of 9-methylguanine (a trough at around 270 nm and a peak at around 255–260 nm in addition to a peak at 291 nm). The site at which protonation occurs is asserted to be the seventh nitrogen (N-7) of the guanine base in the case of both guanosine[12] and

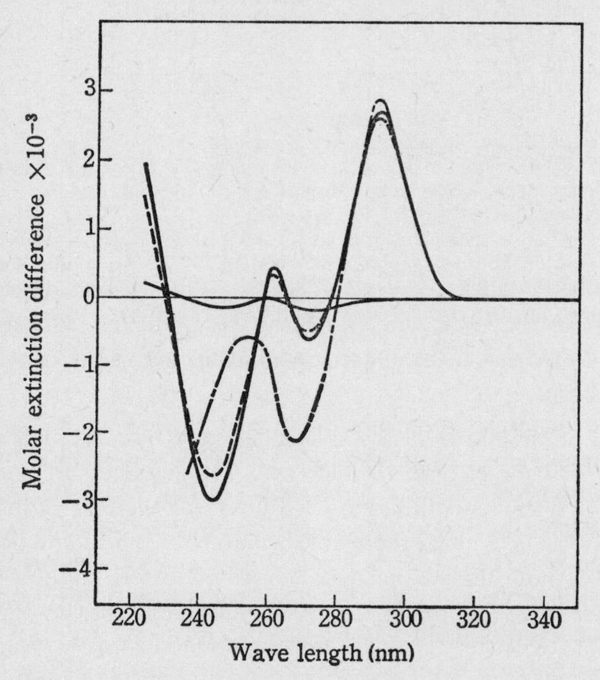

Fig. 3. Difference spectra produced by changing the pH of guanine derivative solutions. ——, Guanosine-3′-phosphate from pH 6.5 to 1.0. ------, Guanosine from pH 6.5 to 1.35. — — —, 9-Methylguanine from pH 6.5 to 1.2. —— (flat curve), Guanosine-3′-phosphate from aqueous to 18% (v/v) dioxane at pH 6.5. Changes in absorbancy of guanosine-2′-phosphate from pH 6.5 to 1.0 were nearly identical with those for guanosine-3′-phosphate.

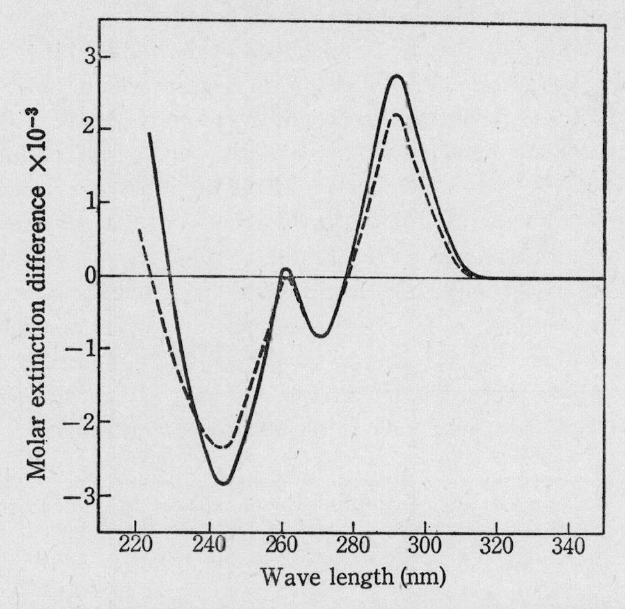

Fig. 4. Difference spectra produced by changing the pH of the solution of guanosine-5′-phosphate from pH 6.5 to 1.1 (——) and of deoxy-guanosine from pH 6.5 to 1.8 (------).

9-methylguanine,[13] when they are in acidic solution. The difference spectra shown in Fig. 1, therefore, could be interpreted as representing the protonation of N-7 of the guanine base at the active site of RNase T_1.

The molar extinction difference in the vicinity of 290 nm between protonated and non-protonated guanylate or guanosine was measured from the spectrophotometric titration curve shown in Fig. 5 and determined to be 3.0×10^3 cm^{-1} M^{-1}. The pK value of guanosine-2′-phosphate was observed spectroscopically to be 2.3 from Fig. 5, which is in good agreement with the literature pK value[14] for the guanine base. The absorbancy difference upon binding was estimated to be 3.1×10^3 cm^{-1} M^{-1} for the complex formed in the mixture, assuming that the RNase T_1 in the mixture was fully saturated with the guanosine-3′-phosphate ligand when 9.5×10^{-5} M of the enzyme was mixed with 2.65×10^{-4} M of the guanine nucleotide at pH 5.6, of which the K_i value has been reported to be about 7.7×10^{-6} M.[7] From a comparison of the spectra shown in Fig. 1 and 2 with those in Fig. 3 and 4, it was suggested that the differ-

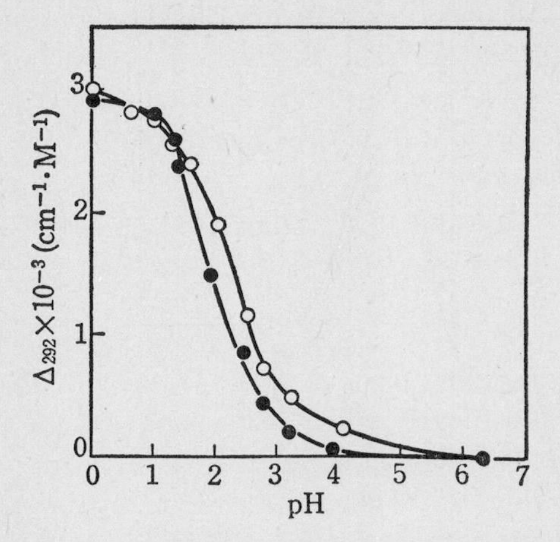

Fig. 5. Spectrophotometric titration of guanosine-2′-phosphate (○) and guanosine (●). Changes in absorbancy at 292 nm are plotted against pH.

ence spectra observed in the region of 250–300 nm upon mixing of RNase T_1 with guanine derivatives such as guanylate or guanosine, can be assigned to the protonation of N-7 of the guanine base in the enzyme-analog complex, since the former spectrum is closely similar both qualitatively and quantitatively to the latter. The conformational change of aromatic residues of RNase T_1 upon the binding would have little or no detectable effect on the difference spectrum.

2.3. Binding of Adenylate to RNase T_1

The difference spectrum observed upon mixing adenosine-2′(3′)-phosphate (a mixture of adenosine-2′-phosphate and adenosine-3′-phosphate) with RNase T_1 failed to manifest the characteristic absorbancy difference observed upon formation of RNase T_1-guanosine-2′ or 3′-phosphate complex, although adenosine-2′ or 3′-phosphate has been shown to be a competitive inhibitor for the enzyme and to form a complex.[7,15] Slight absorbancy differences in the vicinity of 280 and 255 nm were observed upon mixing adenosine-2′(3′)-phosphate with the enzyme, as shown in Fig. 6. In order to assign the difference spectrum shown in this figure, the spectral change of adenylate was investigated by changing the solvent. Fig. 7 gives the difference spectrum of adenosine-2′(3′)-

phosphate resulting from a change of solvent from citrate buffer (pH 6.5) to the buffer containing 18% dioxane, that is, to a partially non-aqueous solvent. The difference spectrum resembles that observed by mixing adenosine-2′(3′)-phosphate with RNase T_1, as shown in Fig. 6. It is concluded that a solution containing dioxane is a good approximation of the environment in the region of the binding site of RNase T_1, suggesting a hydrophobic interaction between the nucleotide and the enzyme.

The enzyme interaction spectrum of adenylate bears no resemblance to the difference spectrum resulting from a change of the pH of the solution of adenylate (Fig. 7), suggesting that the adenine base is not protonated in the complex. The difference spectrum of interaction of the enzyme and adenylate (Fig. 6) cannot be interpreted by a simple "polarization red-shift" of the perturbed adenosine chromophore because the spectrum bears little resemblance to the difference spectrum observed upon addition of ethylene glycol (final concentration 20%) to an adenylate solution, which is similar to the hypothetical difference spectrum originating from a "polarization red-shift"[16] (shown in Fig. 7). A stacking force between the nucleotide base and a certain residue of

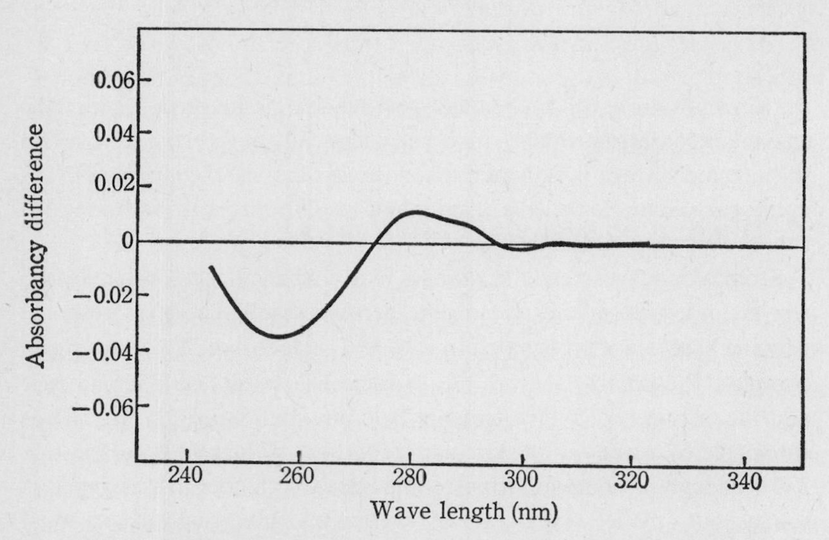

Fig. 6. Difference spectrum from the interaction of 2′(3′)-adenylate and RNase T_1. The enzyme (6.8×10^{-5} M) was mixed with adenosine-2′(3′)-phosphate (2.65×10^{-4} M) in sodium acetate (0.014 M), pH 5.6. Cells of 3 mm path length were used.

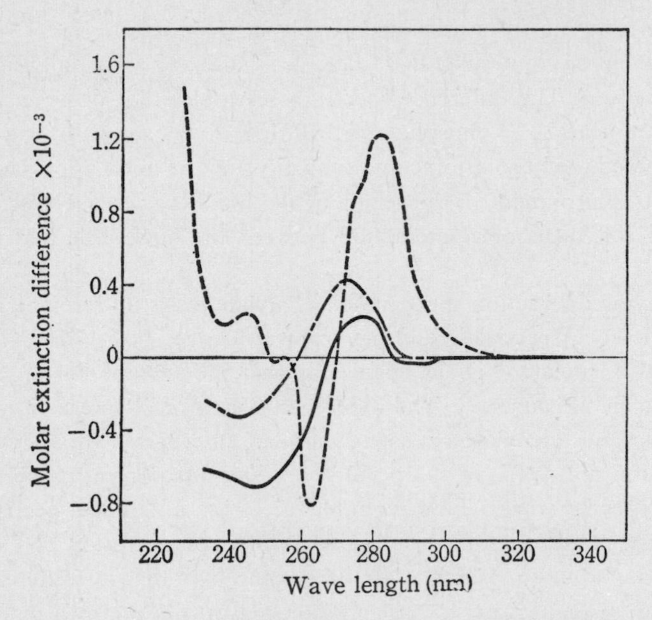

Fig. 7. Perturbation difference spectra of adenosine-2′(3′)-phosphate. The nucleotide was perturbed by dioxane (final, 18%) (——), acid (final pH 1.0) (------), and ethylene glycol (final, 20%) (—-—). The spectra were recorded using an aqueous solution of adenosine-2′(3′)-phosphate, pH 6.5, as reference.

the enzyme, probably a phenylalanine residue, is favored because the enzyme preparation which is modified chemically at tyrosine residues on the enzyme surface is fully active for the hydrolysis of guanosine-2′,3′-cyclic phosphate,[17] and the tryptophan residue seems to be buried inside the enzyme molecule.[18]

From a comparison of the changes in absorbancy of adenylate by the enzyme interaction and by solvent perturbation, it appears that the adenine base is not protonated in RNase T_1-adenosine-2′(3′)-phosphate complex. Presumably there is no interaction between the adenine base and the enzyme except the hydrophobic one mentioned above. This is supported by a study of the circular dichoroism of RNase T_1-adenosine-2′(3′)-phosphate complex, which will be dealt with later in this paper. A similar but smaller trough in the vicinity of 255 nm was observed by mixing a high concentration of cytidine-2′(3′)-phosphate (1.2×10^{-3} M) with RNase T_1 (6.7×10^{-5} M). Cytidine-2′(3′)-phosphate has also been shown to be a competitive inhibitor, for which the K_i value is much higher than in the case of adenosine phosphate.[15,16] Considering the specificity of RNase T_1, it seems likely that the interaction between the enzyme and the nucleotide, which reveals the absorbancy difference shown in Fig. 1 or 2, is essential during the catalytic action.

From the above conclusion that the binding of adenosine-2′(3′)-phosphate to RNase T_1 may be simulated by dissolving the adenylate in 18% dioxane, a solution of guanosine or guanine nucleotide in acidic dioxane may be considered as a better approximation of the guanosine chromophore in the enzyme complex. As shown in Fig. 3, very small troughs at around 270 and 250 nm were observed in the difference spectrum recorded by adding dioxane to a guanosine-3′-phosphate solution at neutral pH. No significant difference was detected when the difference spectrum of guanosine-3′-phosphate resulting from addition of both acid and dioxane (final concentration 18 or 30%) was compared with that resulting from lowering the pH of the solution (Fig. 3). These results suggest that the hydrophobic interaction between the guanine base and RNase T_1, which is presumed from spectral studies of RNase T_1-adenylate complex formation, has little effect on the difference spectrum.

2.4. Dissociation Constants of RNase T_1-Analog Complexes

Assuming that the absorbancy difference in the vicinity of 290 nm is 3.1×10^3 cm^{-1} M^{-1} of the complex formed, the amount of RNase T_1-substrate analog complex in the mixture may be estimated from the difference spectrum observed. From data so obtained, an attempt was made to calculate the binding constants (K_s) of the enzyme-substrate analog complexes, and the results are summarized in Table 1. The values in the table are in fair accord with those obtained by gel-filtration experiments.[7] The dissociation constant of the enzyme-guanosine complex is not affected by the addition of inorganic phosphate ion, suggesting that the interaction between the enzyme and guanosine group and that between the enzyme and phosphate group are not cooperative.

A comparison of the dissociation constants will give an idea of the relative intensities of binding of the analogs to RNase T_1. The affinities of various guanine derivatives to the enzyme decrease in the order, guanosine-2′ or 3′-phosphate> guanosine-5′-phosphate> guanosine> deoxyguanosine> 9-methylguanine. These results demonstrate the ex-

TABLE 1.

Dissociation Constants of Various RNase T_1-Substrate Analog Complexes

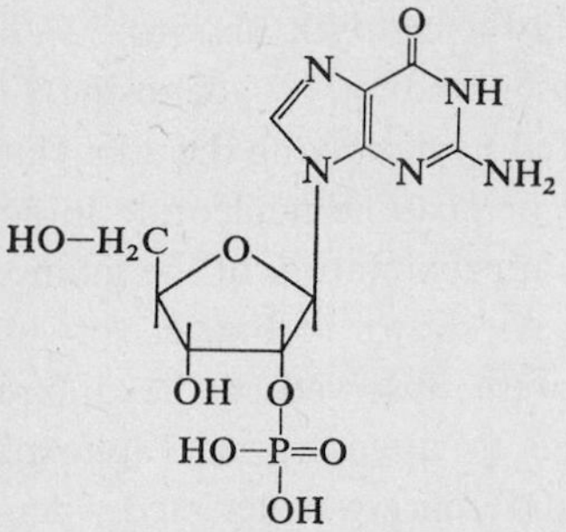

Compound	K_S	ΔG°
guanosine-2′-phosphate	$K_S = 0.9 \times 10^{-5}$M	$\Delta G^\circ = 6.9 \times 10^3$cal†
guanosine-3′-phosphate	$K_S = 1.2 \times 10^{-5}$M	$\Delta G^\circ = 6.7 \times 10^3$cal
guanosine-5′-phosphate	$K_S = 3.0 \times 10^{-5}$M	$\Delta G^\circ = 6.2 \times 10^3$cal
guanosine	$K_S = 3.5 \times 10^{-4}$M	$\Delta G^\circ = 4.7 \times 10^3$cal
deoxyguanosine	$K_S = 1.9 \times 10^{-3}$M	$\Delta G^\circ = 3.7 \times 10^3$cal
9-methylguanine	$K_S = 3.0 \times 10^{-3}$M	$\Delta G^\circ = 3.4 \times 10^3$cal

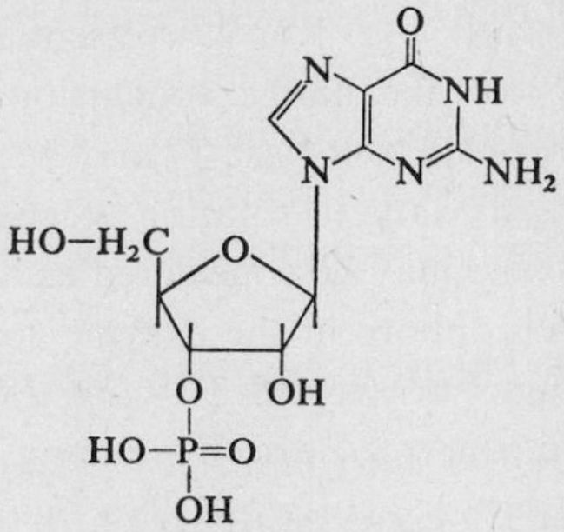

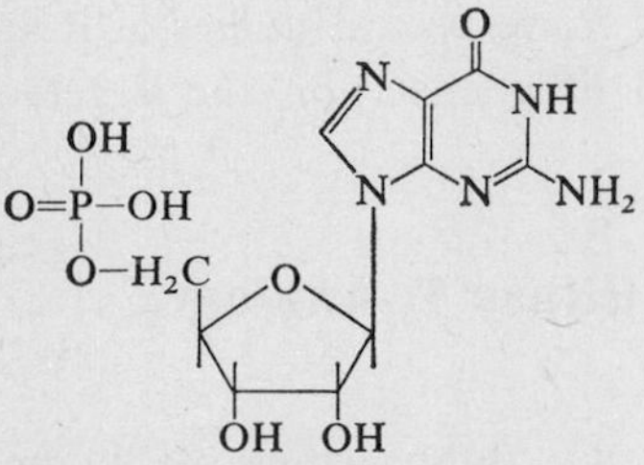

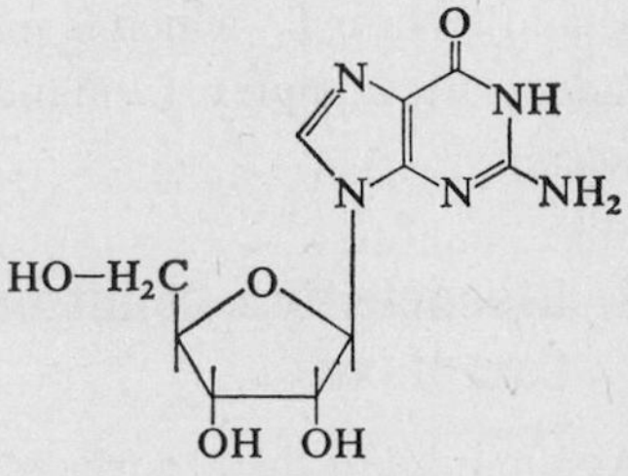

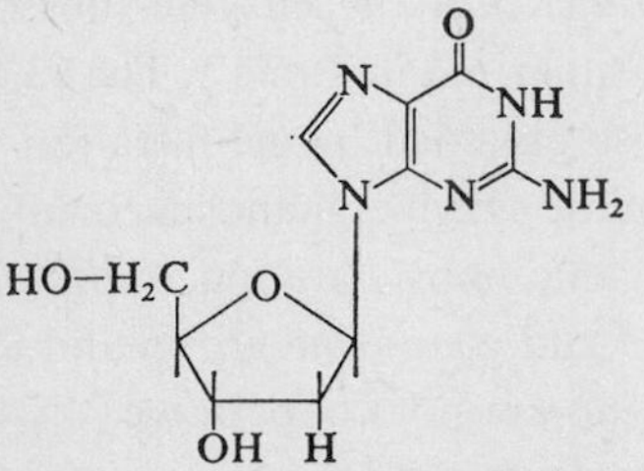

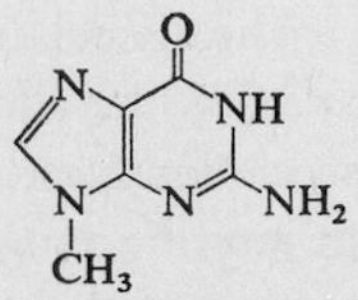

† The free energy changes upon dissociation were estimated according to the relationship, $\Delta G^\circ = -RT \ln K_S$.

istence of interactions between the enzyme and both ribose and phosphate groups in addition to the guanine base, and the requirement of an appropriate geometrical arrangement of guanine, ribose, and phosphate groups for highest affinity to RNase T_1. The changes in free energy upon binding were calculated from the K_s values and are listed in Table 1 for comparison. It can be seen from the table, for instance, that the 2'-phosphate group contributes about 2 kcal to the binding energy.

2.5. Stoichiometric Studies on the Binding

It has been reported that one molecule of guanosine-2'-phosphate binds to one molecule of the enzyme indicating one active site per molecule of RNase T_1.[9] This conclusion was confirmed by a stoichiometric study on the interaction of guanosine and ribonuclease as shown in Fig. 8. To a constant concentration of RNase T_1, different amounts of guanosine were added, and from the absorbency differences at 291 nm the concentrations of the complex formed in the mixtures were determined as above. In Fig. 7, the reciprocal of the concentration of the complex is plotted against the reciprocal of the concentration of unbound guanosine in the mixture. The data indicate a 1 : 1 complex formation and a K_s value of 3.6×10^{-4} M.

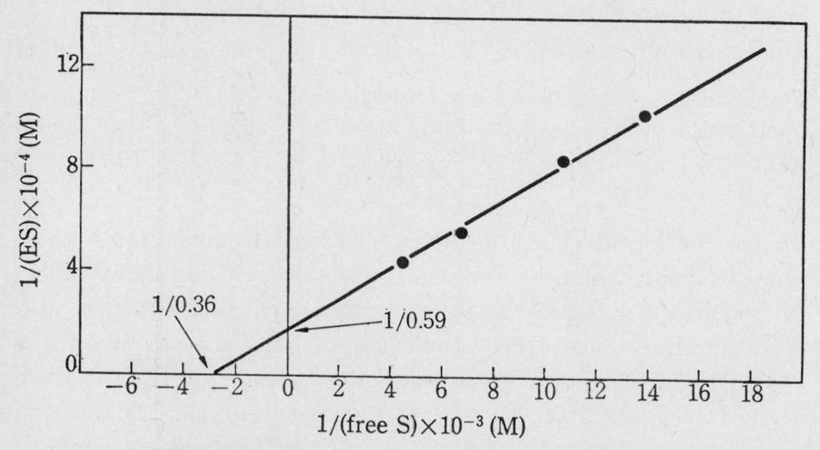

Fig. 8. The dependence of complex formation on guanosine concentration. To 5.45×10^{-5} M RNase T_1, various concentrations of guanosine were added in 0.0125 M sodium acetate, pH 5.6. The reciprocal of the amount of complex formed is plotted against the reciprocal of the concentration of free guanosine in the mixture.

2.6. Binding of Guanosine-3′-phosphate to Carboxymethylated RNase T_1

It has been observed that Glu-58 of the T_1 enzyme is unusually reactive with bromo- or iodoacetate in solution at pH 5.5[19] The product of the reaction was identified to be Glu-58–carboxymethylated RNase T_1. The modified enzyme is catalytically inactive. Substrate analogs such as guanosine-2′-phosphate or guanosine-3′-phosphate protect the enzyme from inactivation by iodoacetate;[9] thus, it is conceivable that the proton donor of the enzyme to the guanine base is the carboxyl group of Glu-58.

To examine the role of Glu-58, carboxymethylated enzyme was prepared by mixing RNase T_1 with iodoacetate and tested spectrophotometrically for its binding activity with guanosine-3′-phosphate. The results of preliminary experiments are presented in Table 2. When

TABLE 2

The Binding of Guanosine-3′-phosphate to Glu-58–carboxymethylated (CM) RNase T_1

RNase T_1 preparation	Concentration	Concentration of 3′-GMP	RNase T_1 bound with 3′-GMP (%)	K_s value
native RNase T_1	7.3×10^{-5}M	1.05×10^{-4}M	95	1.2×10^{-5}M
CM-RNase T_1	$2\ \times10^{-5}$M	$0.7\ \times10^{-4}$M	40	
CM-RNase T_1	5.6×10^{-5}M	1.05×10^{-4}M	50	$9\ \times10^{-5}$M
CM-RNase T_1	3.6×10^{-5}M	1.20×10^{-4}M	55	

the enzyme was alkylated, the apparent affinity to guanosine-3′-phosphate estimated from the absorbancy difference in the vicinity of 290 nm was greatly reduced, suggesting the participation of the glutamic acid residue in the interaction between the enzyme and the nucleotide.[11] The "apparent" dissociation constant was estimated to be 9×10^{-5} M, which is about ten times larger than that for the native enzyme. The dissociation constant estimated from the ultraviolet difference spectrum is labeled "apparent" because the possibility cannot be excluded that carboxymethylated RNase T_1 combines with guanosine-3′-phosphate without the protonation of the N-7 of the guanine base.

Although the role of Glu-58 of the enzyme in the catalytic action cannot be stated definitely from the data presented here, one possible proton donor for the protonation of the N-7 of the guanine base in the enzyme-substrate complex is the carboxyl group of the glutamic acid residue. It is necessary to study in more detail the binding of the modified enzyme with substrate analogs in order to determine the role of Glu-58 in the enzyme action. More elaborate measurements of the interaction between the modified enzyme and the substrate analogs including the circular dichroism of the modified enzyme-analog complexes are now being studied.

3. Circular Dichroism of RNase T_1-Substrate Analog Complexes

3.1. Circular Dichroic Spectra

The circular dichroic spectrum of RNase T_1-guanosine-3′-phosphate complex is quite different from the graphical summation of the spectra of the components as shown in Fig. 9.[20] Circular dichroic spectra of RNase T_1 and guanosine-3′-phosphate are given in the figure along with the dichroic difference spectrum calculated graphically. The difference spectrum is characterized by a peak at around 250 nm and a trough at around 280 nm.

Similar dichroic differences were observed when RNase T_1 was mixed in solution with guanosine-2′ or 5′-phosphate, guanosine or deoxyguanosine.[11] The spectra observed are shown in Fig. 10 and 11.

3.2. Assignment of Dichroic Change

It is generally acceptable to think of the conformation of the guanosine group as fixed into the *syn* or *anti* form in the enzyme-nucleotide complex. The *anti* and *syn* conformations (Fig. 12) correspond to $\phi_{CN} = -30°$ and $+150°$ respectively, using the angle ϕ_{CN} defined by Donohue *et al.*[21] Much effort has been devoted to the correlation of the sign of the principal Cotton effect or dichroic band to the

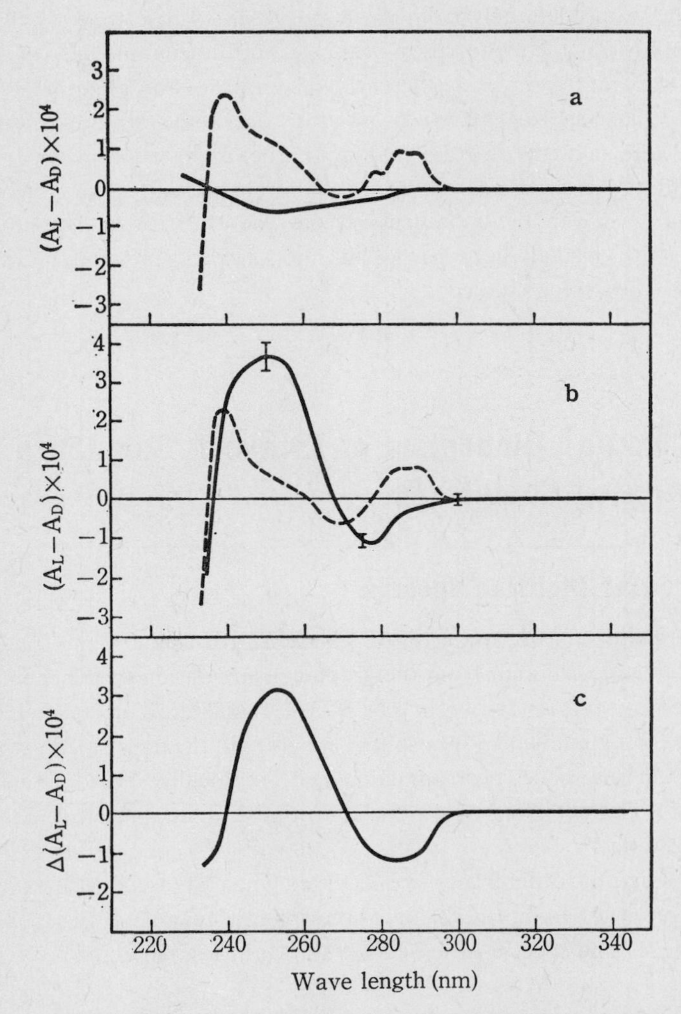

Fig. 9. Circular dichroic difference spectrum of interaction of RNase T_1 and guanosine-3′-phosphate, recorded at 25 °C using cells of 3 mm path length. (a) Dichroic spectra of the enzyme, 9.5×10^{-5} M (------) and nucleotide, 2.65×10^{-4} M (——) in 0.014 M sodium acetate, pH 5.6. (b) Dichroic spectrum of the mixture (——), and the algebraic sum of the spectra of the enzyme and nucleotide (------). (c) Dichroic difference spectrum.

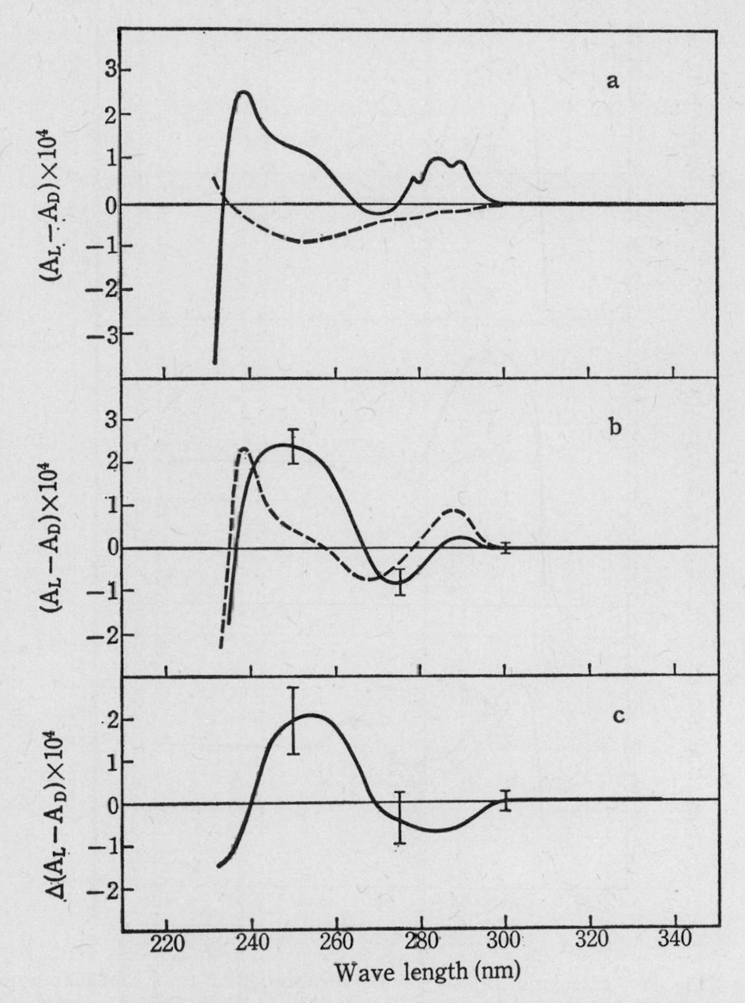

Fig. 10. Circular dichroic difference spectrum of interaction of RNase T_1 and guanosine, recorded at 25°C using cuvettes of 3 mm path length. (a) Dichroic spectra of the enzyme, 9.2×10^{-5} M (——) and nucleoside, 3.26×10^{-4} M (------). (b) Dichroic spectrum of the mixture (——), and the graphical summation of the spectra of the enzyme and nucleoside (------). (c) Dichroic difference spectrum.

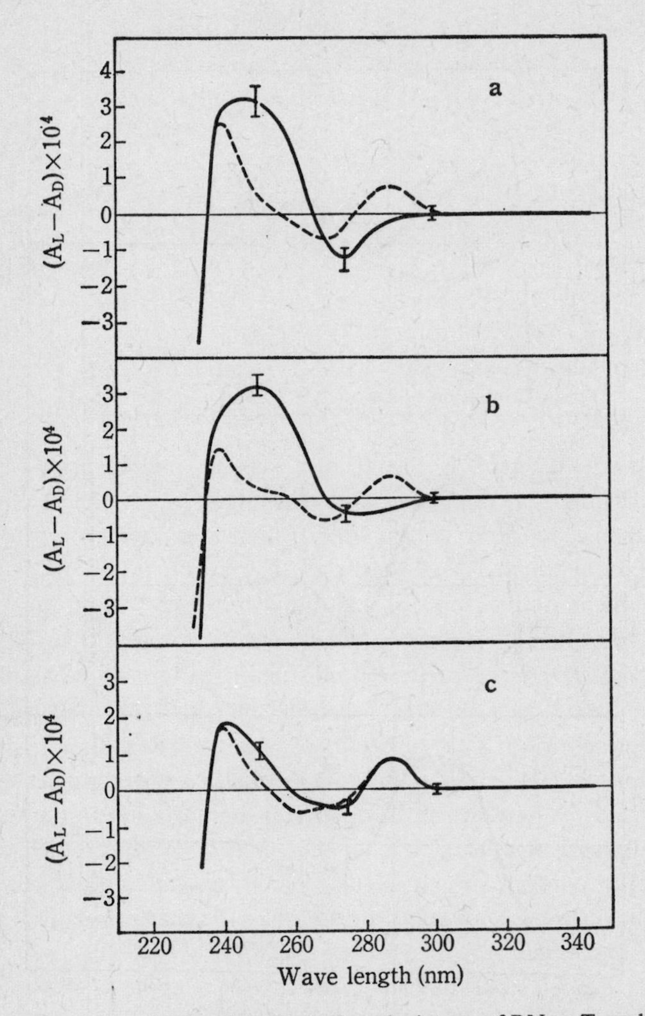

Fig. 11. Circular dichroic spectra of mixtures of RNase T_1 and guanosine-2′ or 5′-phosphate or deoxyguanosine, recorded at 25 °C using cuvettes of 3 mm path length. ——, the dichroism of the mixture; ------, the graphical summations of the dichroic spectra of the constituents measured separately.

(a) The enzyme (9.5×10^{-5} M) was mixed with guanosine-5′-phosphate (2.2×10^{-4} M) in sodium acetate (0.014 M), pH 5.6. (b) The enzyme (7.0×10^{-5} M) was mixed with guanosine-2′-phosphate (2.2×10^{-4} M) in sodium acetate (0.013 M), pH 5.6. (c) The enzyme (9.5×10^{-5} M) was mixed with deoxyguanosine (2.65×10^{-4} M) in sodium acetate (0.013 M), pH 5.6.

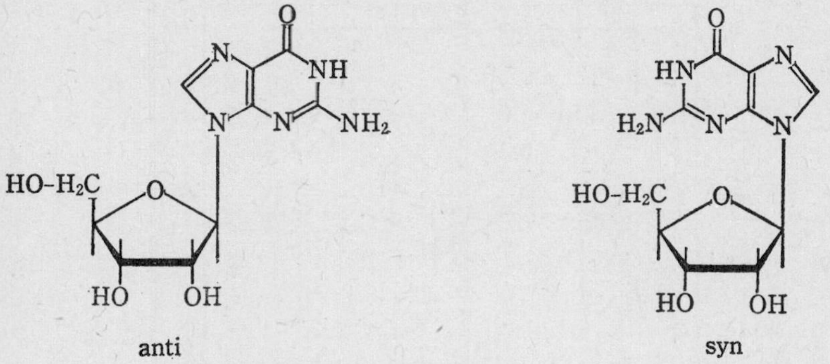

anti syn

Fig. 12. The conformation of guanosine.

conformation.[22-31] Among the studies on the optical activity of guanosine, it has been suggested that the preferred conformation of guanosine may change into the *syn* form from the *anti* upon lowering of the pH of the solution.[32] As shown in Fig. 13, an inversion of the dichroic band in the vicinity of 250 nm is observed when guanine nucleoside or nucleotide is protonated in an acidic solution. The almost exclusive presence of the preferred *anti* conformation for all natural nucleosides including guanosine has been suggested in crystals and neutral solutions from X-ray diffraction[33] and NMR studies,[34,35] although it appears from circular dichroic spectra that purine nucleosides or nucleotides in solution have much freedom to rotate about the glycoside linkage. Realizing that circular dichroism reflects the configurational or conformational features of the molecule, the inversion of sign in the dichroic band may correspond to the change in conformation of the nucleoside from the *anti* to *syn* form or vice versa, i.e. the solid line in Fig. 13 represents the dichroism of guanine nucleoside in the *syn* conformation. For the study of the circular dichroism of guanosine or guanylate fixed in the *syn* conformation, guanosine in acidic solutions might be a better model compound than cyclized model compounds such as *S*-cycloguanosine or 3,5′-cycloguanosine, in which additional asymmetries are present and complicate the interpretation. From the magnitude of the dichroic bands, it appears that guanosine in acidic solutions has a great deal of freedom to rotate about its glycoside linkage. Comparing the spectra in Fig. 9, 10, and 11 with that of guanosine in acidic solutions, the conclusion is that the dichroic changes shown

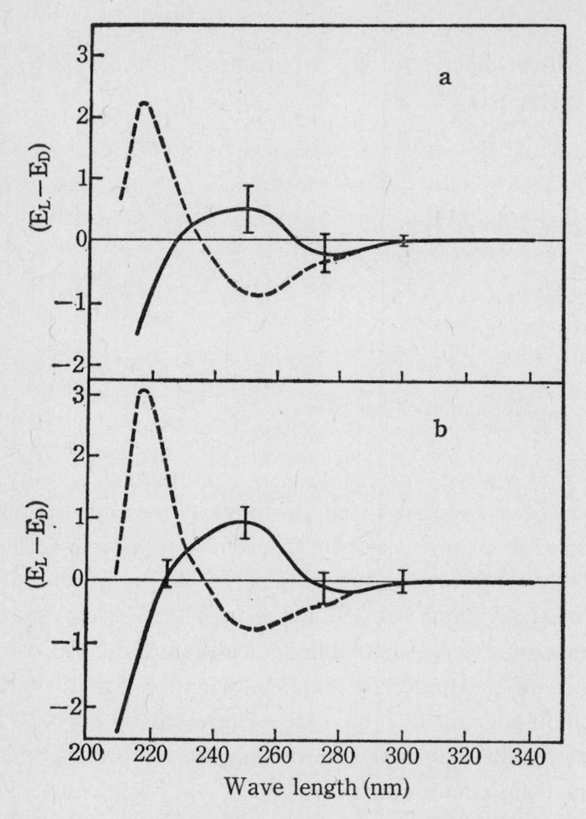

Fig. 13. Circular dichroic spectra of guanosine and guanosine-3′-phosphate recorded at 25 °C. (a) Guanosine in 0.02 M sodium citrate, pH 6.5 (------) and guanosine at pH 0.5 (——). (b) Guanosine-3′-phosphate at pH 6.5 (------) and at pH 1.0 (——).

in these figures are a result of the fixation of the guanosine group into the *syn* conformation in the enzyme-analog complex. From the spectra of protonated guanine derivatives, a peak at around 250 nm and a small trough at around 280 nm appear to be characteristic of the dichroic spectra of these compounds and are qualitatively in good agreement with the dichroic difference spectra shown in the above figures. Addition of dioxane (final concentration 18 %) had no significant effect on the circular dichroism of guanosine in acidic solutions.

3.3. Magnitude of Dichroic Differences upon Mixing Analog with RNase T_1

The magnitudes of the difference observed at 250 or 285 nm were calculated and the results are summarized in Table 3, where the data are expressed using the molar ellipticity difference ($\Delta[\theta]_\lambda$) defined by Eq. (1).

$$\Delta[\theta]_\lambda = \left(A - E - N + N\frac{m}{n} \right) \times 3300/m \text{ (deg cm}^2 \text{ dmole}^{-1}\text{)}, \tag{1}$$

where A, E and N represent, respectively, the magnitude ($A_L - A_R$) of the mixture, the enzyme at a given concentration and the guanine derivative at a given concentration measured at the wave length, λ, using a cuvette of 1 cm light path; n and m denote, respectively, the molar concentrations of the guanine compound and the enzyme-inhibitor complex formed in the mixture.

TABLE 3

Comparison of Magnitudes of the Dichroic Difference of Interaction of RNase T_1 and Guanine Nucleotide or Guanosine

Nucleoside or nucleotide used	Concentration of complex formed in mixture	$\Delta[\theta]_\lambda$ (deg.cm^2.dmole^{-1})	
		at 250 nm	at 285 nm
2′-GMP	7.05×10^{-5}M	4.3×10^4	-1.5×10^4
3′-GMP	9.2	3.4	−1.5
3′-GMP	6.8	3.3	−1.9
3′-GMP	3.2	3.6	−1.1
5′-GMP	8.4	3.0	−1.5
guanosine	4.25	4.4	−1.7
Average		3.7×10^4	-1.5×10^4

As seen in Table 3, the values of the molar dichroic difference calculated from the various RNase T_1-guanine derivative complexes are in good agreement with each other, so that the concentration of complex present in the mixture of enzyme and guanine nucleoside or nucleotide can be determined by the magnitude of the dichroic difference observed at 250 nm assuming a molar ellipticity difference of 3.7×10^4 deg cm^2 dmole^{-1} at 250 nm. The equation for this purpose is

$$m = (A - E - N) / \left(11.2 - \frac{N}{n}\right), \tag{2}$$

where A, E, N, n, and m are as defined above.

Although determination of the enzyme-analog complex by Eq. (2) employing the circular dichroism difference seems to be less accurate than the method using the ultraviolet absorbancy difference at 291 nm, the CD method may be useful for studying the complex formation in

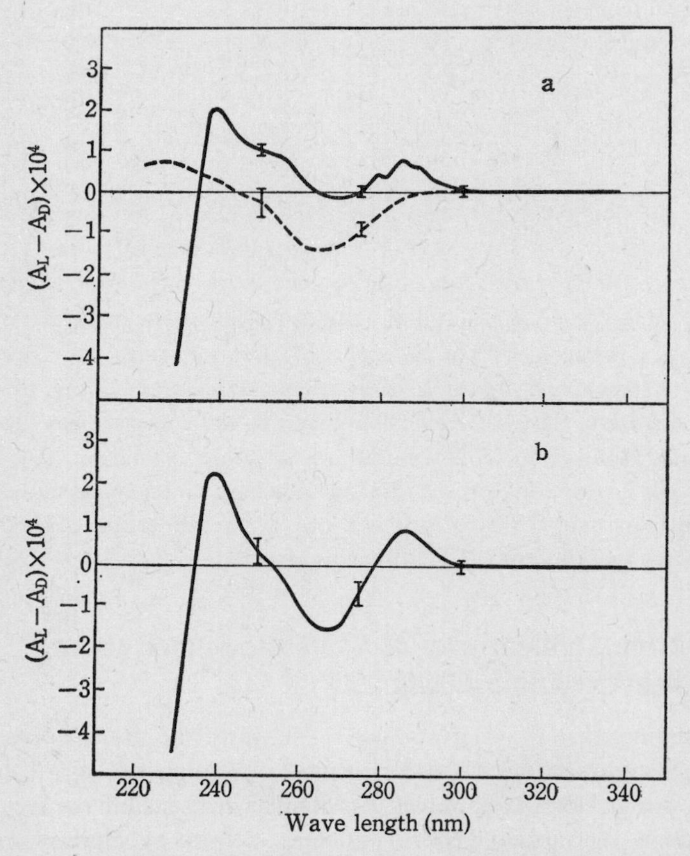

Fig. 14. Circular dichroic spectrum of a mixture of RNase T_1 and adenosine-2′(3′)-phosphate. (a) Dichroic spectra of the enzyme, 6.8×10^{-5} M (——) and nucleotide, 2.65×10^{-4} M (------). (b) Dichroic spectrum of the mixture. No significant difference is seen between the spectrum in (b) and the simple sum of the spectra in (a). The spectra were recorded at 25°C in 0.014 M sodium acetate, pH 5.6, using cells of 3 mm path length.

some cases, especially the binding of guanine nucleosides or nucleotides to chemically modified RNase T_1.

In short, it was suggested from studies of the circular dichroism of the complex that the guanosine group is fixed in the *syn* conformation in the course of RNase T_1-substrate analog complex formation, although this consideration is yet to be confirmed by X-ray crystallography. In the previous section the ultraviolet absorbancy difference observed upon binding of guanylates to RNase T_1 was interpretable assuming no contribution from a conformational change of the enzyme protein. In like manner the conformational change of the enzyme protein is assumed to contribute no significant dichroic difference, although the data presented cannot exclude dichroic difference due to a conformational change of the aromatic residues of the enzyme protein during the course of the binding process.

The circular dichroism of a mixture of RNase T_1 and adenosine-2′(3′)-phosphate was investigated. The dichroic spectrum was almost identical with the simple sum of the spectra of the enzyme and adenylate as shown in Fig. 14. From the K_i value appearing in the literature,[7] one can expect the formation of 5.1×10^{-5} M of enzyme-adenylate complex in the mixture under the conditions outlined in the legend of Fig. 14. The results indicate that the adenosine group in the complex may not be restricted to a specific conformation in the complex, suggesting a lack of certain interactions between the adenine base and enzyme molecule. This is consistent with the interpretation of the ultraviolet absorption difference spectrum of the complex discussed above.

3.4. Circular Dichroism of Nucleoside Fixed in an Enzyme-Analog Complex

Assuming that the conformational change of the protein does not significantly affect the circular dichroism of the mixture and that the guanosine group in the complex is protonated and fixed in the *syn* conformation, the molar ellipticity difference, defined as above, will correspond to the molecular ellipticity coefficient of the protonated guanosine fixed in the *syn* conformation. The circular dichroism of the *syn*, protonated guanosine-3′-phosphate was thus estimated from the dichroic spectra of RNase T_1, guanosine-3′-phosphate and a mixture of both. The proposed spectrum is given in Fig. 15.

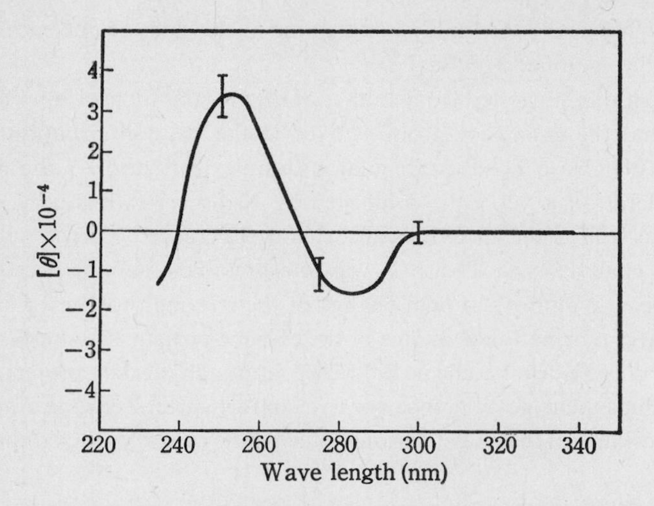

Fig. 15. Proposed circular dichroic spectrum for *syn* fixed guanosine-3′-phosphate.

Studies on the circular dichroism of RNase T_1-substrate analog complexes have opened the way for a new application of the enzyme to optical studies of the conformation of guanosine and related nucleosides or nucleotides. It seems promising to study the correlation of circular dichroism with the conformation of nucleosides and nucleotides through the use of enzyme-substrate analog complexes. It is probable that the conformation of a nucleoside would be rigidly restricted by the architecture of the binding site of the enzyme. It follows then that the 8-bromoguanosine group in 8-bromoguanosine-2′(3′)-phosphate bound to RNase T_1 is fixed into the *syn* form as well. An absorbancy difference is observed upon mixing 8-bromoguanylate with RNase T_1 as shown in Fig. 16. The difference spectrum with a peak at 295 nm is similar to those observed for the binding of guanylates to the enzyme (Fig. 1), indicating binding of the 8-bromoguanosine-2′(3′)-phosphate to the enzyme.[36] This result is consistent with the recent observation[37] that 8-bromoguanosine-2′,3′-cyclic phosphate is hydrolyzed by RNase T_1. The circular dichroism of the RNase T_1-8-bromoguanosine-2′(3′)-phosphate complex (Fig. 16) was studied, and the estimated dichroic spectrum of *syn* fixed 8-bromoguanosine-2′(3′)-phosphate is given in Fig. 17.[36] Similarly, the binding of 8-bromoguanosine to the enzyme was investigated (see Fig. 18) and a spectrum resembling that of 8-

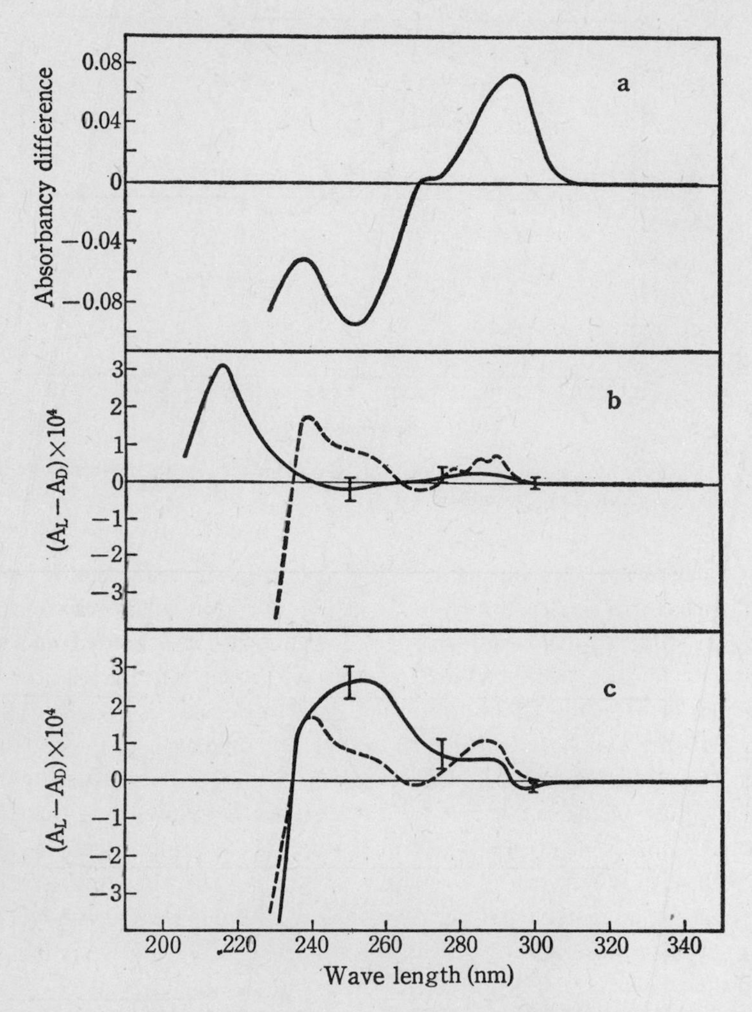

Fig. 16. Interaction of RNase T_1 and 8-bromoguanosine-2′(3′)-phosphate. (a) Ultraviolet absorption difference spectrum produced by the interaction. (b) Dichroic spectra of the enzyme (------) and nucleotide (——). (c) Dichroic spectrum of the mixture (——), and the graphical summation of the spectra of the enzyme and nucleotide (-----). The spectra were recorded at 25 °C using cuvettes of 3 mm path length. The concentrations employed were 6.75×10^{-5} M (enzyme) and 2.6×10^{-4} M (nucleotide) in 0.014 M sodium acetate, pH 5.6.

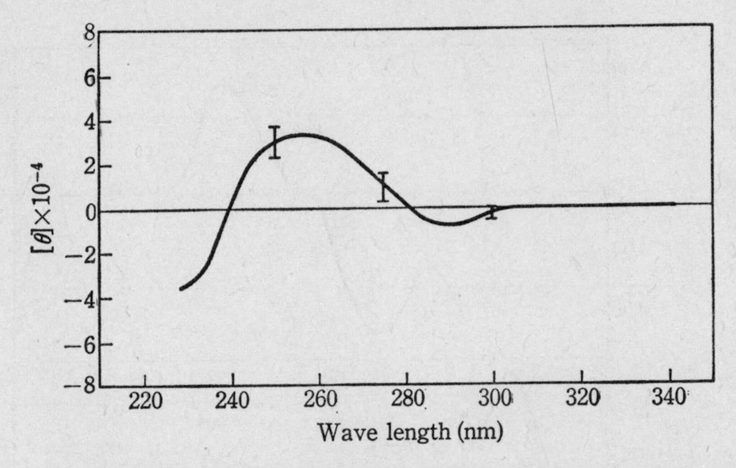

Fig. 17. Proposed circular dichroic spectrum for *syn* fixed 8-bromoguanosine-2′(3′)-phosphate.

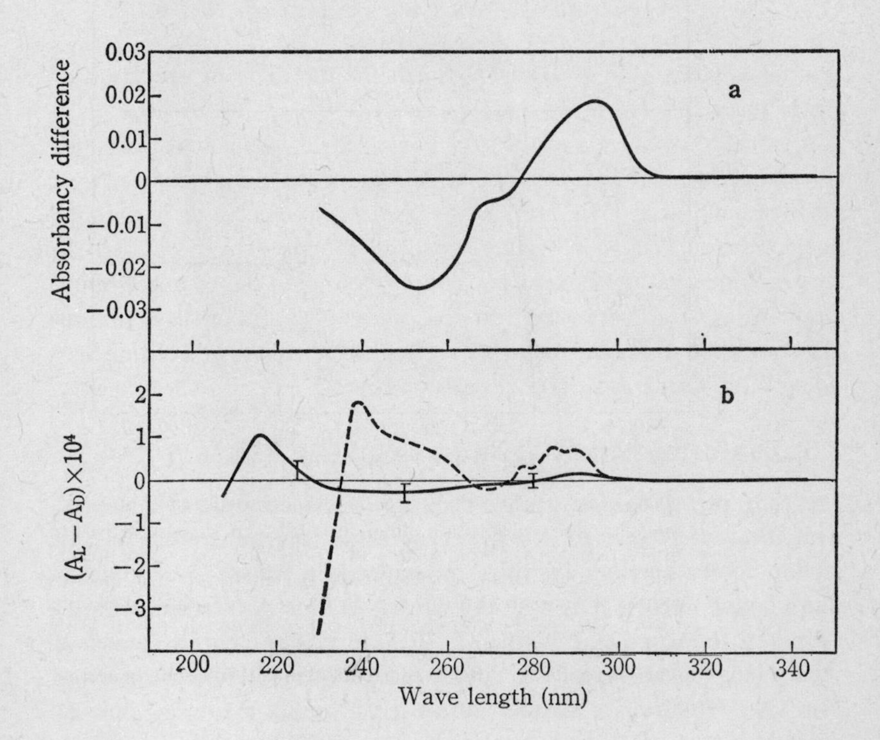

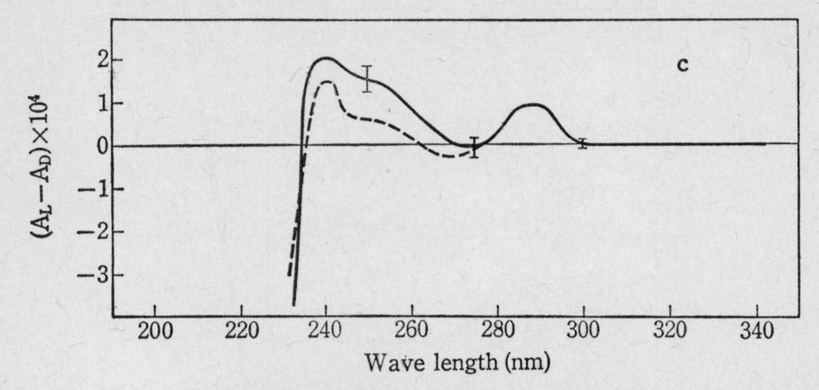

Fig. 18. Interaction of RNase T_1 and 8-bromoguanosine. (a) Ultraviolet absorption difference spectrum produced by the interaction. (b) Dichroic spectra of the enzyme (------) and nucleoside (——). (c) Dichroic spectrum of the mixture (——), and the graphical summation of the spectra of the enzyme and nucleoside (------). The spectra were recorded at 25 °C using cells of 3 mm path length. The concentrations employed were 6.75×10^{-5} M (enzyme) and 1.7×10^{-4} M (nucleoside) in 0.014 M sodium acetate, pH 5.6.

bromoguanylate (Fig. 17) was obtained for the dichroic spectrum of *syn* bromoguanosine using Eq. (1).

A circular dichroic spectrum of *syn* fixed 2′,3′-inosinate was also calculated from similar experiments. The spectrum is characterized by a positive band at around 245 nm. Also, dichroic spectra of pyrimidine nucleotides of fixed conformation can be proposed from circular dichroic studies on pyrimidine nucleotides bound to pancreatic ribonuclease A, as described in the next section. Application of these ribonucleases will be useful in studying the circular dichroism of fixed nucleotides. Work along such lines is now in progress.

3.5. Conformation of Cytidine Bound to RNase A

It is of interest to compare the details of the mechanism of RNase T_1 with those of RNase A. A brief reference is made, therefore, in this section to the circular dichroism of pancreatic RNase A-cytidine-2′-phosphate complex. It is well known that RNase A interacts with 2′-cytidylate to form a 1 : 1 complex.[38-42] The circular dichroic spectra of the enzyme protein, cytidine-2′-phosphate and the mixture are given in Fig. 19.[43] The difference spectrum observed upon mixing cytidine-2′-phosphate with the enzyme is also shown in the figure.

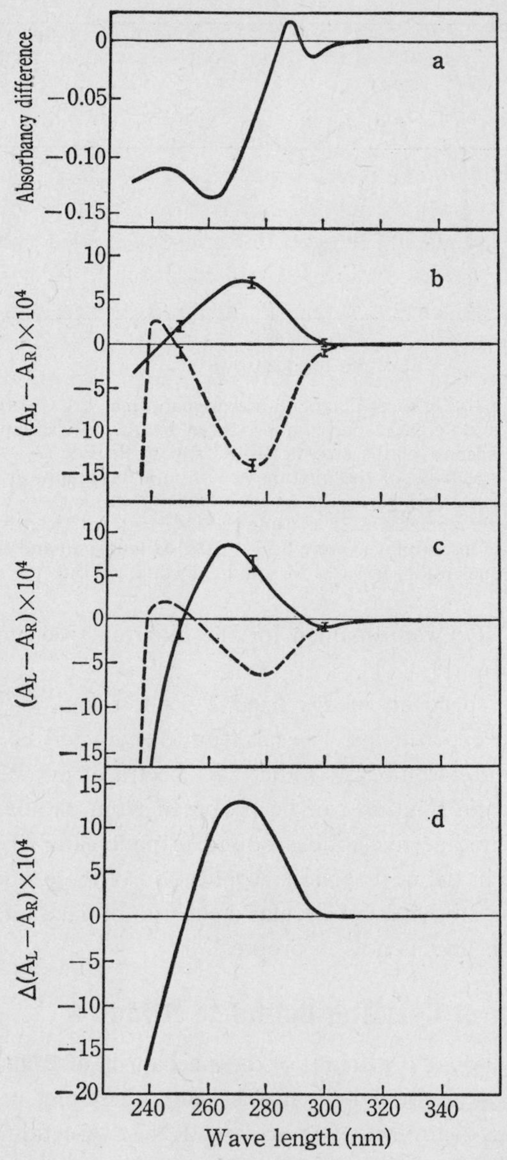

Fig. 19. Interaction of pancreatic RNase A and cytidine-2′-phosphate. (a) Ultraviolet absorption difference spectrum produced by the interaction. (b) Dichroic spectra of the enzyme (------) and nucleotide (——). (c) Dichroic spectrum of the mixture (——), and the algebraic summation of the spectra of the enzyme and nucleotide (------). (b) Dichroic difference spectrum. The spectra were recorded at 25 °C

using cells of 3 mm path length. The concentrations employed were 1.47×10^{-4} M (enzyme) and 2.65×10^{-4} M (nucleotide) in 0.013 M sodium acetate, pH 5.6.

The dichroic difference was calculated graphically from Fig. 19c and is shown in Fig. 19d. It appears from the data in the literature[28-31] that pyrimidine nucleosides fixed in the *anti* conformation have a positive Cotton effect around the major absorption band, that is, a positive dichroic band. It may be concluded that the dichroic difference in Fig. 19d is attributable to the fixation of the cytidine group into the *anti* form on the enzyme surface. The maximum of the dichroic difference spectrum in Fig. 19d is more nearly identical with that of the dichroic band of cytidine-2′-phosphate (270 nm) rather than the minimum (275–280 nm) of the enzyme protein, suggesting the lack of a significant contribution from conformational changes of tyrosine residues to the dichroic difference spectrum. It has been suggested also from an absorbancy difference spectrum study,[44] a chemical modification study on tyrosine residues[45] and a spectropolarimetric study of the enzyme and inhibitor complex[46] that a conformational change of tyrosine residues would not take place during complex formation, although some conflicting opinions have appeared in the literature.[38,42] The view favored here is that a conformational change of tyrosine residues of the nuclease would not contribute significantly to the present difference spectrum even if the conformations of tyrosine residues underwent change during complex formation.

The molar concentration of the complex formed can be determined spectrophotometrically assuming that the molar extinction difference at 260 nm is -2.8×10^3 absorbancy units.[38] Under the conditions described in the legend of Fig. 19, the amount of complex in the enzyme-cytidine-2′-phosphate mixture was nearly identical with that of the RNase A used in the mixture, i.e. the enzyme was saturated with the cytidylate ligand. Assuming that the dichroic difference observed is attributable only to fixation of the cytidine group in the *anti* conformation, the circular dichroism of *anti* fixed cytidine-2′-phosphate can be calculated to be as shown in Fig. 20 according to Eq. (1). The molecular ellipticity coefficient at 270 nm should be 2.6×10^4 deg cm^2 dmole^{-1}. Although the circular dichroism of the pyrimidine nucleoside fixed in the *anti* form has not yet been reported, this value seems to be reasonable when compared with those values reported for cyclopurine nucleosides.[23,24,26-8]

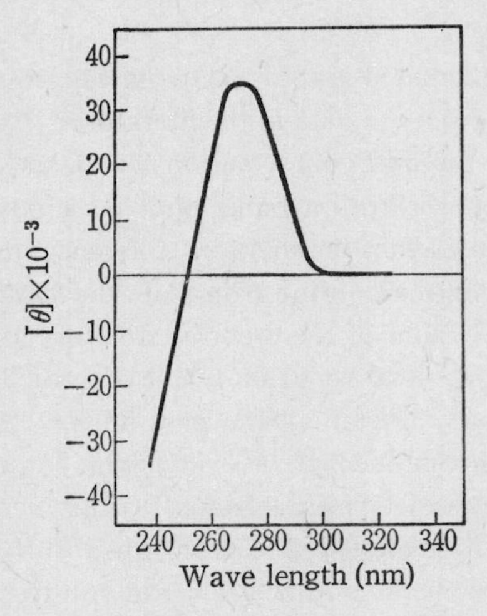

Fig. 20. Proposed circular dichroic spectrum for *anti* fixed cytidine-2′-phosphate.

4. Mechanism of RNase T_1-Substrate Complex Formation

The basic question concerning the mechanism of RNase T_1-substrate complex formation is which parts of the substrate and which parts of the enzyme participate in the interaction. From investigations of the optical properties of RNase T_1-analog complexes, a certain amount of information has been accumulated in relation to the above question.

The protonation of N-7 of the base by some proton-donating group in the enzyme protein must be one of the steps of complex formation. It is also suggested that the guanylate group of the substrate (RNA) is fixed partially in a non-aqueous environment. Presumably, the guanine base is stacked with a phenylalanine residue in the binding site of the enzyme. The stacking of a pyrimidine base with Phe-120 has been indicated from the results of an X-ray diffraction study[47] of RNase S and uridine-3′-phosphate complex. Similarly it has been reported that a tyrosine ring seems to be aligned parallel to the pyrimidine ring of an inhibitor, thymidine-3′,5′-phosphate, upon its binding to staphylococcal nuclease.[48] Using a carboxymethylated enzyme preparation, it was suggested that the carboxyl group of Glu-58 participates in the binding and might be a proton donor to the guanine base. The participation of a carboxyl group of the enzyme in the binding was also suggested by studies of the pH dependence of the absorbancy difference upon mixing guanylate with the enzyme.[9] The dependence of the absorbancy difference upon pH was reported to give a bell-shaped curve with a maximum at around pH 5. These results, which were recently confirmed by the authors' group, suggest the implication of certain groups of pK about 4 and 7 in the binding. A similar conclusion was reported by Irie from detailed kinetic studies on competitive inhibition by guanosine-2′-phosphate.[49] The carboxyl group appears to be the most likely candidate for the group in the binding site with a pK of about 4.

The dissociation constants of these complexes were estimated spectrophotometrically assuming an absorbancy difference at 291 nm of 3.1×10^3 cm^{-1} M^{-1} of the complex formed in the mixture. A comparison of the dissociation constants indicates that the correct assemblage of guanine, ribose, and phosphate groups is required for strong binding, suggesting the presence of at least three subsites on the enzyme molecule which are responsible independently for these three groups.

The physical basis of the binding of the guanylic acid group of the substrate RNA to RNase T_1 can be attributed to an electrostatic interaction between the 3′-phosphate of a guanylic acid group and a positively charged group in the phosphate binding site. The existence of another coulombic force is also presumed between the protonated N-7 of the base and the proton donor. The fact that the affinity of deoxyguanosine is much lower than that of guanosine suggests the importance of the 2′-oxygen for strong binding. It is probable that the interactions which participate in the binding of the ribose group are van der Waals contact and a specific hydrogen bond between the 2′-oxygen and a basic group in the enzyme, which might play an important role in the transphosphorylation step to form 2′,3′-cyclic phosphate. The difference between changes in free energy upon dissociation of RNase T_1-guanosine and -deoxyguanosine complexes, is estimated to be about 1 kcal/mole, which represents the strength of interaction of the 2′-oxygen and the enzyme (see Table 1). This value might be sufficient to presume a hydrogen bond between them. The enzyme does not split DNA whereas it binds with deoxyguanosine. One possible explanation for this is that there is no acceptor available for the phosphate transfer reaction; there may not be space enough for a water molecule near the phosphate group when the deoxyguanylic acid group of the DNA molecule fits into the binding site of RNase T_1. This proposal suggests a strong van der Waals contact at the binding site. From a comparison of the K_s values of enzyme-guanosine and -guanosine-5′-phosphate complexes, it appears that the binding is enhanced by the phosphate group at the 5′ position. This enhancement leads the present authors to conclude that a fourth subsite exists which is responsible for the 5′-phosphate.

The considerations which relate to the mechanism of enzyme-substrate complex formation and which were obtained from optical properties of RNase T_1-substrate analog complexes, can be summarized by the scheme in Fig. 21. In addition to the interactions discussed above, there are many other interactions between the enzyme and substrate molecules. For instance, from the results of detailed studies on the base specificity of RNase T_1 by Egami and his co-workers,[4-6] it is possible to

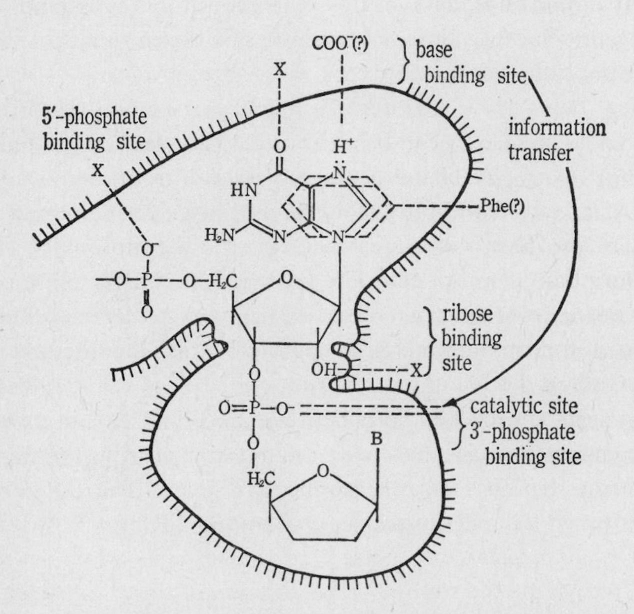

Fig. 21. Proposed mechanism of RNase T_1-substrate complex formation.

conclude a specific interaction between the 6-oxo group of the base and a group from the enzyme, as shown in Fig. 21.

From the circular dichroic studies, it appears that the conformation of the guanosine group in the RNase T_1-substrate complex is restricted rigidly into the *syn* form (see Fig. 21). Circular dichroism of the RNase T_1-adenosine-2′,3′-phosphate complex suggests that the adenosine group is not fixed into one conformation in that complex. The most likely interpretation is the presence of almost equal amounts of the *anti* and *syn* conformations of the adenosine group fixed in the complex. From the absorbancy difference upon complex formation and the base specificity of RNase T_1, the authors favor a mechanism in which the protonation of N-7 of the purine base is closely related both to the restriction of the conformation of the nucleoside group in the enzyme-analog complex and to the catalytic process of the enzyme. To go a step further, the authors would like to present the hypothesis that protonation of the N-7 position of the guanine base by some group of RNase T_1 plays a role as an actuator to form the catalytic site around the phosphate group. This protonation might induce a slight conformational change of the enzyme molecule to form the catalytically active conformation. It should be noted that the phosphorus atom is roughly 7 Å from the seventh nitrogen atom of the base in the guanosine-3′-phosphate molecule.

In contrast to the *syn* conformation of the guanosine group in the RNase T_1-guanylate complex, the *anti* conformation was suggested for the pyrimidine nucleoside ligand in the enzyme-substrate complex during RNase A action. Fixation of the pyrimidine nucleoside group in the *anti* form has also been suggested by the high susceptibility of polyformycin phosphate to pancreatic RNase A action,[50] although a contradictory consideration has been proposed from NMR studies.[51] It will be necessary to confirm the conclusions by X-ray analysis of the enzyme-nucleotide complex.

More elaborate studies on the optical properties of RNase T_1-substrate analog complexes seem to give promise for elucidating the nature of the interaction between the enzyme protein and the substrate, and of the mechanism of the unique specificity of enzyme action.

References

1. K. Sato and F. Egami, *J. Biochem.* (*Tokyo*), **44**, 753 (1957).
2. K. Sato-Asano, *ibid.*, **46**, 31 (1959).
3. K. Takahashi, *J. Biol. Chem.*, **240**, 4117 (1965).
4. F. Egami, K. Takahashi and T. Uchida, *Progr. Nucleic Acid Res. and Mol. Biol.*, **3**, 59 (1964).
5. F. Egami, *Nippon Kagaku Zasshi* (Japanese), **87**, 909 (1966).
6. K. Takahashi, T. Uchida and F. Egami, *Advan. Biophys.* (*Tokyo*), **1**, 53 (1970).
7. K. Takahashi, see p. 285.
8. P. R. Whitfeld and H. Witzel, *Biochim. Biophys. Acta*, **72**, 338 (1963).
9. S. Sato and F. Egami, *Biochem. Z.*, **342**, 437 (1965).
10. T. Oshima and K. Imahori, *J. Biochem.* (*Tokyo*), **69**, 987 (1971).
11. K. Imahori and T. Oshima, *Protein, Nucleic Acid and Enzyme* (Japanese), **16**, 1092 (1971).
12. H. T. Miles, F. B. Howard and J. Frazier, *Science*, **142**, 1458 (1963).
13. H. M. Sobell and K. Tomita, *Acta Cryst.*, **17**, 126 (1964).
14. W. E. Cohn, *Nucleic Acids: Chemistry and Biology* (ed. E. Chargaff and J. N. Davidson), vol. 1, p. 217, Academic Press, 1955.
15. *a.* M. Irie, *J. Biochem.* (*Tokyo*), **56**, 495 (1964); *b.* M. Irie, *ibid.*, **61**, 550 (1967).
16. J. W. Donovan, M. Laskowski Jr. and H. A. Scheraga, *J. Am. Chem. Soc.*, **83**, 2686 (1961).
17. H. Kasai and K. Takahashi, *private communication.*
18. S. Kawashima and T. Ando, *Intern. J. Protein Res.*, **1**, 185 (1969).
19. K. Takahashi, W. H. Stein and S. Moore, *J. Biol. Chem.*, **242**, 4682 (1967).
20. T. Oshima and K. Imahori, *J. Biochem.* (*Tokyo*), **70**, 197 (1971).
21. J. Donohue and K. N. Trueblood, *J. Mol. Biol.*, **2**, 363 (1960).
22. A. Hampton and A. W. Nichol, *J. Org. Chem.*, **32**, 1688 (1967).
23. M. Ikehara, M. Kaneko, K. Muneyama and H. Tanaka, *Tetrahedron Letters*, **1967**, 3977.
24. M. Ikehara, M. Kaneko and M. Sagai, *Chem. Pharm. Bull.* (*Tokyo*), **16**, 1151 (1968).
25. W. A. Klee and S. H. Mudd, *Biochemistry*, **6**, 988 (1967).
26. *a.* D. W. Miles, R. K. Robins and H. Eyring, *Proc. Natl. Acad. Sci. U.S.*, **57**, 1138 (1967); *b.* D. W. Miles, R. K. Robins and H. Eyring, *J. Phys. Chem.*, **71**, 3931 (1967).
27. *a.* T. R. Emerson, R. J. Swan and T. L. V. Ulbricht, *Biochem. Biophys. Res. Commun.*, **22**, 505 (1966); *b.* G. T. Rogers and T. L. V. Ulbricht, *ibid.*, **39**, 419 (1970).
28. D. W. Miles, M. J. Robins, R. K. Robins, M. W. Winkley and H. Eyring, *J. Am. Chem. Soc.*, **91**, 824, 831 (1969).

29. J. T. Yang, T. Samejima and P. K. Sarkar, *Biopolymers*, **4**, 623 (1966).
30. *a*. T. R. Emerson, R. J. Swan and T. L. V. Ulbricht, *Biochemistry*, **6**, 843 (1967); *b*. G. T. Rogers and T. L. V. Ulbricht, *Biochem. Biophys. Res. Commun.*, **39**, 414 (1970).
31. T. Nishimura, B. Shimizu and I. Iwai, *Biochim. Biophys. Acta*, **157**, 221 (1968).
32. W. Guschlbauer and Y. Courtois, *FEBS Letters*, **1**, 183 (1968).
33. A. E. V. Haschemeyer and A. Rich, *J. Mol. Biol.*, **27**, 369 (1967).
34. *a*. M. P. Schweizer, A. D. Broom, P. O. P. Ts'o and D. P. Hollis, *J. Am. Chem. Soc.*, **90**, 1042 (1968); *b*. S. I. Chan and J. H. Nelson, *ibid.*, **91**, 168 (1969).
35. *a*. S. S. Danyluk and F. E. Hruska, *Biochemistry*, **7**, 1038 (1968); *b*. I. C. P. Smith, B. J. Blackburn and T. Yamane, *Can. J. Chem.*, **47**, 513 (1969).
36. T. Oshima and K. Imahori, in preparation.
37. R. Yuki and H. Yoshida, *Biochim. Biophys. Acta,* **246**, 206 (1971).
38. J. P. Hummel, D. A. Ver Ploeg and C. A. Nelson, *J. Biol. Chem.*, **236**, 3168 (1961).
39. *a*. C. A. Nelson and J. P. Hummel, *J. Biol. Chem.*, **236**, 3173 (1961); *b*. J. P. Hummel and W. J. Dreyer, *Biochim. Biophys. Acta*, **63**, 530 (1962).
40. C. A. Ross, A. P. Mathias and B. R. Rabin, *Biochem. J.*, **85**, 145 (1962).
41. E. A. Barnard and A. Ramel, *Nature*, **195**, 243 (1962).
42. *a*. G. G. Hammes and P. R. Schimmel, *J. Am. Chem. Soc.*, **87**, 4665 (1965); *b*. R. E. Cathou, G. G. Hammes and P. R. Schimmel, *Biochemistry*, **4**, 2687 (1965).
43. T. Oshima and K. Imahori, *J. Biochem.* (*Tokyo*), **70**, 193 (1971).
44. M. Irie and F. Sawada, *ibid.*, **62**, 282 (1967).
45. M. E. Friedman and H. A. Scheraga, *Biochim. Biophys. Acta*, **128**, 576 (1966).
46. T. Samejima, M. Kita, M. Saneyoshi and F. Sawada, *ibid.*, **179**, 1 (1969).
47. H. Wyckoff, F. Richards, M. Doscher, D. Tsernoglou, T. Inagami, L. Johnson, K. Hardman and N. Allewell, presented at the 7th Int. Congr. Biochem., Tokyo, 1967. Abstracts I, p. 155.
48. A. Arnone, C. J. Bier, F. A. Cotton, E. E. Hazen Jr., D. C. Richardson and J. S. Richardson, *Proc. Natl. Acad. Sci. U.S.*, **64**, 420 (1969); *J. Biol. Chem.,* **246**, 2302 (1971).
49. M. Irie, *J. Biochem.* (*Tokyo*), **63**, 649 (1968).
50. D. C. Ward, W. Fuller and E. Reich, *Proc. Natl. Acad. Sci. U.S.*, **62**, 581 (1969).
51. D. H. Meadows, G. C. K. Roberts and O. Jardetzky, *J. Mol. Biol.*, **45**, 491 (1969).

Acknowledgments

Permission to reproduce the following copyright material is gratefully acknowledged.

ACADEMIC PRESS INC., *New York*

Source	Used in
Arch. Biochem. Biophys., **115**, 232 (1966). M. Shinitzsky *et al.* Table I.	Table 14 (HAYASHI)
Biochem. Biophys. Res. Commun., **25**, 116 (1966). G. Johannin and J. Yon. Fig. 2.	Fig. 5 (INAGAMI)
Biochem. Biophys. Res. Commun., **28**, 779 (1967). J. J. Pollock *et al.* Table I.	Table 26 (HAYASHI)

ACADEMIC PRESS INC. (LONDON) LTD., *London*

Source	Used in
J. Mol. Biol., **15**, 489 (1966). C. Tanford *et al.* Fig. 3, 5, 6.	Fig. 27, 28, 29 (HAMAGUCHI)
J. Mol. Biol., **50**, 71 (1970). M. Inouye *et al.* Fig. 1.	Fig. 7 (TSUGITA)
J. Mol. Biol., **54**, 219 (1970). Y. Ocada *et al.* Fig. 6, 7; Table 10.	Fig. 15, 17, 19; Table 7 (TSUGITA)

THE AMERICAN CHEMICAL SOCIETY, *Washington*

Source	Used in
Biochemistry, **1**, 844 (1962). M. Caplow and W. P. Jencks. Fig. 3.	Fig. 10 (INAGAMI)
Biochemistry, **2**, 844 (1963). C. G. Trowbridge *et al.* Fig. 1.	Fig. 11 (INAGAMI)
Biochemistry, **4**, 1088 (1965). J. A. Stewart and J. E. Dobson. Fig. 3.	Fig. 12 (INAGAMI)
Biochemistry, **8**, 713 (1969). F. W. Dahlquist and M. A. Raftery. Fig. 13.	Table 13 (HAYASHI)

Biochemistry, **8**, 1110 (1969). B. B. Sykes. Table I. — Table 15 (HAYASHI)
Biochemistry, **8**, 4579 (1969). K. C. Aune and C. Tanford. Fig. 5. — Fig. 31 (HAMAGUCHI)
J. Am. Chem. Soc., **84**, 2251 (1964). M. L. Bender and B. Zerner. Fig. 1. — Fig. 4 (INAGAMI)

THE AMERICAN SOCIETY OF BIOLOGICAL CHEMISTS, INC., *Baltimore*

J. Biol. Chem., **238**, 228 (1963). D. G. Smyth *et al.* Fig. 1. — Fig. 3 (TAKAHASHI)
J. Biol. Chem., **239**, 2516 (1964). A. J. Sophianopoulos and K. E. Van Holde. Fig. 1. — Fig. 22 (HAMAGUCHI)
J. Biol. Chem., **240**, 4117 (1965). K. Takahashi. Fig. 2. — Fig. 2 (TAKAHASHI)
J. Biol. Chem., **242**, 4388 (1967). D. M. Chipman *et al.* Fig. 1, 3; Table II, IV. — Fig. 46, 51; Table 18, 19 (HAYASHI)
J. Biol. Chem., **242**, 4644 (1967). S. S. Lehrer and G. D. Fasman. Fig. 2. — Fig. 21 (HAMAGUCHI)
J. Biol. Chem., **242**, 4688 (1967). K. Takahashi *et al.* Fig. 1, 3, 4; Table III. — Fig. 4, 7, 8; Table 1 (TAKAHASHI)
J. Biol. Chem., **243**, 391 (1968). A. Tsugita *et al.*, Fig. 5; Table I, II, III. — Fig. 11; Table 2, 4, 5 (TSUGITA)

ELSEVIER PUBLISHING CO., *Amsterdam*

Biochim. Biophys. Acta, **122**, 116 (1966). J. Chevallier and J. Yon. Fig. 3, 4. — Fig. 7, 8 (INAGAMI)
Biochim. Biophys. Acta, **140**, 439 (1967). A. D'Albis and J.-J. Bechet. Fig. 1. — Fig. 13 (INAGAMI)
Biochim. Biophys. Acta, **167**, 150 (1968). P. Jolles *et al.* Table I. — Table 21 (HAYASHI)

MACMILLAN (JOURNALS) LTD., *London*

Nature, **221**, 337 (1967). D. M. Blow *et al.* Fig. 2–4. — Fig. 2 (INAGAMI)
Ibid. **221**, 339 (1967). — Eq. 26, 27, 28 (INAGAMI)

NATIONAL ACADEMY OF SCIENCES, *Washington*

Proc. Natl. Acad. Sci., **55**, 498 (1966). R. S. Edger *et al.* Fig. 1. — Fig. 1 (TSUGITA)
Proc. Natl. Acad. Sci., **56**, 26 (1966). F. W. Dahlquist *et al.* Table I. — Table 17 (HAYASHI)

Proc. Natl. Acad. Sci., **57**, 496 (1967). J. A. Rupley and V. Gates. Table 3. — Table 23 (HAYASHI)

NATIONAL RESEARCH COUNCIL OF CANADA, *Ottawa*

Can. J. Chem., **37**, 750 (1959). J. A. Stewart and J. Ouellet. Fig. 7. — Fig. 3 (INAGAMI)

Index

A

B

D

E

F

G

M

N

O

P

Q

R

S

T

U

V

W

X